D1539063

*Dedicated to the memory
of my father and mother*

Jeremy J. Gray

Linear Differential Equations and Group Theory from Riemann to Poincaré

Second Edition

Birkhäuser
Boston • Basel • Berlin

Jeremy J. Gray
Faculty of Mathematics and Computing
The Open University
Milton Keynes MK7 6AA
United Kingdom

Library of Congress Cataloging-in-Publication Data
Gray, Jeremy, 1947-
    Linear differential equations and group theory from Riemann to
Poincaré / Jeremy Gray. — 2nd ed.
      p.   cm.
    Includes bibliographical references and index.
    ISBN 0-8176-3837-7 (alk. paper). — ISBN 3-7643-3837-7 (alk.
paper)
    1. Differential equations, Linear–History.   2. Group theory–
History.   I. Title.
QA372.G68   1999
515'.354' 09—dc21   99-14284
                                       CIP

AMS Subject Classifications: 01A55, 33Cxx, 33C80

Printed on acid-free paper.
© 2000 Birkhäuser Boston, 2nd edition.
© 1986 Birkhäuser Boston, 1st edition.

Birkhäuser

ISBN 0-8176-3837-7   SPIN 10533762
ISBN 3-7643-3837-7

Reformatted from author's files in LATEX 2e by TEXniques, Inc., Cambridge, MA.
Printed and bound by Hamilton Printing Company, Rensselaer, NY.
Printed in the United States of America.

9 8 7 6 5 4 3 2 1

Contents

Chapter I: Hypergeometric Equations and Modular Equations

Chapter II: Lazarus Fuchs

Chapter III: Algebraic Solutions to a Differential Equation

Chapter IV: Modular Equations

Chapter V: Some Algebraic Curves

Chapter VI: Automorphic Functions

Notes

Four passages which contain technical details which may not be to everyone's taste have been marked with a star*.

Introduction to the
Second Edition

I have added some new material, most of it on the period from the mid-1880s (the notional terminus of the first edition) to the early 1900s. The Riemann–Hilbert problem, as it is called, which concerns the existence of linear ordinary differential equations whose solutions have prescribed behaviour at the singular points, was taken to be solved when I wrote the first edition. Affirmative answers were attributed to Plemelj [1908] and Birkhoff [1913], and these seemed to lie outside the period under consideration. Then around 1990 it emerged that there was a significant gap in the proofs, and Anosov and Bolibruch showed that in fact the Riemann–Hilbert problem cannot be solved in general. I have therefore added a few pages on the history of this topic; the details of the achievement of Anosov and Bolibruch can be found in their book [1994]. Another account of the modern state-of-the-art can be found in the paper by Varadarajan [1991], in which the wider class of meromorphic differential equations is also considered.

My eyes were opened to the tradition of Picard–Vessiot theory, otherwise known as the Galois theory of linear ordinary differential equations, when I was instructed on the history of this topic by some of its current practitioners. I have tried to catch up, as an historian, by bringing this story up to the early 1900s; the current scene is described in Magid [1994] and several papers of Singer (e.g., [1992]). Once this work of Picard's was taken on board, it seemed capricious not to embrace the work he, Appell, and Goursat did in the 1880s on the hypergeometric equation in two variables, which turned out to have interesting connections with partial differential equations studied by Euler, and the telegraphists equation. Again, there is a lively modern interest in these topics, see Holzapfel [1986a, b] and Yoshida [1987].

I took the opportunity to round out the account of Poincaré's work in 1883/84 by amplifying the discussion of his work on solving differential equations. This culminated in his sketchy explanation of how every linear ordinary differential equation can be solved, which has implications for the Riemann–Hilbert problem and which in turn made use of a no more than plausible uniformisation theorem. His analysis breaks into two parts: one establishing the result in the generic case, and one using a continuity argument to establish it in general. Schlesinger made several contributions in this area, which were punctured by Plemelj. I therefore took the story of this approach to differential equations up to 1908, by which time the uniformisation theorem had been solved (by Poincaré and Koebe, independently) but the implications of the continuity method for were in doubt. Amusingly, I gather that the modern theory of moduli spaces should be capable of fully rigorising such continuity arguments.

New energy for the study of Poincaré emanates from his home town of Nancy, where a *Centre de Recherche Henri Poincaré* opened in 1994. Editions are planned of his scientific correspondence, and I am particularly pleased that the *Centre* has already published Poincaré's three supplements on differential equations of 1880 (Poincaré [1997]).

Apart from the new material, the text is largely unaltered from the first edition. I have put right numerous minor errors, and would like to thank the critics who pointed them out. I also thank the critics who found kind things to say. I have made some specific acknowledgements at places in the text, and I have updated the references to current historical literature.

I am very happy to be able to thank David Fowler and Ian Stewart for all their help over the years. I am also very pleased to thank Toni Cokayne for scanning the whole text of the first edition and editing the scanned version. I am equally pleased to thank Karen Lemmon and Sharon Powell for their fine job of preparing the T_EXed version of the material.

Jeremy Gray
The Open University
Milton Keynes
MK7 6AA
England

Introduction to the First Edition[1]

Mathematicians often speak of the unity of their subject, whether to praise it or lament its passing. This book traces the emergence of such a unity from its nineteenth century origins in the history of linear ordinary differential equations, especially those closely connected with elliptic and modular functions. In later chapters the book is concerned with the impact of group theoretical and geometrical ideas upon the problem of understanding the nature of the solutions to a differential equation. So far as is possible, the mathematics is developed from scratch, and no more than an undergraduate knowledge of the subject is presumed. It is my hope that this historical treatment will reacquaint many mathematicians with a rich and important area of mathematics perhaps known today only to specialists.

The story begins with the hypergeometric equation

$$x(1-x)\frac{d^2y}{dx^2} + (c - (a+b+1)x)\frac{dy}{dx} - aby = 0$$

studied by Gauss in 1812. This equation is important in its own right as a linear ordinary differential equation for which explicit power-series solutions can be given and, more importantly, their inter-relations examined. Gauss' work in this direction was extended by Riemann in a paper of 1857, in which the crucial idea of analytically continuing the solutions around their singularities in the complex domain was first truly understood. This approach, which may be termed the monodromy approach, gave a thorough global understanding of the solutions of the hypergeometric equation. It was extended in 1865 by Fuchs to those nth order linear ordinary differential equations, none of whose solutions have essential singularities, a class he was able to characterise. The work of Fuchs revealed that, for technical reasons, equations other than the hypergeometric would be less easy to understand globally. He suggested, however, that a sub-class could be isolated, consisting of differential equations all of whose solutions were algebraic functions, and this problem was tackled in the 1870s by many mathematicians: Schwarz (for the hypergeometric equation), Fuchs, Gordan, and Klein (for the second order equation), Jordan (for the nth order). The methods may be described as geometric (Schwarz and Klein), invariant-theoretic (Fuchs and Gordan), and group-theoretic (Jordan), and of these, the group-theoretic was the most strikingly successful. It arose from the monodromy approach by taking all the monodromy transformations, i.e., all analytic transformations of a basis of solutions under analytic continuations around all paths in the domain, and considering this set in totality as a group. In the case at hand this group, which is evidently composed of linear transformations, is finite if and only if the solutions are all algebraic.

The hypergeometric equation is also important because it contains many interesting equations as special cases, notably Legendre's equation

$$(1 - k^2)\frac{d^2y}{dk^2} + \frac{1 - 3k^2}{k}\frac{dy}{dk} - y = 0,$$

which is satisfied by the periods

$$K = \int_0^1 \frac{dx}{\sqrt{[(1 - x^2)(1 - k^2x^2)]}} \quad \text{and} \quad K' = \int_1^{1/k} \frac{dx}{\sqrt{[(1 - x^2)(1 - k^2x^2)]}}$$

of elliptic integrals considered as functions of the modulus k^2.

This equation was studied by Legendre, Gauss, Abel, Jacobi, and Kummer. It is useful in the study of the transformation problem of elliptic integrals: given a prime p and a modulus k^2, find a second modulus λ^2 for which the associated periods L and L' satisfy

$$\frac{L}{L'} = p\frac{K}{K'}.$$

The transformation problem also yields a polynomial relation between k^2 and λ^2 of degree $p + 1$, but Galois showed that, when $p = 5$, 7, and 11, the degree can be reduced to p. This mysterious observation was discussed by Galois in 1832 in terms of what would now be called the Galois group $PSL(2; \mathbf{Z}/p\mathbf{Z})$ and shown to depend on the existence of sub-groups of small index (in fact, p) in these groups. Galois' work was taken up by Betti, Hermite, Kronecker, and Jordan in the 1850s and 1860s, and a connection was discovered between the general quintic equation and the transformation problem at the prime $p = 5$, which enabled Hermite to solve the quintic equation by modular functions. Analogous questions at higher primes remained unsolved until, in 1879, Klein explained the special significance of the Galois group $PSL(2; \mathbf{Z}/7\mathbf{Z})$.

Klein's work grew out of an attempt to reformulate the theory of modular functions and modular transformations without using the theory of elliptic functions. This had been almost completely achieved by Dedekind in 1878, when, inspired by some notes of Riemann's and Fuchs' considerations of the monodromy transformations of the Legendre equation, he constructed a theory of modular transformations based on the idea of the lattice of periods of an elliptic function. Klein enriched Dedekind's approach with an explicit use of group-theoretic ideas and an essentially Galois-theoretic approach to the fields of rational and modular functions. Both men relied on the theory of the hypergeometric equation at technical points in their arguments. Klein's theory divided into the case when the genus of a certain Riemann surface was zero, when it connected with his earlier work on differential equations and with Hermite's theory of the quintic equation, and when the genus was greater than zero.

The latter cases related to the study of higher plane curves, and in particular the case of the group $PSL(2; \mathbf{Z}/7\mathbf{Z})$ (when the genus is 3) related to plane quartics and their 28 bitangents. These curves had been studied projectively by Plücker and Hesse, and function-theoretically by Riemann, Roch, Clebsch, and Weber. Klein was able to give an account of their work in the spirit of his new theoretical formulations, and began to develop a systematic theory of modular functions.

Meanwhile, and at first quite independently, Poincaré began in 1880 to develop a more general theory of Riemann surfaces, discontinuous groups (such as $PSL(2; \mathbf{Z})$), and differential equations. At first he seems not to have known of the work of Schwarz or Klein, but to have picked up the initial idea only from a paper of Fuchs. Soon he learned, in correspondence with Klein, of what had already been done, but there is nonetheless a marked contrast between the novelty of Poincaré's work and the impeccably educated approach of Klein. The group-theoretic and Riemann-surface theoretic aspects of the theory were developed jointly, but the connection with differential equations remained Poincaré's concern.

This presentation ends in 1882 with the publication of Poincaré's and Klein's papers on the Fuchsian theory (a choice of name that Klein hotly denounced) and with Klein's collapse from nervous exhaustion. The less complete Kleinian theory of 1883–1884 is scarcely discussed.

It can be argued that the history of mathematical ideas is a history of problems, methods, and results. These terms are, of course, not precise — a problem may be to find the method leading to an already known or intuited result, or to explain why a method works. But still one may sit gingerly on the edge of this Procrustean bed and argue that results in mathematics are, more or less, truths. Such-and-such numbers are prime, such-and-such geometries are possible, such-and-such functions exist. Problems are a more varied class of objects. They may be existence questions, or they may be more theoretical or methodological. They may arise outside mathematics, or so deep inside it that only specialists can raise them. Finally, the methods are the means employed to formulate the theory. These methods are often quite personal, and may be subjected to two kinds of test. The first is their critical scrutiny by other mathematicians as to their mathematical validity, itself a matter having an historical dimension. The second is a matter of taste or style, leading mathematicians to adopt or reject methods for their own use.

If this trichotomy is applied to the subject at hand, it suggests that the hypergeometric question raised the problem: understand the global relationships of the solutions given locally by power-series. Power-series and their convergence were dealt with carefully by Gauss and did not pose a problem, but the global question did. Riemann gave results in terms of monodromy, and in terms of the associated Riemann surfaces defined by the solutions. The monodromy method was not itself problematic, and could be used to formulate problems about, for example, Legendre's equation. This led to the successful elaboration of new results about modular functions. But the Riemann surface approach constructed functions transcendentally and by means of a doubtful use of Dirichlet's principle, so it raised methodological problems. These problems were avoided by Dedekind, Klein, and Poincaré, but tackled directly by Schwarz and C. A. Neumann. The transcendental approach was also rejected for a while in the study of algebraic curves, and Clebsch, Brill, and M. Noether developed a more strictly algebraic approach to the theorems of Abel, Riemann, and Roch.

The study of the transformation problem was explicitly raised as a methodological problem, that of emancipating the theory of modular equations from the theory of elliptic functions. As such, it was tackled by Dedekind. That solution was reformulated by Klein and completed by Hurwitz in 1881, who freed the theory of modular functions from the embrace of elliptic functions. Poincaré and Klein chose to base their theories of automorphic functions firmly on Riemannian ideas.

This brief sketch is intended to illuminate the importance, for the conduct of the history of mathematics, of the state of the subject as a theoretically organised body of knowledge. Mathematics is not just a body of results, each one attached to a technical argument, but an intricate system of theories and a historical process. Histories which proceed from problems to results and leave out the methods by which those results were achieved omit a crucial aspect of the process of this theoretical development. The gain in space, essential if a broad period is to be described, is made only by risking making the reasons for the development unintelligible. The vogue for histories of modern mathematics has, with notable exceptions, been too willing to leave out the details of how it was all done, and thus leave unexplored the question of why it was done. Similarly, all but the best histories of foundational topics have concentrated naively on the foundations without appreciating the status of foundational enquiry within the broader picture of mathematical discovery. It is probably not true that mathematics can be built from the bottom up, but is it certainly false that its history can be told in that way.

For reasons of space and ignorance I have made no serious attempt to ground this account in questions of a social-historical kind, although I believe such inquiries are most important. However, the trichotomy of problems, methods, and results would accommodate itself to such an approach. Problems may evidently be socially determined, either from outside the subject altogether or within the development of competing mathematical schools. Methods are naturally historically and socially specific in large part, but results are more objective. The formulation here adopted at least suggests what aspects of the history of modular and automorphic functions may be treated more sociologically while still securing an objectivity for the mathematical results.

I do not claim that this study establishes any conclusions that can be stated at an abstract level. Rather, it represents an attempt to try out a methodology for exploring certain past events.

One striking observation can, however, be made at once. Only Gauss was concerned with finding scientific applications for his results in the theory of differential equations. Riemann, who was deeply interested in physics, sought no role there for his P-functions. Klein, who made great claims for the importance of mathematics in physics, and Poincaré, who later did work of the greatest importance in astronomy and electromagnetism, all confined their research to the domain of pure mathematics. The Berlin school, led by Weierstrass and Fuchs, showed less interest in physics anyway, but during the period 1850–1880 it is clear that none of the works here discussed were inspired by scientific concerns. It may be a fastidious contemporary preference for not claiming that large purposes stand behind narrow papers, and it may be that a real concern for physics is shown by some of these mathematicians elsewhere in their work, although their published work is largely free of such claims. It is more likely that we are confronted with an emerging speciality — pure mathematics — perhaps being practised, in Germany at least, by a professeriat disdainful of applications. Several of Klein's remarks about physics suggest such a divorce was taking place, and an examination of physicists' work during the period might suggest a comparable lack of interest in mathematics. On the other hand, this period marks the introduction of the tools of elliptic function theory into that curious hybrid — applied mathematics (notably into the study of heat diffusion and Lamé's equation, and the motion of the top). It would be quite important to examine the literature with an eye to the distinction between pure and applied science, say between dynamics and electromagnetic theory, and to see what the role of mathematics was in each of them.

I might add that the unity of mathematics, which is often invoked in a rather imprecise way, was brought home to me very vividly as I worked on the various aspects of this story. It is not just the striking harmony of group theory and geometry in geometric function theory which is here on display, but the many parallels between the theory of the differential equations and the polynomial equations derived from problems in elliptic functions. Nor is this a static unity imposed, as it were, after the discoveries have been made. In particular, the different ways in which transformations may be formed, or groups act, is a guiding thread for many mathematicians. Klein justifiably took pleasure in seeing how the regular solids could be found to lie so often at the heart of different problems, as indeed they still do, and I hope some of that pleasure has been conveyed here. But perhaps the reader will find the continuation of Klein's work into the world of non-Euclidean geometry an even greater delight.

Unity would be a terrible thing if it did not respect individual detail, and many specific items have been presented here because, finally, I could not keep them out. This is particularly true of the algebraic curves in Chapter V. Sometimes details have been included so that I may try to explain the mathematics. Perhaps, historically, this was unwise, and the jaded reader should skip such passages, I have in mind much of Chapter I, II.2, the end of II.3 and III.3. On the other hand, mathematics should not be made unduly mysterious. I have written numerous exercises to help elucidate the mathematics. Two books may serve as excellent companions: on linear differential equations, Poole's book of that name [1960]; and on modular functions, the last chapter of Serre's *A Course of Arithmetic* [1978]. Two good books on automorphic functions and Fuchsian groups are Lehner's *Short Course ...* [1966] and

Ford's classic *Automorphic Functions* [1929] which is certainly still worth reading. A good modern introduction is provided by Beardon [1983]. The reader seeking more information on elliptic functions can still scarcely do better than to consult the famous book by Whittaker and Watson [1973].

Observations on the text

I conclude with some observations.

References have generally been given in the following forms: Shakespeare [1603] refers to the entry in the bibliography under Shakespeare for that date, which is usually the date of first publication. The symbols $a, b, \ldots$ distinguish between works published in the same year. Well known works are referred to by name, so the above would be given as *Hamlet*.

Notation is largely as it appears in the original works under discussion, so x and y are generally complex variables, but some simplifications have been introduced. Mathematical comments of an anachronistic kind, which I have felt it necessary to make on occasion, are usually set off in square brackets [   ].

References to classic works, such as the German *Encyklopädie der Mathematischen Wissenschaften*, are relatively few. This is because the classics are *Berichte* (reports) not *Geschichte* (histories) a useful distinction carefully observed by their authors, but which has reasonably restricted them to brief historical remarks. I have tended to use these extensive reference works as guides, but to report directly on what they had led me to find.

Conjectures and opinions of my own are usually stated as such; the personal pronoun always indicates that a personal view is being expressed.

I have followed recent German practice in using the symbol $:=$, as in $X := Y$ or $Y =: X$, to mean that 'X is defined to be Y'.

Acknowledgements

I would like to thank my supervisors David Fowler and Ian Stewart for all their help and encouragement during the years. An earlier incarnation of this work formed my doctoral thesis at the University of Warwick, and I would also like to thank my examiners, Ivor Grattan-Guinness and Rolph Schwarzenberger, for their helpful comments. I thank the Niedersächsische Staatsbibliothek in Göttingen for permission to quote from the unpublished letter from Jordan to Klein, and the Académie des Sciences for permission to quote from the unpublished papers of Poincaré in their possession.

Last but certainly not least, my thanks to the Open University typists for their excellent job of typing this: Christine, Frances, Kim, Michelle, Norma, Shirley and Sue. My apologies to them, and the reader, for the mistakes my poor proof reading have allowed to slip through.

Chapter I

Hypergeometric Equations

This chapter does three things. It gives a short account of the work of Euler, Gauss, Kummer, and Riemann on the hypergeometric equation, with some indication of its immediate antecedents and consequences. It therefore looks very briefly at some of the work of Gauss, Legendre, Abel and Jacobi on elliptic functions, in particular at their work on modular functions and modular transformations. It concludes with a description of the general theory of linear differential equations supplied by Cauchy and Weierstrass. There are many omissions, some of which are rectified elsewhere in the literature.[1] The sole aim of this chapter is to provide a setting for the work of Fuchs on linear ordinary differential equations, to be discussed in Chapter II, and for later work on modular functions, discussed in Chapters IV and V.

1.1 Euler and Gauss

Euler gave two accounts of the differential equation:

$$x(1-x)\frac{d^2y}{dx^2} + [\gamma - (\alpha + \beta + 1)x]\frac{dy}{dx} - \alpha\beta y = 0 \tag{1}$$

and the power series which represents one solution of it

$$y = 1 + \frac{\alpha\beta}{1\cdot\gamma}x + \frac{\alpha(\alpha+1)\beta(\beta+1)}{1\cdot 2\cdot\gamma\cdot(\gamma+1)}x^2 + \cdots, \tag{2}$$

now known as the hypergeometric equation and the hypergeometric series, respectively. Here y is a real valued function of a real variable x and α, β, and γ are real constants. The earlier account occupies four chapters of the *Institutiones Calculi Integralis* [1769, Vol. II, Part I, Chs. 8–11]; the later one is a paper [1794] presented to the St. Petersburg Academy of Science in 1778 and published posthumously in 1794.

In the paper Euler demonstrated that the power series satisfies the differential equation, and conversely, that the method of undetermined coefficients yields the power series as a solution to the differential equation. In the *Institutiones* he also considered the slightly more general equation

$$x^2(a + bx^n)\frac{d^2y}{dx^2} + x(c + ex^n)\frac{dy}{dx} + (f + gx^n)y = 0, \tag{3}$$

which has two solutions of the form

$$y = Ax^\lambda + Bx^{\lambda+n} + Cx^{\lambda+2n} + \cdots$$

where λ satisfies $\lambda(\lambda + 1)a + \lambda c + f = 0$. When the two values of λ obtained from this equation differ by an integer, Euler derived a second solution containing a logarithmic term.[2] The substitution $x^n = u$ reduces Euler's equation to

$$n^2u^2(a + bu)\frac{d^2y}{du^2} + [(a + bu)n(n - 1)u + nu(c + eu)]\frac{dy}{du} + (f + gu)y = 0, \tag{4}$$

which would be of the hypergeometric type if f were zero and one could divide throughout by u.

Euler, following Wallis [1655], used the term "hypergeometric" to refer to a power series in which the nth term is $a(a + b) \ldots (a + (n - 1)b)$. The modern use of the term derives from Johann Friedrich Pfaff, who devoted his *Disquisitiones Analyticae* (Vol. I, 1797) to functions expressible by means of the series. Pfaff's purpose was to solve the hypergeometric equation and various transforms of it in closed form, and he gave many examples of how this could be done. His view of power series expansions of a function seems to have been the typical view of his day; namely, that they were a means to the end of describing the function in terms of other, better understood functions. The view that a large class of functions can be understood using power series, without there being a reduction to closed form, is one of the characteristic advances made by Pfaff's student Gauss.

Gauss

Much has been made of Gauss' legendary ability to calculate with large numbers, and mention of this will be made below, but Gauss was also a prodigious manipulator with series of all kinds. He himself said that many of his best discoveries were made at the end of lengthy calculations, and much of his work on elliptic functions moves in a sea of formulae with an uncanny sense of direction. Of course, Gauss' skills as a reckoner with numbers are unusual, even among mathematicians, whereas the ability to think in formulae is much more common; one is struck as often by the technical power of great mathematicians as by their profundity. Nonetheless, some mathematicians have the ability more than others. It is present in a high degree in Euler, Gauss, and Kummer, but much less in Klein or Poincaré. Certainly it enabled Gauss to leave the circumscribed eighteenth century domain of functions and move with ease into the large class of functions known only indirectly. This move confronts all who take it with the question: when is a function 'known'?, to which there were broadly speaking two answers. One was to develop a theory of functions in terms of some characteristic traits which can be used to mark certain functions as having particular properties. Thus one might seek integral representations for the functions, and be able to characterise those functions for which a such-and-such kind of representation is possible. The second answer is to side-step the question and to regard the inter-relation of functions given in power series as itself the answer. Of course, most mathematicians adopted a mixture of the two approaches, depending on their own success with a given problem. We shall see that Gauss was happy to publish a work of the second kind on the hypergeometric series, and that the work of the Berlin school led by Weierstrass regarded the study of series as a cornerstone of their theory of functions, as did Lagrange earlier. On the other hand, Riemann and later workers, chiefly Klein and Poincaré, sought more geometric answers. Fuchs, although at Berlin, adopted an interestingly ambiguous approach.

It is almost impossible to describe Gauss. Gifted beyond all his contemporaries, he was doubly isolated: by the startling novelty of his vision, and by a quirk of history which produced no immediate successors to the French and German (rather Swiss) mathematicians who had dominated the eighteenth century. In 1800 Lagrange was 64, Laplace 51, Legendre 48, Monge 54. Contact with them would have been difficult for Gauss because of the Napoleonic war, and perhaps distasteful, given his conservative disposition. His teachers, Pfaff (then 35) and Kaestner (81) were not of the first rank, nor, unsurprisingly, were his contemporaries Bartels and W. Bolyai. By the time there were young mathematicians around with whom he could have conversed (Jacobi, Abel, or the generation of Cauchy and Fourier) he had become confirmed in a life-long avoidance of mathematicians. His contact was with astronomers, notably Bessel, in whose subject he worked increasingly. Only the tragic figure of Eisenstein caught Gauss' imagination towards the end of his life.[3]

Gauss, it is said, wrote much but published little. However, May [1972, 300] found that Gauss published 323 works in his lifetime, many on astronomy — *matura sed non pauca?* Nonetheless, it is true that he published almost nothing of his work on elliptic functions; the vast store of discoveries left in the *Nachlass* account for most of the disparity between what he knew and what he saw fit to print. There is no space here for an adequate account of Gauss' approach to elliptic functions, and only one aspect can be discussed, which bears most closely on the themes of this chapter.[4]

Gauss' elliptic functions

Gauss discovered for himself the arithmetico-geometric mean (agm) when he was 15. It is defined as follows for positive numbers a_0 and b_0:

$$a_1 = \tfrac{1}{2}(a_0 + b_0), \quad \text{their arithmetic mean, and}$$
$$b_1 = \sqrt{(a_0 b_0)}, \quad \text{their geometric mean.}$$

The iteration of this process, defining

$$a_{n+1} = \tfrac{1}{2}(a_n + b_n),$$
$$b_{n+1} = \sqrt{(a_n b_n)}, \qquad \text{for } n \geq 1,$$

produces two sequences $\{a_n\}$ and $\{b_n\}$, which in fact converge to the same limit, α, called the agm of a_0 and b_0. Convergence follows from the inequality $a_{n+1} - b_{n+1} < \tfrac{1}{2}(a_n - b_n)$. Gauss wrote $M(a, b)$ for the agm of a and b. Plainly, $M(\lambda a, \lambda b) = \lambda M(a, b)$, and Gauss considered various functions of the form $M(1, x)$. For example $M(1, 1 + x) = M\left(1 + \tfrac{x}{2}, \sqrt{(1 + x)}\right)$, so setting $x = 2t + t^2$, he obtained power series expansions with undetermined coefficients for M in terms of x and then in terms of t, from which the coefficients could be calculated. They display no particular pattern, but various manipulations led Gauss to the dramatic series for the reciprocal of $M(1 + x, 1 - x)$:

$$y := M(1 + x, 1 - x)^{-1} = 1 + \frac{1}{4}x^2 + \frac{9}{64}x^4 + \frac{25}{256}x^6 + \cdots$$

$$= 1 + \left(\frac{1}{2}\right)^2 x^2 + \left(\frac{1 \cdot 3}{2 \cdot 4}\right)^2 x^4 + \left(\frac{1 \cdot 3 \cdot 5}{2 \cdot 4 \cdot 6}\right)^2 x^6 + \cdots$$

(5)

As a function of x, y satisfies the differential equation

$$(x^3 - x)\frac{d^2 y}{dx^2} + (3x^2 - 1)\frac{dy}{dx} + xy = 0,$$

and Gauss found another linearly independent solution $M(1, x)^{-1}$.

The connection with elliptic integrals was discovered by Gauss on 30 May 1799, as he tells us in his diary [1917, entry 98]. He considered the lemniscatic integral

$$\int \frac{dx}{(1 - x^4)^{1/2}}, \qquad \text{for which} \qquad \int_0^1 \frac{dx}{(1 - x^4)^{1/2}} := \frac{\omega}{2},$$

and showed that $M(1, \sqrt{2}) = \frac{\pi}{\omega}$ "to eleven places".[5] This alerted him to the possibility of making a much more general discovery applicable to any complete elliptic integral, and he was able to claim such a result almost exactly a year later [diary entries 105, 106 May 1800]. He took

$$\int_0^\pi \frac{d\phi}{(1 - k^2 \cos^2 \phi)^{1/2}},$$

expanded the denominator as a power series in $\cos^2 \phi$, used the known integrals

$$\int_0^\pi \cos^{2n} \phi \, d\phi = \frac{1 \cdot 3 \dots (2n - 1)\pi}{2 \cdot 4 \dots 2n},$$

and found

$$\int_0^\pi \frac{d\phi}{(1 - k^2 \cos^2 \phi)^{1/2}} = \frac{\pi}{M(1 + k, 1 - k)} = \frac{\pi}{M(1, (1 - k^2)^{1/2})}. \tag{6}$$

A hypergeometric equation is readily obtained from the equation for $M(1 + x, 1 - x)^{-1}$ by the substitution $x^2 = z$ when it becomes

$$z(1 - z)\frac{d^2 y}{dz^2} + (1 - 2z)\frac{dy}{dz} - \frac{1}{4}y = 0. \tag{7}$$

This is a form of Legendre's equation and as will be seen, is a special case of the general hypergeometric equation, in which $\alpha = \beta = \frac{1}{2}, \gamma = 1$. Gauss obtained an expression for a period of the elliptic integral

$$\int_0^\pi \frac{d\phi}{(1 - k^2 \cos^2 \phi)^{1/2}}$$

as a function of the modulus k. The term 'period' derives from the analogy with the integral

$$\int_0^1 \frac{dx}{(1 - x^2)^{1/2}}$$

which is further discussed below.

In 1827, Gauss made a study of a function more or less inverse to M, expressing the modulus as a function of the quotient of the periods. These later discoveries will be described in due course, for they remained unpublished until 1866 [Gauss *Werke* III, 470–480], by which time others had made them independently.

The hypergeometric equation

Gauss only published the first part of his study of the hypergeometric equation, [1812a] in 1812. The second part, [1812b], found among the extensive *Nachlass*, follows on from the first, in numbered paragraphs (§§38–57).

Gauss' published paper is not remarkable by Gauss' own standards, but even so it has several claims to fame: it considers x as a complex variable; and it contains the earliest rigorous argument for the convergence of a power series and a study of the behaviour of the function at a point on the boundary of the circle of convergence, as well as a thorough examination of continued fraction expansions for certain quotients of hypergeometric functions. Part two is given to finding several solutions of the hypergeometric equation and the relationships between them, and will be of more concern to us in the sequel.

In Part I Gauss observed that the series

$$1 + \frac{\alpha\beta}{1 \cdot \gamma}x + \frac{\alpha(\alpha + 1)\beta(\beta + 1)}{1 \cdot 2\gamma(\gamma + 1)}x^2 + \cdots$$

is a polynomial if either $\alpha - 1$ or $\beta - 1$ is a negative integer, and is not defined at all if γ is a negative integer or zero (this case he excluded). In all other cases, the series is convergent for $x = a + bi$ by the ratio test, provided that $a^2 + b^2 < 1$.

He gave, as Pfaff [1797] had before him, a list of functions which can be represented by means of hypergeometric functions, and then introduced the idea of contiguous functions[6] (Section 2, §7): $F(\alpha, \beta, \gamma, x)$ is contiguous to any of the six functions $F(\alpha \pm 1, \beta \pm 1, \gamma \pm 1, x)$ obtained from it by increasing or decreasing one coefficient by 1. He obtained (§14) 15 equations connecting $F(\alpha, \beta, \gamma, x)$ with each of the fifteen pairs of its different contiguous functions by systematically permuting the α's, β's, γ's, $(\gamma - 1)$'s, etc. and comparing coefficients. As an example, the fifteenth equation is

$$0 = \gamma(\gamma - 1 - (2\gamma - \alpha - \beta - 1)x)F(\alpha, \beta, \gamma; x)$$
$$+ (\gamma - \alpha)(\gamma - \beta)xF(\alpha, \beta, \gamma + 1; x)$$
$$- \gamma(\gamma - 1)(1 - x)F(\alpha, \beta, \gamma - 1; x).$$

As Klein remarked [1894, 16], these establish that any three contiguous functions satisfy a linear relationship with rational functions for coefficients. They can be worked up to give linear relationships over the rational functions between any three functions of the form $F(\alpha \pm m, \beta \pm n, \gamma \pm p, x)$, where m, n, and p are integers. Gauss' purpose in introducing contiguous functions was to obtain continued fractions for quotients of hypergeometric functions, e.g.,

$$\frac{F(\alpha, \beta + 1, \gamma + 1; x)}{F(\alpha, \beta, \gamma; x)},$$

and hence for several familiar elementary functions. Observe that

$$F = F(\alpha, \beta, \gamma; x), \frac{dF}{dx} = \frac{\alpha\beta}{\gamma}F(\alpha + 1, \beta + 1, \gamma + 1; x)$$

and

$$\frac{d^2F}{dx^2} = \frac{\alpha(\alpha + 1)\beta(\beta + 1)}{\gamma(\gamma + 1)}F(\alpha + 2, \beta + 2, \gamma + 2; x)$$

are contiguous in the obvious generalised sense, the relationship between them being, essentially, the differential equation itself. Gauss made this observation at the start of the second unpublished paper.

In the third and final section of the published paper, Gauss considered the question of the value of $F(\alpha, \beta, \gamma, 1)$, i.e., of $\lim_{x \to 1} F(\alpha, \beta, \gamma, x)$, at least for real α, β and γ. He introduced the gamma function in a somewhat modified form which is, perhaps, more intuitively acceptable, by defining

$$\Pi(k, z) := \frac{1 \cdot 2 \ldots k \cdot k^z}{(z + 1)(z + 2) \ldots (z + k)},$$

where k is a positive integer, and $\Pi(z) := \lim_{k \to \infty} \Pi(k, z)$. The limit certainly exists for $\text{Re}(z) \geq 0$, and Π satisfies the functional equation $\Pi(z + 1) = (z + 1)\Pi(z)$, with

$\Pi(0) = 1$, from which it follows that $\Pi(n) = n!$ for positive integral n; Π may be called (Gauss') factorial function; it is infinite at all negative integers. (In the usual notation, due to Legendre [1814], $\Pi(z) = \Gamma(z+1)$.) The factorial function enabled Gauss to write

$$F(\alpha, \beta, \gamma, 1) = \frac{\Pi(k, \gamma - 1)\Pi(k, \gamma - \alpha - \beta - 1)}{\Pi(k, \gamma - \alpha - 1)\Pi(k, \gamma - \beta - 1)} F(\alpha, \beta, \gamma + k, 1),$$

and since $\lim_{k \to \infty} F(\alpha, \beta, \gamma + k, 1) = 1$, he obtained (§23)

$$F(\alpha, \beta, \gamma, 1) = \frac{\Pi(\gamma - 1)\Pi(\gamma - \alpha - \beta - 1)}{\Pi(\gamma - \alpha - 1)\Pi(\gamma - \beta - 1)}.$$

This expression is meaningful provided $\alpha + \beta - \gamma < 0$.

The gamma function also enabled him to attend to certain integrals, for example, Euler integrals of the first kind, in Legendre's terminology:

$$\int_0^x z^{\lambda-1}(1 - z^\mu)^\nu dz,$$

which vanishes at $x = 0$. This Gauss expressed as

$$\frac{x^\lambda}{\lambda} F\left(-\nu, \frac{\lambda}{\nu}, \frac{\lambda}{\mu} + 1, x^\mu\right).$$

When $x = 1$ the definite integral

$$\int_0^1 z^{\lambda-1}(1 - z^\mu)^\nu dz = \frac{\Pi\left(\frac{\lambda}{\mu}\right)\Pi(\nu)}{\lambda\Pi\left(\frac{\lambda}{\mu} + \nu\right)}.$$

Gauss commented (§27) "Whence many relations, which the illustrious Euler could only get with difficulty, fall out at once". As examples of spontaneous results Gauss considered the lemniscatic integrals:[7]

$$A = \int_0^1 \frac{dx}{(1 - x^4)^{1/2}}, \qquad B = \int_0^1 \frac{x^2 dx}{(1 - x^4)^{1/2}}$$

and showed that

$$A = \frac{\Pi\left(\frac{1}{4}\right)\Pi\left(-\frac{1}{2}\right)}{\Pi\left(-\frac{1}{4}\right)}, \quad B = \frac{\Pi\left(\frac{3}{4}\right)\Pi\left(-\frac{1}{2}\right)}{3\Pi\left(\frac{1}{4}\right)} = \frac{\Pi\left(-\frac{1}{4}\right)\Pi\left(-\frac{1}{2}\right)}{4\Pi\left(\frac{1}{4}\right)}, \quad \text{so } AB = \frac{\pi}{4}.$$

The published paper virtually concludes with a study of a function Ψ. Gauss attributed to Euler the equation (now usually called Stirling's series)

$$\log \Pi(z) = \left(z + \frac{1}{2}\right)\log z - z + \frac{1}{2}\log 2\pi + \frac{B_1}{1 \cdot 2 \cdot z} - \frac{B_2}{3 \cdot 4 \cdot z^3} + \frac{B_3}{5 \cdot 6 \cdot z^5} + \cdots,$$

where B_1, B_2, B_3, etc. are the Bernoulli numbers, which are defined by the expansion

$$\frac{x}{e^x - 1} = 1 - \frac{x}{2} + \sum \frac{(-1)^{k+1} B_k x^{2k}}{(2k)!}$$

(so $B_1 = \frac{1}{6}$, $B_2 = \frac{1}{30}$, $B_3 = \frac{1}{42}$). He introduced

$$\Psi(z) := \log z + \frac{1}{2z} - \frac{B_1}{2z^2} + \frac{B_2}{4z^4} - \frac{B_3}{6z^6} + \cdots$$

and showed that

$$\frac{d}{dz} \Pi(z) = \Pi(z) \cdot \Psi(z),$$

so Ψ is the logarithmic derivative of Π. The last paragraph is a two-page tabulation of z, $\log \Pi(z)$, and $\Psi(z)$ where $0 \leq z \leq 1$, z increases in steps of 0.01, and the tabulated values are given to 18 decimal places, thus surpassing Legendre's 12-place tables [1814] of $\log \Gamma(z)$ over the same range. The paper frequently carries calculations of specific values of functions to over 20 decimal places; such calculations were both easy and congenial to Gauss.

Gauss began the second and unpublished part of the paper, *Determinatio series nostrae per Aequationem Differentialem Secundi Ordinis*, by observing that $P := F(\alpha, \beta, \gamma; x)$ is a solution of the differential equation

$$(x - x^2)\frac{d^2 P}{dx^2} + (\gamma - (\alpha + \beta + 1)x)\frac{dP}{dx} - \alpha\beta P = 0. \tag{8}$$

To find a second linearly independent solution, he set $1 - y = x$, when the equation becomes

$$(y - y^2)\frac{d^2 P}{dy^2} + (\alpha + \beta + 1 - \gamma - (\alpha + \beta + 1)y)\frac{dP}{dy} - \alpha\beta P = 0,$$

which is the first equation with γ replaced by $\alpha + \beta + 1 - \gamma$. It, therefore, has a solution $F(\alpha, \beta, \alpha + \beta + 1 - \gamma, 1 - x)$, and the differential equation in general has solutions of the form

$$MF(\alpha, \beta, \gamma, x) + NF(\alpha, \beta, \alpha + \beta + 1 - \gamma, 1 - x), \tag{§39}$$

where M and N are constants.

Other solutions may arise which do not at first appear to be of this type, but, he remarked, any three solutions must satisfy a linear relationship with constant coefficients. This fact was of most use to him when transforming the differential equation by means of a change of variable. For instance, the substitution $P = x^{1-\gamma} P'$ transforms the differential equation into

$$(x - x^2)\frac{d^2 P'}{dx^2} + (2 - \gamma - (\alpha + \beta + 3 - 2\gamma)x)\frac{dP'}{dx} - (\alpha + 1 - \gamma)(\beta + 1 - \gamma)P' = 0,$$

which has the general solution

$$P' = MF(\alpha+1-\gamma, \beta+1-\gamma, 2-\gamma; x)+NF(\alpha+1-\gamma, \beta+1-\gamma, \alpha+\beta+1-\gamma; 1-x).$$

But Gauss was able to show that

$$F(\alpha, \beta, \alpha+\beta+1-\gamma; 1-x) = \frac{\Pi(\alpha+\beta-\gamma)\Pi(-\gamma)}{\Pi(\alpha-\gamma)\Pi(\beta-\gamma)}F(\alpha, \beta, \gamma; x)$$

$$+\frac{\Pi(\alpha+\beta-\gamma)\Pi(\gamma-2)}{\Pi(\alpha-1)\Pi(\beta-1)}x^{1-\gamma}(1-x)^{\gamma-\alpha-\beta}F(1-\alpha, 1-\beta, 2-\gamma; x).$$

He considered various substitutions to obtain a variety of equations relating the hypergeometric functions. For instance (§47), setting $x = \dfrac{y}{y-1}$ gives a new equation for P and y, whence, setting $P = (1-y)^\mu P'$, an equation for P' and y. From this equation he deduced, when $\mu = \alpha$,

$$F(\alpha, \beta, \gamma, x) = (1-y)^\alpha F(\alpha, \gamma-\beta, \gamma, y)$$

$$= (1-x)^{-\alpha} F\left(\alpha, \gamma-\beta, \gamma, \frac{-x}{1-x}\right),$$

and when $\mu = \beta$,

$$F(\alpha, \beta, \gamma, x) = (1-x)^{-\beta} F\left(\beta, \gamma-\alpha, \gamma, \frac{-x}{1-x}\right).$$

The substitutions he considered are of two types: the transformations of x:

$$x = 1-y, \quad x = \frac{1}{y}, \quad x = \frac{y}{y-1}, \quad x = \frac{y-1}{y},$$

and these transformations of P : $P = x^\mu P'$, $P = (1-x)^\mu P'$ for particular values of μ. These gave him several solutions to the original equation in terms of functions like $F(-, -, -, x)$ and $F(-, -, -, 1-x)$, etc., possibly multiplied by powers of x and $1-x$, and also some linear identities between these solutions. He also gave an impressive calculation to illustrate the linear dependence of the three solutions

$$P = F(\alpha, \beta, \gamma; x), \quad Q = x^{1-\gamma} F(\alpha+1-\gamma, \beta+1-\gamma, 2-\gamma; x)$$

and

$$R = F(\alpha, \beta, \alpha+\beta+1-\gamma, 1-x):$$
$$R = F(\alpha, \beta, \gamma)P + F(\alpha+1-\gamma, \beta+1-\gamma, 2-\gamma)Q,$$

where

$$F(\alpha, \beta, \gamma) = \frac{\Pi(\alpha+\beta-\gamma)\Pi(-\gamma)}{\Pi(\alpha-\gamma)\Pi(\beta-\gamma)}.$$

The paper concluded with a discussion of certain special cases that can arise when α, β and γ are not independent, for example, when $\beta = \alpha + 1 - \gamma$, and the quadratic change of variable $x = 4y - 4y^2$ can be made.

Gauss made a very interesting observation at this point. The equation has as one solution in this case

$$F\left(\alpha, \beta, \alpha + \beta + \frac{1}{2}, 4y - 4y^2\right) = F\left(2\alpha, 2\beta, \alpha + \beta + \frac{1}{2}, y\right).$$

If, he said, y is replaced by $1 - y$, this produces

$$F\left(\alpha, \beta, \alpha + \beta + \frac{1}{2}, 4y - 4y^2\right) = F\left(2\alpha, 2\beta, \alpha + \beta + \frac{1}{2}, 1 - y\right),$$

and one is led to the seeming paradox

$$F\left(2\alpha, 2\beta, \alpha + \beta + \frac{1}{2}, y\right) = F\left(2\alpha, 2\beta, \alpha + \beta + \frac{1}{2}, 1 - y\right)$$

which equation is certainly false (§55). To resolve the paradox he distinguished between F as a function, which satisfies the hypergeometric equation, and F as the sum of an infinite series. The sum is only defined within its circle of convergence, but the function is to be understood for all continuous changes in its fourth term, whether real or imaginary, provided the values 0 and 1 are avoided. This being so, he argued that one would no more be misled than one would infer from arcsin $\frac{1}{2} = 30°$ and $\sin 150° = \frac{1}{2}$ that $30° = 150°$, for a (many-valued) function may have different values even though its variable has taken the same value, whereas a series may not.

Gauss here confronted the question of analytically continuing a function outside its circle of convergence. It was his view that the solutions of the differential equation exist everywhere but at 0, 1, (and ∞, although he avoided the expression). However, their representation in power series is a local question, and the same function may be represented in different ways. In particular, the series expression may not be recaptured if the variable is taken continuously along some path and restored to its original value.

Because he talked here of continuous change in the variable in the complex number plane, one may thus infer that Gauss here was truly discussing analytic continuation, and not merely the plurality of series solutions at a given point. In later terminology, used by Cauchy and Riemann in the 1850s, a function is *monodromic* if its analytic continuations always yield a unique value for the function at each point, and such considerations are called "monodromy questions". Gauss is therefore the first to raise the monodromy problem in the context of differential equations, albeit in the unpublished part of his paper. One may reasonably speculate that it was connected in his mind with the linear relations that exist between any three solutions at a point, but he does not say so explicitly.[8]

In these papers, Gauss introduced a large class of functions of a complex variable which were defined by the hypergeometric equation and were capable of various expressions in series. The main direction of his research was in studying relationships between the series, which in turn provided information about the nature of the functions under consideration.

1.2 Jacobi and Kummer

Gauss published only a small part of his work on elliptic functions, and made no mention of it in his paper on the hypergeometric series. Kummer, the next author to discuss the series significantly, did so with a view to using them to explore the new functions, which by then had been announced publicly. So before discussing his work, it will be necessary to look briefly at the theory of the elliptic functions as begun by Euler and Legendre and developed by Abel and, in particular, by Jacobi.[9]

Elliptic integrals

The equation

$$k(1 - k^2)\frac{d^2y}{dk^2} + (1 - k^2)\frac{dy}{dk} + ky = 0,$$

now called Legendre's equation for elliptic integrals of the second kind, was studied by Euler in [1733], because of its connection with the rectification of the ellipse, and again in [1750], when he found the solution for which $y = 1$ when $k = 0$ to be

$$y = 1 + \alpha k^2 + \beta k^4 \cdots + \log k(\gamma k^2 + \delta k^4 + \cdots)$$

(for suitable $\alpha, \beta, \gamma, \delta, \ldots$). This is interesting, for it shows that logarithmic terms may be expected near a singular point of the equation, a matter to be discussed more fully in Chapter II.

The equation

$$\frac{\partial}{\partial \mu}(1 - \mu^2)\frac{\partial U}{\partial \mu} + \frac{1}{(1 - \mu^2)}\frac{\partial^2 U}{\partial \phi^2} + n(n + 1)U = 0$$

was obtained by Laplace [1782] in a study of potential theory. When ϕ is absent, the equation becomes Legendre's equation for elliptic integrals of the first kind, (9) below. Legendre used this equation (with ϕ absent) in the course of his own work on potential theory, [1793], and introduced the differential equations for elliptic integrals of the first and second kinds in [1785], and, more influentially, in his *Exercises de calcul intégral*, [1814] and *Traité des fonctions elliptiques et des intégrales eulériennes* [1825, Vol. 1, Ch. 13] where he derived the equation

$$k(1 - k^2)\frac{d^2Q}{dk^2} + (1 - 3k^2)\frac{dQ}{dk} - kQ = 0 \tag{9}$$

for the periods $4K$ and $2iK'$ as functions of the modulus k, by differentiating the complete integral

$$\int_0^{\pi/2} \frac{d\phi}{\sqrt{(1 - k^2 \sin^2 \phi)}}$$

with respect to k. Here

$$K := \int_0^1 \frac{dx}{\sqrt{[(1-x^2)(1-k^2x^2)]}} \quad \text{and} \quad K' := \int_0^1 \frac{dx}{\sqrt{[(1-x^2)(1-k'^2x^2)]}}$$

and $k'^2 := 1 - k^2$ is the so-called complementary modulus.

Legendre's elliptic integrals are real, and the modulus $c = \sin\theta$ lies between 0 and 1, [*Traité*, Ch. 5]. They define single-valued functions of their upper end points which have single-valued inverse functions. Legendre's tables and his accompanying comments make it clear that he regarded the problem of inversion as solved (in this case). In the *Exercices de calcul intégral* ... [1811] he remarked (p. 380): "The same formulae serve to solve the inverse problem, that is to say, to determine the amplitude ϕ when one knows the function $F(c,\phi)$." In the *Traité* (p. 383) he said, "Thus being given an arbitrary value of the angle ψ, one can find the corresponding value of the time t and reciprocally." (Quoted in Krazer [1909, 55n].)

Once the upper end point is allowed to vary in the complex plane, the presence of the periods means that the integral

$$u = \int_0^\phi \frac{d\phi}{\sqrt{(1-k^2\sin^2\phi)}} = \int_0^x \frac{dx}{\sqrt{[(1-x^2)(1-k^2x)]}},$$

where $x = \sin\phi$, defines only an infinitely many-valued "function" of the upper end point, since the path of integration may loop several times around ± 1 and $\pm 1/k$. Jacobi (and Abel independently) had the idea of studying instead the inverse functions $\phi = am(u)$, $x = x(u) = \sin am\, u$. Jacobi was from the first particularly interested in the transformation problem for this function, the change of modulus from k^2 to λ^2 given by

$$\frac{dy}{\sqrt{[(1-y^2)(1-\lambda^2y^2)]}} = \frac{dx}{M\sqrt{[(1-x^2)(1-k^2x^2)]}}. \tag{10}$$

This is connected to the problem of relating the inverse functions $\sin am\, u$ and $\sin am\, nu$ for odd integers n, which was suggested to him by the evident analogy with

$$u = \int_0^{\sin u} \frac{dx}{\sqrt{(1-x^2)}}.$$

But whereas the equation connecting $\sin n\theta$ and $\sin\theta$ has n roots, the equation between $\sin am\, nu$ and $\sin am\, u$ has n^2 when n is odd (the case n even is a little more complicated). Jacobi explained this in terms of the double periodicity of $\sin am$, which, he showed, satisfied $\sin am(u + 4K) = \sin am(u + 2iK') = \sin am\, u$, whence $4K$ and $2iK'$ are the periods.

In his *Fundamenta Nova* [1829] Jacobi, inspired by a paper of Abel's, gave explicit rational transformations of the kind he sought, $y = \frac{U(x)}{V(x)}$, and when U was a polynomial of order p, he said the transformation was of order p. To take one of his examples, corresponding to the transformation of order 3, if the periods of the integral taken with modulus λ are Λ and Λ', then one asks for $\frac{\Lambda'}{\Lambda} = 3\frac{K'}{K}$. Jacobi found that the substitution

$$y = \frac{x(a + a'x^2)}{1 + b'x^2} = \frac{U(x)}{V(x)}$$

where $a = 1 + 2\alpha$, $a' = \alpha^2$, and $b' = \alpha(2 + \alpha)$ produces a complicated expression for

$$\frac{dy}{\sqrt{[(1 - y^2)(1 - \lambda^2 y^2)]}}$$

of the form

$$\frac{P(x)dx}{\sqrt{Q(x)}},$$

where $P(x)$ is of degree 4 and $Q(x)$ of degree 6. But, for suitable λ, $P(x)$ occurs squared as a factor of $Q(x)$ and the expression in y reduces to

$$\frac{dx}{M\sqrt{[(1 - x^2)(1 - k^2 x^2)]}}$$

for some constant M. In this case

$$a' = \sqrt{\left(\frac{k^3}{\lambda}\right)};$$

so, setting $4\sqrt{k} = u$ and $4\sqrt{\lambda} = v$, he obtained this equation connecting the moduli k and λ:

$$u^4 - v^4 + 2uv(1 - u^2 v^2) = 0. \qquad \text{[1829 §13 in 1969, I, 74]}$$

This equation, and others like it for transformations of higher order, will be discussed in Chapter IV where their significance as polynomial equations will be analysed in the context of the emerging Galois theory. Jacobi also derived differential equations connecting k and λ, and since it was these equations which interested Kummer, they will be presented here. Jacobi argued [*Fund. Nova* §32–34 in 1969, I, 129–138] that if $Q = aK + bK'$ and $Q' = a'K + b'K'$ are two solutions of Legendre's equations, then their quotient satisfies

$$d\left(\frac{Q'}{Q}\right) = -\frac{\pi}{2}\left(\frac{ab' - a'b}{k(1 - k^2)}\right)\frac{dk}{Q^2}.$$

The same equations hold for the periods Λ and Λ' taken with respect to a different modulus λ:

$$(\lambda - \lambda^3)\frac{d^2 L}{d\lambda^2} + (1 - 3\lambda^2)\frac{dL}{d\lambda} - \lambda L = 0.$$

So, if λ is obtained from k by a transformation of order n, i.e., $\frac{\Lambda'}{\Lambda} = \frac{nK'}{K}$, then $\Lambda = \frac{K}{M}$, and the corollary of Legendre's equation implies that

$$\frac{ndk}{k(1 - k^2)K^2} = \frac{d\lambda}{\lambda(1 - \lambda^2)\Lambda^2},$$

from which it follows that

$$M^2 = \frac{1}{n} \frac{\lambda(1-\lambda^2)dk}{k(1-k^2)d\lambda}.$$

If n is fixed and M is regarded as a function of k, then Legendre's equations can be used to eliminate M from this equation, and the result is[10]

$$3\left(\frac{d^2\lambda}{dk^2}\right)^2 - 2\frac{d\lambda}{dk}\cdot\frac{d^3\lambda}{dk^3} + \left(\frac{d\lambda}{dk}\right)^2 \cdot \left\{\left[\frac{1+k^2}{k-k^3}\right]^2 - \left[\frac{1+\lambda^2}{\lambda-\lambda^3}\right]^2\left(\frac{d\lambda}{dk}\right)^2\right\} = 0.$$

(Jacobi's equation 12)

This equation, which may be called Jacobi's differential equation for the moduli, was one of the targets of Kummer's work. It has, among its particular integrals, the quotients of solutions of Legendre's equations:

$$\frac{a'K + b'K'}{aK + bK'} + \frac{\alpha'\Lambda + \beta'\Lambda'}{\alpha\Lambda + \beta\Lambda'}.$$

The second part of Gauss' paper on the hypergeometric series raises two main types of question. First, it would be useful to have a systematic account of the solutions obtained by the various substitutions, and of the nature of the substitutions themselves. Second, it would be instructive to connect the hypergeometric functions with the newer functions in analysis, especially in complex analysis, such as the elliptic functions. It is striking that Kummer's 1836 paper [Kummer 1836 = *Coll. Papers*, II] sets itself both these tasks and resolves them while, moreover, observing Gauss' restrictions where the work would otherwise be too difficult (for example, by considering only real coefficients).

Kummer

Ernst Eduard Kummer had studied mathematics at Halle, after first intending to study Protestant theology — a common enough false start — and in 1836 was a lecturer at the Liegnitz Gymnasium. His earliest work was in function theory, but from the mid-1840s onwards he concerned himself with algebraic number theory, which he came to dominate; he is also remembered for his quartic surface with 16 nodal points. Leopold Kronecker was one of his students at Liegnitz, and with Kronecker and Weierstrass, Kummer dominated the Berlin school of mathematics from 1856 until his retirement in 1883. He was a gifted teacher and organiser of seminars; he concerned himself greatly with the fortunes of his many students, and his students were correspondingly devoted to him. He was also a man of great charm, and he had a great appetite for administration, being dean of the University of Berlin twice, rector once, and perpetual secretary of the physics-mathematics section of the Berlin Academy from 1863 to 1878. Although he never attended a lecture by Dirichlet, he considered him to have been his real teacher, and this is perhaps reflected in the topics which he came to study most closely.[11]

Kummer alluded briefly to other discussions of the hypergeometric equation at the start of his long paper [1836].[12] Of Gauss' paper he remarked: "But this work is only the first part of a greater work as yet unpublished, and wants comparison of hypergeometric series in which the last element x is different. This will therefore be the principal purpose of the present work; the numerical application of the discovered formulae will preferably be made to elliptic transcendents, to which in great part the general series corresponds." To elucidate the first problem, Kummer sought the most general transformation there could be between two hypergeometric equations.

He considered two hypergeometric equations:

$$\frac{d^2y}{dx^2} + p\frac{dy}{dx} + qy = 0,\tag{11}$$

and

$$\frac{d^2v}{dz^2} + P\frac{dv}{dz} + Qv = 0,\tag{12}$$

where x and z are real, $z = z(x)$ is a function of x, and $y = w \cdot v$ where $w = w(x)$ is a function to be determined. Eliminating y and v he showed that

$$w^2 = c \cdot \left(e^{\int Pdz - \int pdx}\right)\frac{dx}{dz}.\tag{13}$$

He went on to remark that if z was known as a function of x, then w would also be known as a function of x; so he eliminated w and found that[13]

$$2\frac{d^3z}{dx^3}\left(\frac{dz}{dx}\right)^{-1} - 3\left(\frac{d^2z}{dx^2}\right)^2\left(\frac{dz}{dx}\right)^{-2} - \left(2\frac{dP}{dz} + P - 4Q\right)\frac{dz}{dx}$$
$$- \left(2\frac{dp}{dx} + p - 4q\right) = 0,\tag{14}$$

and so, he said, the only difficulty was to solve this equation and determine z as a function of x. Since Kummer stipulated that the transformations between the transcendental functions should be algebraic, he required algebraic solutions of (14). The general problem of finding all algebraic solutions to a differential equation appeared to him to be impossible, but in this case he had shown in an earlier paper [1834] that the general solution of (14) was of the form

$$A\phi(x)\psi(z) + B\phi(x)\psi_1(z) + C\phi_1(x)\psi(z) + D\phi_1(x)\psi_1(z) = 0,$$

where $\phi(x)$ and $\phi_1(x)$ are independent solutions of (11) and $\psi(z)$ and $\psi_1(z)$ are independent solutions of (12), so

$$A\phi(x) + B\phi_1(x) = A'w\psi(z) + B'w\psi_1(z).\tag{15}$$

Indeed, equation (13) implies that

$$\frac{ce^{-\int pdx}dx}{(A\phi(x)+B\phi_1(x))^2}=\frac{e^{-\int Pdz}dz}{(A'\psi(z)+B'\psi_1(z))^2},$$

whence

$$\frac{A\phi(x)+B\phi_1(x)}{C\phi(x)+D\phi_1(x)}=\frac{A'\psi(z)+B'\psi_1(z)}{C'\psi(z)+D'\psi_1(z)} \tag{16}$$

is the general solution of (14).

Kummer's analysis of the cases when certain relationships exist between α, β, and γ, and accordingly quadratic changes of variable are possible, must be looked at only briefly. It led him in fact to the same paradox as Gauss, a seemingly impossible equation between

$$F(-,-,-,x) \quad \text{and a sum of two} \quad F\left(-,-,-,\frac{1}{x}\right)$$

for certain α's, β's and γ's. In this case either one side converges or the other, but not both. Kummer took more or less Gauss' (unpublished) view that the equality meant the same function was being represented in two ways, one valid when $x < 1$, the other when $x > 1$. But Kummer's discussion was confined to real x, and so lacks the concept of continuous change in x connecting the two branches. For Kummer, $x = 1$ is a genuine barrier; the series fail to converge and are, so to speak, kept apart. When in Chapter VII he allowed x to become complex, he did not return to this problem, so one cannot infer that he was aware of monodromy considerations. This claim is made for him by Klein [1967, 267] and Biermann, [1973, 523] and several workers risk implying it when they connect Kummer's 24 solutions with the question of monodromy. While they are making an entirely permissible interpretation of Kummer's results, it is not one made by Kummer himself. The honour of discovery must go to Gauss, and the first to grasp the significance of the idea was Riemann.

Kummer's 24 solutions

In Section II of his paper Kummer produced a set of 24 solutions to the hypergeometric equation which, in some sense, are the complete solutions to the equation.[14] To be precise, if the variable is allowed to be complex, they provide not only sets of bases for the solutions everywhere, but a description of their analytic continuation on the complex sphere. For this reason they are central to the insights of Riemann and Schwarz, as we shall see. However, in Kummer's work, the variable is only real, and he regarded them as the best way to obtain solutions valid near the singular points in a variety of convenient forms. Let the coefficients in (11) and (12) be α, β, γ and α', β', γ'.

Kummer imposed the following simplifying requirements: z is to be a function of x alone, there are to be no relationships between α, β and γ or α', β', and γ', but

α', β', and γ' are to be linear combinations α, β, and γ with constant coefficients.[15] Under these restrictions Kummer found

$$z = \frac{ax + b}{cx + d} \quad \text{(48 in \emph{Coll. Papers} II 84)}.$$

In §6 Kummer noted that for such a function z of x,

$$2\left(\frac{d^3z}{dx^3}\right)\left(\frac{dx}{dz}\right) - 3\left(\frac{d^2z}{dx^2}\right)^2\left(\frac{dx}{dz}\right)^2 = 0,$$

simplifying (14) considerably. It becomes

$$\frac{Ax^2 + Bx + C}{x^2(1-x)^2} = \frac{(ad - bc)^2(A'(ax + b)^2 + B'(ax + b)(cx + d) + C'(cx + d)^2)}{(ax + b)^2(cx + d)^2((c - a)x + d - b)^2}. \tag{17}$$

There can be no common factors either side since α, β, γ are arbitrary so the two numerators can only differ by a constant factor, say m. There are precisely six solutions to the two equations that arise for m by equating the numerators and denominators separately:

$$
\begin{array}{lll}
c = 0 & = b = a - d, & m = a^6 \\
c = 0 & = d - b = a + b, & m = a^6 \\
a = 0 & = d = c - b, & m = b^6 \\
a = 0 & = d - b = c + d, & m = b^6 \\
c - a = 0 = b = c + d, & & m = a^6 \\
c - a = 0 = d = a + b, & & m = b^6
\end{array}
$$

for which the substitutions are precisely

$$z = x, \ z = 1 - x, \ z = \frac{1}{x}, \ z = \frac{1}{1 - x}, \ z = \frac{x}{x - 1}, \ z = \frac{x - 1}{x},$$

respectively.

In the first case, there are four possible substitutions for α', β', γ', in place of α, β, γ:

$$
\begin{aligned}
(\alpha', \beta', \gamma') &= (\alpha, \beta, \gamma) \\
&= (\gamma - \alpha, \gamma - \beta, \gamma) \\
&= (\alpha - \gamma + 1, \beta - \gamma + 1, 2 - \gamma) \\
&= (1 - \alpha, 1 - \beta, 2 - \gamma), \quad \text{(50 in \emph{Coll. Papers} II, 86)}
\end{aligned}
$$

as can be seen from the equation $Ax^2 + Bx + C = A'x^2 + B'x + C'$ upon replacing A by $(\alpha - \beta)^2$, A' by $(\alpha' - \beta')^2$, and so on.

Furthermore, there are 4 substitutions in α, β, γ for each of the other substitutions of z for x giving rise to 24 solutions to the differential equations, which

Kummer listed in §8. They have become known as Kummer's 24 solutions to the hypergeometric equation, and they are displayed in Table 1.1 [52, 53 in *Coll. Papers*, II, 88, 89].

Independently of Gauss, Kummer pointed out that several of these substitutions could be spotted without going through the argument above. For example (14) is unaltered by the substitution of $\gamma' - \alpha'$ for α', $\gamma' - \beta'$ for β', so from the solution $F(\alpha', \beta', \gamma', z)$, the solution $(1 - z)^{\gamma' - \alpha' - \beta'} F(\gamma' - \alpha', \gamma' - \beta', \gamma', z)$ is obtained.

Kummer's analysis presents a complete answer to the first problem: what are the allowable changes of variable for the general hypergeometric equation?

It immediately raises the second question: what are the solutions themselves? As Kummer pointed out (§9), there are many relationships between the solutions. Some, in any case, are equal to others, for example

$$F(\alpha, \beta, \gamma, x) = (1 - x)^{\gamma - \alpha - \beta} F(\gamma - \alpha, \gamma - \beta, \gamma, x)$$

$$= (1 - x)^{-\alpha} F\left(\alpha, \gamma - \beta, \gamma, \frac{x}{x - 1}\right).$$

In fact, the six families of four solutions rearrange themselves into six different families of four equal solutions thus: 1, 2, 17, and 18; 3, 4, 19 and 20; 5, 6, 21, 22; 7, 8, 23, 24; 9, 12, 13, 15; and 10, 11, 14, 16. So, to find all the linear relations between the twenty-four solutions is enough to consider the six different ones 1, 3, 5, 7, 13, 14. Of these, 5 and 7 converge or diverge exactly when 13 and 14 diverge or converge, respectively. Kummer here restricted x to be real, but the observation is valid for complex x. The problem is thus reduced to finding the relations between the following triples: 1, 3, 5; 1, 3, 7; 1, 3, 13; 1, 3, 14. As an example of Kummer's results, the relationship between 1, 3, and 5 is

$$F(\alpha, \beta, \gamma, x) = \frac{\Pi(\gamma - 1)\Pi(\alpha - \gamma)\Pi(\beta - \gamma)}{\Pi(1 - \gamma)\Pi(\alpha - 1)\Pi(\beta - 1)} F(\alpha - \gamma + 1, \beta - \gamma + 1, 2 - \gamma, x) x^{1 - \gamma}$$

$$+ \frac{\Pi(\alpha - \gamma)\Pi(\beta - \gamma)}{\Pi(\alpha + \beta - \gamma)\Pi(-\gamma)} F(\alpha, \beta, \alpha + \beta - \gamma + 1, 1 - x),$$

where Π stands for Gauss' factorial function. He listed the relationships that arise in §11 (see Table 1.2). They all arise from evaluating $F(-, -, -, x)$ at 0 or 1, and hence obtaining equations for the A's and B's.

Chapter 6 of Kummer's paper is his analysis of the transcendental functions that can be represented by the hypergeometric functions. He found that $F\left(\frac{1}{2}, \frac{1}{2}, 1, c^2\right)$ was already well known in analysis, for

$$F\left(\frac{1}{2}, \frac{1}{2}, 1, c^2\right) = \frac{1}{\pi} \int_0^1 u^{-1/2}(1 - u)^{-1/2}(1 - c^2 u)^{-1/2} du$$

$$= \frac{2}{\pi} \int_0^{\pi/2} \frac{d\phi}{(1 - c^2 \sin \phi^2)^{1/2}}$$

$$= F^1(c),$$

Table 1.2

21.
$$\left\{\begin{aligned}
&F = A\,x^{1-\gamma}F(\alpha-\gamma+1,\ \beta-\gamma+1,\ 2-\gamma,\ x) + B\,F(\alpha,\ \beta,\ \alpha+\beta-\gamma+1,\ 1-x),\\
&A = \frac{\Pi(\gamma-1)\,\Pi(\alpha-\gamma)\,\Pi(\beta-\gamma)}{\Pi(1-\gamma)\,\Pi(\alpha-1)\,\Pi(\beta-1)},\qquad
B = \frac{\Pi(\alpha-\gamma)\,\Pi(\beta-\gamma)}{\Pi(\alpha+\beta-\gamma)\,\Pi(-\gamma)}.
\end{aligned}\right.$$

22.
$$\left\{\begin{aligned}
&F = A_2\,x^{1-\gamma}F(\alpha-\gamma+1,\ \beta-\gamma+1,\ 2-\gamma,\ x)\\
&\qquad + B_2\,(1-x)^{\gamma-\alpha-\beta}F(\gamma-\alpha,\ \gamma-\beta,\ \gamma-\beta-1),\\
&A_2 = \frac{\Pi(\gamma-1)\,\Pi(-\alpha)\,\Pi(-\beta)}{\Pi(1-\gamma)\,\Pi(\gamma-\alpha-1)\,\Pi(\gamma-\beta-1)},\qquad
B_2 = \frac{\Pi(-\alpha)\,\Pi(-\beta)}{\Pi(\gamma-\alpha-\beta)\,\Pi(-\gamma)}.
\end{aligned}\right.$$

23.
$$\left\{\begin{aligned}
&F = A_3\,F(\alpha,\ \beta,\ \alpha+\beta-\gamma+1,\ 1-x)\\
&\qquad + B_3\,(1-x)^{\gamma-\alpha-\beta}F(\gamma-\alpha,\ \gamma-\beta,\ \gamma-\alpha-\beta+1,\ 1-x),\\
&A_3 = \frac{\Pi(\gamma-1)\,\Pi(\gamma-\alpha-\beta)}{\Pi(\gamma-\alpha-1)\,\Pi(\gamma-\beta-1)},\qquad
B_3 = \frac{\Pi(\gamma-1)\,\Pi(\alpha+\beta-\gamma-1)}{\Pi(\alpha-1)\,\Pi(\beta-1)}.
\end{aligned}\right.$$

24.
$$\left\{\begin{aligned}
&F = A_3\,(1-x)^{1-\gamma}F(\alpha-\gamma+1,\ \beta-\gamma+1,\ 2-\gamma,\ x)\\
&\qquad + B_3\,(1-x)^{-\alpha}F\!\left(\alpha,\ \gamma-\beta,\ \alpha-\beta+1,\ \tfrac{1}{1-x}\right),\\
&A_3 = \frac{\Pi(-\beta)\,\Pi(\alpha-\gamma)\,\Pi(-\beta)}{\Pi(1-\gamma)\,\Pi(\alpha-1)\,\Pi(\gamma-\beta-1)},\qquad
B_3 = \frac{\Pi(-\beta)\,\Pi(\alpha-\gamma)}{\Pi(\alpha-\beta)\,\Pi(-\gamma)}.
\end{aligned}\right.$$

25.
$$\left\{\begin{aligned}
&F = A_4\,(1-x)^{1-\gamma}F(\alpha-\gamma+1,\ \beta-\gamma+1,\ 2-\gamma,\ x)\\
&\qquad + B_4\,(1-x)^{-\beta}F\!\left(\beta,\ \gamma-\alpha,\ \beta-\alpha+1,\ \tfrac{1}{1-x}\right),\\
&A_4 = \frac{\Pi(\gamma-1)\,\Pi(\beta-\gamma)\,\Pi(-\alpha)}{\Pi(1-\gamma)\,\Pi(\beta-1)\,\Pi(\gamma-\alpha-1)},\qquad
B_4 = \frac{\Pi(-\alpha)\,\Pi(\beta-\gamma)}{\Pi(\beta-\alpha)\,\Pi(-\gamma)}.
\end{aligned}\right.$$

26.
$$\left\{\begin{aligned}
&F = A_5\,(1-x)^{-\alpha}F\!\left(\alpha,\ \gamma-\beta,\ \alpha-\beta+1,\ \tfrac{1}{1-x}\right)\\
&\qquad + B_5\,(1-x)^{-\beta}F\!\left(\beta,\ \gamma-\alpha,\ \beta-\alpha+1,\ \tfrac{1}{1-x}\right),\\
&A_5 = \frac{\Pi(\gamma-1)\,\Pi(\beta-\alpha-1)}{\Pi(\beta-1)\,\Pi(\gamma-\alpha-1)},\qquad
B_5 = \frac{\Pi(\gamma-1)\,\Pi(\alpha-\beta-1)}{\Pi(\alpha-1)\,\Pi(\gamma-\beta-1)}.
\end{aligned}\right.$$

Table 1.1.

1) $F(\alpha,\ \beta,\ \gamma,\ x),$

2) $(1-x)^{\gamma-\alpha-\beta}F(\gamma-\alpha,\ \gamma-\beta,\ \gamma,\ x),$

3) $x^{1-\gamma}F(\alpha-\gamma+1,\ \beta-\gamma+1,\ 2-\gamma,\ x),$

4) $x^{1-\gamma}(1-x)^{\gamma-\alpha-\beta}F(1-\alpha,\ 1-\beta,\ 2-\gamma,\ x),$

5) $F(\alpha,\ \beta,\ \alpha+\beta-\gamma+1,\ 1-x),$

6) $x^{1-\gamma}F(\alpha-\gamma+1,\ \beta-\gamma+1,\ \alpha+\beta-\gamma+1,\ 1-x),$

7) $(1-x)^{\gamma-\alpha-\beta}F(\gamma-\alpha,\ \gamma-\beta,\ \gamma-\alpha-\beta+1,\ 1-x),$

8) $x^{1-\gamma}(1-x)^{\gamma-\alpha-\beta}F(1-\alpha,\ 1-\beta,\ \gamma-\alpha-\beta+1,\ 1-x),$

9) $x^{-\alpha}F\!\left(\alpha,\ \alpha-\gamma+1,\ \alpha-\beta+1,\ \tfrac{1}{x}\right),$

10) $x^{-\beta}F\!\left(\beta,\ \beta-\gamma+1,\ \beta-\alpha+1,\ \tfrac{1}{x}\right),$

11) $x^{\alpha-\gamma}(1-x)^{\gamma-\alpha-\beta}F\!\left((1-\alpha,\ \gamma-\alpha,\ \beta-\alpha+1,\ \tfrac{1}{x}\right),$

12) $x^{\beta-\gamma}(1-x)^{\gamma-\alpha-\beta}F\!\left(1-\beta,\ \gamma-\beta,\ \alpha-\beta+1,\ \tfrac{1}{x}\right),$

13) $(1-x)^{-\alpha}F\!\left(\alpha,\ \gamma-\beta,\ \alpha-\beta+1,\ \tfrac{1}{1-x}\right),$

14) $(1-x)^{-\beta}F\!\left(\beta,\ \gamma-\alpha,\ \beta-\alpha+1,\ \tfrac{1}{1-x}\right),$

15) $x^{1-\gamma}(1-x)^{\gamma-\alpha-1}F\!\left(\alpha-\gamma+1,\ 1-\beta,\ \alpha-\beta+1,\ \tfrac{1}{1-x}\right),$

16) $x^{1-\gamma}(1-x)^{\gamma-\beta-1}F\!\left(\beta-\gamma+1,\ 1-\alpha,\ \beta-\alpha+1,\ \tfrac{1}{1-x}\right),$

17) $(1-x)^{-\alpha}F\!\left(\alpha,\ \gamma-\beta,\ \gamma,\ \tfrac{x}{x-1}\right),$

18) $(1-x)^{-\beta}F\!\left(\beta,\ \gamma-\alpha,\ \gamma,\ \tfrac{x}{x-1}\right),$

19) $x^{1-\gamma}(1-x)^{\gamma-\alpha-1}F\!\left(\alpha-\gamma+1,\ 1-\beta,\ 2-\gamma,\ \tfrac{x}{x-1}\right),$

20) $x^{1-\gamma}(1-x)^{\gamma-\beta-1}F\!\left(\beta-\gamma+1,\ 1-\alpha,\ 2-\gamma,\ \tfrac{x}{x-1}\right),$

21) $x^{-\alpha}F\!\left(\alpha,\ \alpha-\gamma+1,\ \alpha+\beta-\gamma+1,\ \tfrac{x-1}{x}\right),$

22) $x^{-\beta}F\!\left(\beta,\ \beta-\gamma+1,\ \alpha+\beta-\gamma+1,\ \tfrac{x-1}{x}\right),$

23) $x^{\alpha-\gamma}(1-x)^{\gamma-\alpha-\beta}F\!\left(1-\alpha,\ \gamma-\alpha,\ \gamma-\alpha-\beta+1,\ \tfrac{x-1}{x}\right),$

24) $x^{\beta-\gamma}(1-x)^{\gamma-\alpha-\beta}F\!\left(1-\beta,\ \gamma-\beta,\ \gamma-\alpha-\beta+1,\ \tfrac{x-1}{x}\right).$

an elliptic integral of the first kind, in Legendre's terminology [1825, 11], and

$$F\left(-\frac{1}{2}, \frac{1}{2}, 1, c^2\right) = \frac{2}{\pi} \int_0^{\pi/2} (1 - c^2 \sin \phi^2)^{1/2} d\phi$$
$$= E^1(c),$$

an elliptic integral of the second kind. Legendre had shown that $F^1(c)$ satisfies

$$(1 - c^2)\frac{d^2 y}{dc^2} + \frac{1 - 3c^2}{c}\frac{dy}{dc} y = 0,$$

and that $E^1(c)$ satisfies

$$(1 - c^2)\frac{d^2 y}{dc^2} + \frac{1 - c^2}{c}\frac{dy}{dc} + y = 0.$$

These equations were now identified by Kummer as special cases of the hypergeometric equation (§29). In the course of deriving them Legendre had deduced his famous relationship between the periods of elliptic integrals:

$$F^1(c)E^1(b) + F^1(b)E^1(c) - F^1(b)F^1(c) = \frac{\pi}{2} \quad \text{(where } b^2 = 1 - c^2\text{)}.$$

Kummer derived this result as a consequence of his theory (§30). As has been remarked, his interest in such matters had been awakened by the related equations for the transformations of elliptic functions found by Jacobi, which had been the subject of his [1834] work.

Kummer concluded the paper with an unremarkable study of what happens when x is allowed to be complex, but α, β, and γ stay real, and is largely devoted to the study of special functions. The first significant advance on Kummer's work was to be made by Riemann, who went at once to a general discussion of the complex case. This will be discussed next, after a brief sketch of its context, for Riemann's ideas about complex function theory form a coherent whole and some of his ideas about Riemann surfaces will be encountered again later on. The contemporary theory of differential equations themselves is described at the end of the chapter. Material, which Riemann expounded in lectures and which seems only to have reached a significant audience well after his death, is discussed in Appendix 2.

1.3   Riemann's approach to complex analysis

The central theme of both Riemann's mathematics and his physics is that of the complex function, understood geometrically. Riemann sought to prise the independent variable of a complex function off the complex plane and free it to roam over a more general surface. This removed an unnecessary constraint upon mathematicians' attitudes to such functions and opened the way to a topological study of their properties. To accomplish this, Riemann gave a purely local definition of a complex function, so that it may be equally well regarded as defined on a patch of surface or on a part of the plane. He then sought global restrictions determining the nature of function under certain given conditions. The same dialectic between local and global properties can be found in other parts of his works not particularly concerned with complex variables, for example, in his work on the foundations of geometry. It differs considerably from the emphasis on convergent power series and the theory of analytic continuation[16] of his influential contemporary Karl Weierstrass. Where Weierstrass worked outwards from a function defined by an infinite series on a disc towards the complete function, and used analytic tools, Riemann sought to anchor his global ideas in the specifics of a given problem. There is an interesting contrast within the study of differential equations between Riemann and Lazarus Fuchs, an exponent of the Berlin school whose work is considered in the next chapter. First, however, Riemann's general view of the theory of functions of a complex variable will be considered.

The papers [1851, 1857c] have become famous for several reasons. They introduced what are now called Riemann surfaces, in the form of domains spread out over the complex plane. They presented enough tools to classify all compact orientable surfaces, and so gave a great impetus to topology.[17] They provided a topological meaning for an otherwise unexplained constant which entered into Abel's work on Abelian integrals (see Chapter V), and more generally gave a geometric framework for all of complex analysis. They are thus the first mature, obscure, papers in the study of topology of manifolds and are equally decisive for the development of algebraic geometry and the geometric treatment of complex analysis. Riemann's own use of his ideas in his study of Abelian functions and integrals on algebraic curves is perhaps the greatest indication of their profundity, and will be described below (Chapter V). This chapter concentrates on the implications of his idea of a complex function for the theory of differential equations, which Riemann presented in an earlier paper of 1857, the *Beiträge zur Theorie der durch Gauss'-che Reihe $F(\alpha, \beta, \gamma, x)$ darstellbaren Functionen* [1857a]. His theory of functions will be expounded in this section, and its use in [1857a] in Section 1.4.

In his inaugural dissertation, [1851], Riemann defined a function of a complex variable in this way (§1): "A complex variable w is called a function of another complex variable z if it varies with the other in such a way that the value of the derivative $\dfrac{dw}{dz}$ is independent of the value of the differential dz." This is equivalent to the modern definition of an analytic function w as being differentiable at z_0 (as a function of z) in such a way that $\dfrac{dw}{dz}$ at z_0 is independent of the path of z as it tends to z_0. The function defines a surface in $\mathbf{C}^2$, which Riemann said may be considered as spread out over the complex z-plane. It follows from the definition that infinitesimal neighbourhoods of z and $w(z)$ are conformally equivalent, as Riemann showed in §3, unless the corresponding variations in z and w cease having a finite ratio to one another, which he said it was tacitly assumed they did not. Riemann observed in a footnote that this matter had been thoroughly discussed by Gauss in [Gauss 1822].

Furthermore, if $w = u + vi$, the functions u and v satisfy the equations

$$\frac{\partial u}{\partial x} = \frac{\partial v}{\partial y}, \frac{\partial v}{\partial x} = -\frac{\partial u}{\partial y}$$

(the Cauchy–Riemann equations) and consequently separately satisfy

$$\frac{\partial^2 u}{\partial x^2} + \frac{\partial^2 u}{\partial y^2} = 0, \frac{\partial^2 v}{\partial x^2} + \frac{\partial^2 v}{\partial y^2} = 0.$$

Riemann remarked (§§2–4) that these equations can be used to study the individual properties of u and v and thus the complex function $w = u + vi$.

However, he said, it is not necessary to assume that z lies in **C**. It may lie in some finite domain T, having a boundary, spread out over **C**, and covering the plane several times. The different parts of surface, covering each region of the plane are joined together at points, but not along lines. Under these conditions the number of times a point is covered is completely determined with the boundary and interior specified, but the form of the covering easy be different. More precisely, the surface winds around various branch points at which the various 'leaves' (*Flächentheile*, literally 'pieces of surface') are interchanged in cycles. These may be considered as copies of parts of the plane (cut by a line emanating from the branch point) and joined up according to a certain rule.

A point at which m leaves are interchanged was said by Riemann to have order $m - 1$. Once these branch points are determined, so is the surface (up to a finite number of different shapes deriving from the arbitrariness in the choice of original leaf in each cycle). Functions can be defined on T provided there is not a line of exceptional points (in which case the function would not be differentiable). In a footnote, Riemann observed that this restriction on the set of singular points did not derive from the idea of a function, but from the conditions under which the integral calculus can be applied. He gave as an example of a function discontinuous everywhere in the (x, y) plane the function which takes the value 1 when x and y are commensurable and the value 2 otherwise. Dirichlet [1829] had given the example of a function which takes the value c and $d \neq c$ on the irrationals.

The connectivity of T is important and Riemann proceeded as follows. Two parts of a surface were said to be connected if any point of one can be joined to any point of the other by a curve lying entirely in the surface. Strictly, this defines the modern concept of path connected, but the notions of connectedness and path connectedness agree here, since Riemann's surfaces are manifolds. A boundary cut (*Querschnitt*) is a curve which joins two boundary points without cutting itself. A (bounded) connected surfaces was said by Riemann to be simply connected if any boundary cut makes it disconnected; it then falls into two simply connected pieces. It was said to be n-fold connected if it can be made simply connected by $n - 1$ suitably chosen boundary cuts; the number n is well-defined and independent of the choice of cuts [§6]. The purpose of introducing these concepts was to generalise Cauchy's theorem on contour integration to T. In general the integral of complex function taken around a closed curve in T which contains no poles of the function does not vanish, and Cauchy's theorem is true only when T is simply connected.[18] From this observation the whole of the elementary theory of analytic functions can be generalised to functions satisfying the Cauchy–Riemann equations locally on T. For instance, such functions, w, are infinitely differentiable, and locally one-to-one except near a branch point. Near a branch point z' of order $(n - 1)$ the function becomes one-to-one if $(z - z')^{1/n}$ is taken as the new variable. The image of T is again a surface, S, and the inverse of w is an analytice function $z = z(w)$ [§15].

Riemann was prepared to use power series or Fourier series methods to express a function locally. He described such methods in his [1857c] as standard techniques. Where he differed from Weierstrass was in his emphasis on geometrical reasoning, which was avoided in Berlin. In particular, Riemann was prepared to use Dirichlet's principle to guarantee the existence of functions without seeking them necessarily in any other form. For him a function was known once a certain topological property was known about it (the connectivity of the surface it defined) and once certain discrete facts were known (the net of its singular points). The careful separation of these two kinds of data is characteristic of Riemann, and even more so is his brilliant yoking together of the two. It occurs again, for example, in his study of the zeta function $\zeta(s) = \sum n^{-s} (\mathrm{Re}(s) > 2)$ [1859]. In a more diffuse form, the polarity of continuous and discrete haunts nearly all his work. It also manifests itself as an interplay between global and local properties, and it is in this guise that it appears in his work on differential equations, where it will be seen that he immediately looked for the number of leaves and for the branch points of the solution functions.

1.4   Riemann's P-functions

The paper discussed in this section presents Riemann's analysis, [1857a], of the hypergeometric functions as P-functions (defined below). Riemann's remarkable extension of the theory of the hypergeometric equation as given in his lecture notes of 1858/1859 will be discussed in Appendix 2.

It will be recalled that Kummer had found 24 solutions to the hypergeometric equation in the form of a hypergeometric series in x, $\frac{1}{x}$, $1 - x$, etc., possibly multiplied by some powers of x or $(1 - x)$. Each solution was therefore presented in a form which restricted it to a certain domain and the relationship between overlapping solutions was given. In [1857a] Riemann observed that this method of passing from the series to the function it represents depends on the differential equation itself. It would be possible, he said, to study the solutions expressed as definite integrals, although the theory was not yet sufficiently developed (a task soon to be taken up by Schläfli [1870] and by Riemann himself in lectures [1858/59, *Nachträge* 69– 94]). However, he proposed to study the hypergeometric equation according to his new, geometric methods, which were essentially applicable to all linear differential equations with algebraic coefficients — a rather dramatic claim.

Riemann began by specifying geometrically the functions he intended to study. Any such function P is to satisfy three properties:

1. It has three distinct branch points at a, b, and c, but each branch is finite at all other points;

2. A linear relation with constant coefficients exists between any three branches P', P'', P''', of the function: $c'P' + c''P'' + c'''P''' = 0$;

3. There are constants α and α', called the exponents, associated with the branch point a, such that P can be written as a linear combination of two branches $P^{(\alpha)}$ and $P^{(\alpha')}$ near a, $(z - a)^{-\alpha} P^{(\alpha)}$ and $(z - \alpha)^{-\alpha'} P^{(\alpha')}$ are single-valued, and neither zero nor infinite at a. Similar conditions hold at b and c with constants β, β' and γ, γ', respectively.

To eliminate troublesome special cases Riemann further assumed that none of $\alpha - \alpha'$, $\beta - \beta'$, $\gamma - \gamma'$ are integers, and that furthermore the sum $\alpha + \alpha' + \beta + \beta' + \gamma + \gamma' = 1$. He denoted such a function of z

$$P \left\{ \begin{matrix} a & b & c \\ \alpha & \beta & \gamma & z \\ \alpha' & \beta' & \gamma' \end{matrix} \right\} \quad \text{or} \quad P \left(\begin{matrix} \alpha & \beta & \gamma \\ \alpha' & \beta' & \gamma' \end{matrix} \right) \quad \text{when} \quad (a, b, c) = (0, \infty, 1).$$

The first and third conditions express the nature of a P-function, as Riemann called them, in terms of the singularities at a, b, and c: for example, $P^{(\alpha)}$ is branched like $(x - a)^{\alpha}$. The second condition expresses the global relationship between the leaves and say are at most two linearly independent determinations of the function under analytic continuation of the various separate branches. It is not immediately clear that this information specifies a function exactly; in fact it turns out that it defines P up to a constant multiple.

When a, b, and c take the value 0, ∞, and 1, respectively, as they may be assumed to do without loss of generality, the analogy between P-functions and hypergeometric functions becomes clear. There are two linearly independent solutions of the hypergeometric equation at each singular point; they are branched according to certain expressions in α, β, and γ, and any three solutions are linearly dependent. Riemann showed that information of this kind about the solutions determines the differential equation completely. This goes some way to explain the great significance of the equation, and to illuminate the difficulties we shall see others were to experience in generalizing from it to other equations.

To investigate the global behaviour of P-functions under analytic continuation, Riemann argued that it is enough to specify their behaviour under circuits of the branch points. When two linearly independent branches P' and P'', say, are continued analytically in a loop around the branch point a in the positive (anti-clockwise) direction, they return as two other branches, $\tilde{P}'$ and $\tilde{P}''$. But then

$$\tilde{P}' = a_1 P' + a_2 P''$$
$$\tilde{P}'' = a_3 P' + a_4 P''$$

for some constants a_1, a_2, a_3, a_4, so in some sense the matrix

$$A = \begin{pmatrix} a_1 & a_2 \\ a_3 & a_4 \end{pmatrix}$$

describes what happens at a. Let B and C be the matrices which describe the behaviour of P', P'' under analytic continuation around b and c, respectively.[19] A circuit of a and b can be regarded as a circuit of c in the opposite direction, so

$$CBA = \begin{pmatrix} 1 & 0 \\ 0 & 1 \end{pmatrix}.$$

Any closed path can be written as a product of loops around a, b, or c in the same order, or, as Riemann remarked "the coefficients of A, B, and C completely determine the periodicity of the function." The choice of the word "periodicity" (*Periodicität*) is interesting, suggesting a connection with doubly-periodic functions and the moduli (*Periodicitätsmoduln*) of Abelian functions via the use of closed loops.

The idea of using such a matrix to describe how an algebraic function is branched had been introduced by Hermite [1851], but it seems that Riemann is the first to have considered products of such matrices. We shall see that he effectively determined the monodromy group of the hypergeometric equation; i.e., the group generated by the matrices A, B, C. The term "monodromy group" was first used by Jordan [*Traité*, 278] and its subsequent popularity derives from its successful use by Jordan and Klein, to be discussed in Chapter III.

For definiteness Riemann supposed $a = 0$, $b = \infty$, $c = 1$, and chose branches P^α, $P^{\alpha'}$, P^β, $P^{\beta'}$, P^γ, $P^{\gamma'}$ as in (3) above. A circuit around a in the positive direction returns P^α as $e^{2\pi i \alpha} P^\alpha$, and $P^{\alpha'}$ as $e^{2\pi i \alpha'} P^{\alpha'}$, so

$$A = \begin{pmatrix} e^{2\pi i \alpha} & 0 \\ 0 & e^{2\pi i \alpha'} \end{pmatrix}.$$

To express the effect on P^α and $P^{\alpha'}$ of a circuit around $b = \infty$, he replaced them by their expressions in terms of P^β and $P^{\beta'}$, conducted the new expressions around ∞, and then changed them back into P^α and $P^{\alpha'}$, by writing

$$P^\alpha = \alpha_\beta P^\beta + \alpha_{\beta'} P^{\beta'}$$
$$P^{\alpha'} = \alpha'_\beta P^\beta + \alpha'_{\beta'} P^{\beta'},$$

or more briefly[20]

$$\begin{pmatrix} P^\alpha \\ P^{\alpha'} \end{pmatrix} = B' \begin{pmatrix} P^\beta \\ P^{\beta'} \end{pmatrix}.$$

Then

$$B = B' \begin{pmatrix} e^{2\pi i \beta} & 0 \\ 0 & e^{2\pi i \beta'} \end{pmatrix} B'^{-1}.$$

By similarly conducting P^α and $P^{\alpha'}$ around $c = 1$, Riemann found

$$C = C' \begin{pmatrix} e^{2\pi i \gamma} & 0 \\ 0 & e^{2\pi i \gamma'} \end{pmatrix} C'^{-1},$$

where C' is the matrix relating P^α and $P^{\alpha'}$ to P^γ and $P^{\gamma'}$:

$$\begin{pmatrix} P^\alpha \\ P^{\alpha'} \end{pmatrix} = C' \begin{pmatrix} P^\gamma \\ P^{\gamma'} \end{pmatrix}.$$

Since $CBA = I$, it follows on taking determinants that

$$\det(C)\det(B)\det(A) = 1 = e^{2\pi i(\alpha+\alpha'+\beta+\beta'+\gamma+\gamma')},$$

so $\alpha + \alpha' + \beta + \beta' + \gamma + \gamma'$ must be an integer. Riemann assumed it was in fact 1.

Riemann showed that the entries in B' and C' could be expressed in terms of the six coefficients $\alpha, \ldots, \gamma'$. In the published paper he contented himself with expressions relating the various ratios, $\dfrac{\alpha_\gamma}{\alpha'_\gamma}, \dfrac{\alpha_\beta}{\alpha'_\beta}$, etc., but the precise values of $\alpha_\beta, \ldots, \alpha'_{\gamma'}$ can also be determined; Riemann had obtained them himself in July of the previous year. It is most convenient to do so by obtaining the differential equation which the P-function satisfies and thereafter comparing its various branches with various branches of the 24 solutions obtained by Kummer. To do this, one must first see that the nine quantities $a, b, c, \alpha, \ldots, \gamma'$ define P up to a constant multiple.

Riemann did this in §4 of his paper. The method involves showing from the definition of a P-function that any two with the same $a, b, c, \alpha, \ldots, \gamma'$ have a constant quotient.

The same argument also shows that given two branches P' and P'' of a P-function, whose quotient is not a constant, any other P-function having the same branch points and the same exponents can be expressed linearly as a sum $C'P' + C''P''$.

Riemann went on to study P-functions whose exponents $\alpha, \ldots, \gamma'$ and $\tilde{\alpha}, \ldots, \tilde{\gamma}'$ say, differ, by integers. These are Gauss' contiguous functions transformed to Riemann's setting.

In this way he could determine which differential equation the P-function satisfies, since,

$$P = y, \quad P_1 = \frac{dy}{dx}, \quad \text{and} \quad P_2 = \frac{d^2y}{dx^2}$$

are three such P-functions and, Riemann showed, they satisfy a linear relationship with certain rational functions in x as coefficients. When $\gamma = 0$ he found explicitly that P satisfies the hypergeometric equation in this form:

$$(1-z)\frac{d^2y}{d\log z^2} - (A+Bz)\frac{dy}{d\log z} + (A'-B'z)y = 0.$$

Riemann could therefore connect his P-functions with the functions $F(\alpha, \beta, \gamma, x)$ of Gauss:

$$F(a, b, c, z) = \text{const. } P^\alpha \begin{pmatrix} 0 & a & 0 \\ 1-c & b & c-a-b \end{pmatrix} z \end{pmatrix}.$$

So Riemann had shown that the branching data of the P-function determined its monodromy relations (later generations would say determine the monodromy group of the equation explicitly), and moreover, he had established that the hypergeometric equation is the only second order linear equation whose solutions satisfy the geometric conditions of his three postulates.

Riemann gave neither the general form of the differential equation satisfied by

$$P \left\{ \begin{matrix} a & b & c \\ \alpha & \beta & \gamma & z \\ \alpha' & \beta' & \gamma' \end{matrix} \right\}$$

— a task later accomplished by Papperitz [1889] — nor a full discussion of the P-function as an Euler integral with suitable paths for integration – matters taken up by Pochhammer [1889], Jordan [1915, 251], and Schläfli [1870].

Riemann concluded by illuminating the relationship between P and F, its hypergeometric series representation. Since α and α' may be interchanged,

$$P \left(\begin{matrix} \alpha & \beta & \gamma \\ \alpha' & \beta' & \gamma' \end{matrix} z \right) = P \left(\begin{matrix} \alpha' & \beta & \gamma \\ \alpha & \beta' & \gamma' \end{matrix} z \right);$$

there are 8 P-functions for each hypergeometric series in z, say

$$P \left(\begin{matrix} \alpha & \beta & \gamma \\ \alpha' & \beta' & \gamma' \end{matrix} z \right) = z^\alpha (1-z)^\gamma F(\beta + \alpha + \gamma, \beta' + \alpha + \gamma, \alpha - \alpha' + 1, z).$$

There are six choices of variable, so 48 representations of a function as a P-function.

Concluding comments

The most immediate difference between Riemann and his predecessors is the relative lack of computation. Rather than starting from a hypergeometric series, he began with a P-function having ∞^2 branches and 3 branch-points. To be sure, any two linearly independent branches have expansions as hypergeometric series, and any three branches are linearly dependent, but the argument employed by Riemann inverted that of Gauss and Kummer. His starting point was the set of solutions, functions that are shown to satisfy a certain type of equation. Their starting point was the equation from which a range of solutions are derived. Riemann showed that a very small amount of information, the six exponents at the branch points, entirely characterises the equation and defines the behaviour of the solutions. The hypergeometric equation is special in this respect, as Fuchs [1865] was able to explain, and consequently the task of generalizing the theory to cope with other differential equations was to be quite difficult.

The crucial step in Riemann's argument is his use of matrices to capture the behaviour of the solutions in the neighbourhood of a branch point. These monodromy matrices enable one to study the global nature of the solutions on analytic continuation around arbitrary paths, and were introduced by Hermite in response to Puiseux's work [1850, 1851]. Puiseux had considered the effect on a branch of an algebraic function of analytically continuing it around one of its branch-points,

and also the effect of integration an algebraic function over a closed path containing a branch point. Cauchy also reported on this work [Cauchy, 1851]. Naturally,
the question of influence arises. Riemann was sparing with references especially to
contemporaries. Quick to mention Gauss, especially in works Gauss was to examine, and always willing to acknowledge his mentor Dirichlet, he otherwise usually
contented himself with general remarks of a historical kind which might set the
scene mathematically. Yet, as Brill and Noether said [1892–93, 283]: Riemann
"... everywhere betrayed an exact knowledge of the literature..." and, they continued, (p. 286) "... evidently the fundamental researches of Abel, Cauchy, Jacobi,
Puiseux provide the starting point" for his research on algebraic curves. We must
suppose Riemann was one of those mathematicians who absorbed the work of others and then re-derived it in his own way. He did in fact mention Cauchy fleetingly
when discussing the development of a function in power series, but never Puiseux
by name.[21] However, he tells us that his first work on Abelian functions was carried
out in 1851–52 [1857c, 102], and that would fit well with a contemporary study of
related French work.

The connection between Riemann's study of the hypergeometric equation itself
and those of Gauss and Kummer was, fortunately, a matter on which he chose to be
explicit. In the Personal Report, [1857b], on his [1857a], which he submitted to the
Göttinger Nachrichten, he wrote. "The unpublished part of the Gauss' study on this
series, which has been found in his *Nachlass*, was already supplemented in 1835
by the work of Kummer contained in the 15th volume of *Crelle's Journal*". This
also makes clear that Riemann had spent some of 1856 looking at the Gauss treasure
trove, while working on his own ideas. However, we know from his lectures, see for
example [*Werke* 379–390], that Riemann considered the hypergeometric equation
from the standpoint of the linear differential equations with algebraic coefficients,
which gave the level of generality he most wanted to attain (see Appendix 2).

1.5   Cauchy's theory of the existence of solutions of a differential equation

Cauchy's study of differential equations has been considered frequently by historians of mathematics, notably Freudenthal [1971], Kline [1972] and Dieudonné
[1978]. I follow here the thorough discussion of C. Gilain ([1977] and [1991])
and in his introduction to Cauchy [1981].

Cauchy was critical of his contemporaries' use of power series methods to solve
differential equations, on the grounds that a series so obtained may not necessarily
converge, and if it does converge, it need not represent the solution function. He
offered two methods for solving such equations rigorously. The first, given in lectures at the École Polytechnique in 1823 and published in 1841, [Cauchy 1841] is
a method of approximation by difference equations. It is nowadays known as the

Cauchy–Lipschitz method, and it applies to equations

$$\frac{dy}{dx} = f(x, y)$$

where f and $\frac{\partial f}{\partial y}$ are bounded, continuous functions on some rectangle. He suggested that this theory might be extended to systems of such equations.

The second method [1835] treats a system of first order equations by passing to a certain partial differential equation. Candidates for the solution arise as power series, and Cauchy established their convergence by his "calcul des limites" or, in modern terminology, the method of majorants. In this method a series is shown to be convergent if its nth term is less in modulus than the nth term of a series known to be convergent; Cauchy usually took a geometric series for the majorizing series.

Gilain [1977, 14] notes that the second method supplanted the first in the minds of nineteenth-century mathematicians, and that both were taken to establish the same theorem, although in fact they apply to different types of equations (only the second is analytic). This is in keeping with a tendency throughout the later nineteenth century to regard variables as complex rather than real. Cauchy's preference in this matter is well known; indeed he is the principal founder of the theory of functions of a complex variable, but it was often, if tacitly, understood that function theory really meant complex function theory. Since the eighteenth century had regarded functions as (piecewise) analytic, the critical spirit of the nineteenth century at first provided a rigorous theory of analytic functions. Cauchy's own example of e^{-1/t^2}, for which the Maclaurin series at $t = 0$ is identically zero, showed that such a theory would necessarily be complex analytic (solving his second objection to the older solution methods for differential equation referred to above).

Cauchy's approach was extended by Briot and Bouquet [1856b] to consider the singularities of the solution functions. These can arise when the coefficients of the equation are singular, but may also occur elsewhere if the equation is non-linear. French mathematicians of the 1850s were much concerned with the nature of an analytic function near its singular points; Puiseux and Cauchy had already studied the singularities of algebraic functions in 1850 and 1851. Briot and Bouquet began by simplifying Cauchy's existence proof using the method of majorants, and then considered cases where it breaks down because $f(y, z)$ becomes undetermined in the equation $\frac{dy}{dz} = f(y, z)$. At such points the solution function may have a branch point and also a pole, even an essential singularity. If it has a branch point it is not single-valued, or "monodrome" in their terminology. But unless it has an essential singularity at the point in question (which may be taken to be $z = 0$) it still has a power series expansion of the form

$$z^{\alpha} \sum_{-k}^{\infty} a_n z^n, \, a^{-k} \neq 0.$$

Briot and Bouquet made a particular study [1856c] of equations of the form

$$F\left(u, \frac{du}{dz}\right) = 0,$$

where F is a polynomial, which have elliptic functions as their solutions. For such equations, the solutions have finite poles but no branch points. The nature of the singular points of an analytic function was to remain obscure for a long time. Neuenschwander [1978a, 143] established that the point $z = \infty$ was treated ambiguously by Briot and Bouquet, and the nature of an essential singularity remained obscure until the Casorati–Weierstrass theorem was published. Near an essential singularity the function has a series expansion

$$\sum_{n=-\infty}^{\infty} a_n z^n,$$

but neither f nor $1/f$ can be defined at the singular point; for example, e^z near $z = \infty$, can be regarded as

$$e^{1/t} = \sum_{n=-\infty}^{0} \frac{t^n}{(-n)!}$$

near $t = 0$ and has an essential singularity at $t = 0$. In France, Laurent [1843] had drawn attention to these points; however they had by then received a much more thorough treatment in Germany at the hands of Karl Weierstrass.

Weierstrass developed his theory independently of both Laurent and Cauchy. In his [1842] he studied the system of differential equations

$$\frac{dx_i}{dt} = G_i(x_1, \ldots, x_n) \qquad 1 \le i \le n,$$

where the G_i are rational functions and obtained solutions in the form of power series convergent on a certain domain. To show convergence he took power series expansions of functions $\tilde{G}_i$, which are obtained from the G_i by replacing the coefficients a_{ij} of G_i by positive numbers $a_{ij} \ge |a_{ij}|$, and dominated the $\tilde{G}_i$ by suitably chosen geometric progression. Similar methods are still used today. This paper of Weierstrass was left unpublished until 1894, although he lectured on related topics involving Abelian functions at Berlin in 1863. In these lectures Weierstrass connected the singular points with the domain of validity of a power series expansions and introduced the idea of the analytic continuation of a function outside its circle of convergence. The matter is discussed further in Chapters II and V.

Exercises

The theory of elliptic functions is a vast topic which provides an origin for many of the topics discussed in this book. However, it is not necessary to know anything about that theory to appreciate the story, nor is it difficult to find it out. These exercises cover most of the salient features of the elementary theory of elliptic integrals and functions, and outline the accompanying historical developments, which are discussed in more detail in Houzel [1978].

1. Show that the formula for the element of an arc of the ellipse

$$\frac{x^2}{a^2} + \frac{y^2}{b^2} = 1$$

as a function of x, obtained by using the formula

$$ds^2 = dx^2 + dy^2, \text{ is } ds^2 = \left(\frac{a^2 - e^2 x^2}{a^2 - x^2}\right) dx^2, \text{ where } e^2 = \frac{a^2 - b^2}{a^2}.$$

This integrand cannot in fact be integrated in terms of elementary functions. Failure to do so drove mathematicians to try various other solutions. Newton (1669) gave a power series expansion for the arc-length in terms of x, Euler (1733) and MacLaurin (1742) expansions in terms of e^2.

2. Suppose $x = a \sin \phi$, $y = b \cos \phi$. Show that

$$ds^2 = a^2(1 - e^2 \sin^2 \phi)d\phi^2.$$

The curious term "lemniscatic" derives from the study of a curve, introduced by Jakob Bernoulli in 1694, which can be written in polar coordinates as $r^2 = a^2 \cos 2\phi$.

3. Show that the Cartesian equation of the lemniscate is

$$(x^2 + y^2)^2 = a^2(x^2 - y^2),$$

and that its element of arc in terms of r, using the formula

$$ds^2 = dr^2 + r^2 d\phi^2, \text{ is } ds^2 = \left(\frac{a^4}{a^4 - r^4}\right) dr^2.$$

4. The expression for the arc-length of the lemniscate recalls that for the circle,

$$ds^2 = \frac{a^2}{(a^2 - r^2)} dr^2.$$

In 1752, Euler considered the evident integration of the differential equation

$$\frac{dx}{(1 - x^2)^{1/2}} = \frac{dy}{(1 - y^2)^{1/2}},$$

which yields

$$x^2 + y^2 = c^2 + 2xy(1 - c^2)^{1/2},$$

where c is an arbitrary constant. Prove that this is a solution of the differential equation and show that when it is interpreted in terms of the integrals

$$\int_0^{\sin\phi} \frac{dx}{(1-x^2)^{1/2}} = \phi,$$

etc., for which

$$\int_0^y \frac{dy}{(1-y^2)^{1/2}} = \int_0^x \frac{dx}{(1-x^2)^{1/2}} + \int_0^c \frac{dc}{(1-c^2)^{1/2}},$$

one obtains

$$\sin(\phi + \theta) = \sin\phi\cos\theta + \cos\phi\sin\theta.$$

5. The formulae in (4) are an *algebraic addition theorem* for the integrals, since c is presented as an algebraic function of x and y, and since $\sin(\theta + \phi)$ as an algebraic function of $\sin\theta$ and $\sin\phi$. Euler went on to consider the lemniscatic integrals, for which a duplication formula had been found earlier by Fagnano. Fagnano considered the substitutions

$$r^2 = \frac{2u^2}{1+u^4} \quad \text{and} \quad u^2 = \frac{2v^2}{1-v^4}.$$

Show that they lead to

$$\frac{dr^2}{1-r^4} = \frac{2du^2}{1+u^4} \quad \text{and} \quad \frac{du^2}{1+u^4} = \frac{2dv^2}{1-v^4}.$$

So

$$\frac{dr}{(1-r^4)^{1/2}} = \frac{2dv}{(1-v^4)^{1/2}},$$

from which it follows that if

$$\int_0^R \frac{dr}{(1-r^4)^{1/2}} = S,$$

then

$$\int_0^V \frac{dv}{(1-v^4)^{1/2}} = S/2,$$

so V is half-way along the arc from 0 to R, where V is a known (algebraic) function of R.

6. Euler found that the general solution of the differential equation

$$\frac{dx}{(1-x^4)^{1/2}} = \frac{dy}{(1-y^4)^{1/2}}$$

was $x^2 + y^2 = c^2 + 2xy(1-c^4)^{1/2} - c^2x^2y^2$ for an arbitrary constant c. Verify this. Euler was not satisfied with this rather artificial method of solving the equation, and was much impressed by the young Lagrange's direct method, which I omit.

7. Deduce from (4) that sin is periodic, with period

$$2\pi = 4 \int_0^1 \frac{dt}{(1-t^2)^{1/2}}.$$

8. The conclusions to be drawn from the above exercises are that

$$u = \int_0^x \frac{dt}{(1-t^2)^{1/2}}$$

does not define a single-valued function of x, since $u(x)$ is only well-defined up to a multiple of 2π, but that $x = x(u)$ is a well defined function (the sine function), which moreover is periodic; $\sin(u) = \sin(u + 2\pi n)$, and satisfies an algebraic addition theorem.

What about the similar integral

$$v = \int_0^x \frac{dt}{(1-t^4)^{1/2}}?$$

Deduce from Euler's algebraic addition theorem that it is also periodic, with period

$$2w = 4 \int_0^1 \frac{dt}{(1-t^4)^{1/2}}.$$

Whenever x is real and the path of integration is real, it defines a single-valued function of v, as Legendre was clearly aware. The decisive step is to let x become complex.

Write down the equation for $\sin 2u$ as a function of $\sin u$, and for $\sin nu$, say when n is odd.

9. Deduce from (5) or (6) that there are 8 values of $x(u)$ defined by

$$u = \int_0^x \frac{dt}{(1-t^4)^{1/2}}$$

for a given value of $x(2u)$.

It may be shown that when n is odd, there are either n^2 or $2n^2$ solutions, and so the problem arises: explain the "extra" solutions. Gauss seems to have seen at once that the answer was to admit complex periods.

10. Set $t = it$ in

$$v = \int_0^x \frac{dt}{(1-t^4)^{1/2}},$$

and deduce a connection between $x(v)$ and $x(iv)$: $x(iv) = ix(v)$.

Gauss called the function $x(v)$ the sinus lemniscaticus, $sl(v)$. Since he had shown $sl(u) = sl(u + m2w)$, $m \in \mathbf{Z}$, and $sl(iv) = isl(v)$, he could, by the addition theorem, define $sl(u + iv)$, and deduce that it was doubly periodic, indeed $sl(z + 2(m + ni)w) = sl(z)$ for any complex z and any "Gaussian integer" $m + ni$. The periods are $2w$ and $2iw$. The n values of z such that $\sin nx$ has a given value lie along the interval $[0, 2\pi]$. The n^2 values of z such that $sl(z)$ has a given value are distributed regularly in the square $[0, 2w] \times [0, 2w]$.

11. Show, by differentiating under the integral with respect to (k^2), that

$$K = \int_0^{\pi/2} (1 - k^2 \sin^2 \theta)^{-1/2} d\theta = \frac{\pi}{2} F\left(\frac{1}{2}, \frac{1}{2}, 1, k^2\right)$$

and

$$K' = \int_0^{\pi/2} (1 - k'^2 \sin^2 \theta)^{-1/2} d\theta = \frac{\pi}{2} F\left(\frac{1}{2}, \frac{1}{2}, 1, k'^2\right),$$

where $1 = k^2 + k'^2$ both satisfy the same hypergeometric equation.

12. Show that the substitution $t^2 = \sin^2 \theta + k^2 \cos^2 \theta$ establishes

$$K' = \int_1^{1/k} [(t^2 - 1)(1 - k^2 t^2)]^{-1/2} dt.$$

13. Show that

$$E := \int_0^{\pi/2} (1 - k^2 \sin^2 \theta)^{1/2} d\theta = \frac{\pi}{2} F\left(\frac{1}{2}, -\frac{1}{2}, 1, k^2\right),$$

and

$$E' := \int_0^{\pi/2} (1 - k'^2 \sin^2 \theta)^{1/2} d\theta = \frac{\pi}{2} F\left(\frac{1}{2}, -\frac{1}{2}, 1, k'^2\right)$$

where $k^2 + k'^2 = 1$, satisfy

$$EK' + E'K - KK' = \left(K'\frac{dK}{dk^2} - K\frac{dK'}{dk^2}\right).$$

Show that

$$\lim_{x \to 1} 2x(1 - x)\left(K'\frac{dK}{dk^2} - K\frac{dK'}{dk^2}\right) = \pi \lim_{x \to 1} x(1 - x)\frac{dK}{dk} = \frac{\pi}{2},$$

and deduce Legendre's relation

$$EK' + E'K - KK' = \frac{\pi}{2}.$$

14. Deduce some of Gauss' realizations of the hypergeometric series, e.g.,

(i) $(t + u)^n = t^n F\left(-n, \beta, \beta, \dfrac{-u}{t}\right)$

(ii) $\log(1 + t) = t F(1, 1, 2, -t)$

(iii)

$$e^t = \lim_{k \to \infty} F\left(1, k, 1, \frac{t}{k}\right) = \lim_{k \to \infty} \left\{1 + t F\left(1, k, 2, \frac{t}{k}\right)\right\}$$

$$= \lim_{k \to \infty} \left\{1 + t + \frac{1}{2}t^2 F\left(1, k, 3, \frac{t}{k}\right)\right\} \text{ etc.,}$$

so

(iv) $e^t + e^{-t} = \displaystyle\lim_{k, k' \to \infty} 2F\left(k, k', \frac{1}{2}, \frac{t^2}{4kk'}\right)$

(v) $\sin t = \displaystyle\lim_{k, k' \to \infty} t F\left(k, k', \frac{3}{2}, -\frac{t^2}{4kk'}\right)$

(vi) $t = \sin t \, F\left(\frac{1}{2}, \frac{1}{2}, \frac{3}{2}, \sin^2 t\right)$

[Gauss wrote: k, k' "numeros infinite magnos," infinitely great numbers.]

15. Show that Gauss' function $M(1, x)$ is a continuous and indeed a differentiable function of x.

16. The Bernoulli numbers were defined on p. 14 by an expansion like

$$\frac{x}{e^x - 1} = \sum \frac{b_n}{n!} x^n.$$

Show that $b_0 = 1$, $b_1 = -\frac{1}{2}$, and other $b_n = 0$ if n is odd and greater than 1, by showing that

$$\frac{x}{e^x - 1} = \frac{-x}{2} + \frac{x}{2} \cdot \frac{e^x + 1}{e^x - 1}$$

$$= \frac{-x}{2} - \frac{x}{2} \cdot \left(\frac{e^{-x} + 1}{e^{-x} - 1}\right).$$

Use the expansion

$$x = \left(\sum \frac{b_n}{n!} x^n\right) \left(x + \frac{x^2}{2!} + \frac{x^3}{3!} + \cdots\right)$$

to express b_n in terms of $b_0, \ldots, b_{n-1}$. Deduce that b_n is rational.

17. There is an infinite product expansion for sine, due to Euler:

$$\sin \pi z = \pi z \Pi \left(1 - \frac{z^2}{n^2} \right).$$

Find its logarithmic derivative, which is an infinite series for $\pi z \cot \pi z$, and show from the definition of cot in terms of exponentials, that

$$\pi z \cot \pi z = \sum_{0}^{\infty} \frac{(2\pi)}{(2m)!} B_{2m} z^{2m}.$$

Deduce that

$$\zeta(2k) := \sum_{n} \frac{1}{n^{2k}} = -\frac{(2\pi)^{2k}}{2(2k)!} B_{2k},$$

which gives the exciting result that

$$\frac{1}{(2\pi)^{2k}} \zeta(2k)$$

is rational.

18. The only meromorphic maps of the Riemann sphere to itself are of the form

$$z \rightarrow \frac{az + b}{cz + d}.$$

These maps are called Möbius, or fractional linear, transformations. Show that they form a group.

19. Show that Möbius transformations preserve the cross-ratio of four points; i.e., if $z_i \rightarrow z_i'$, $i = 1, 2, 3, 4$, then

$$(z_1, z_2, z_3, z_4) = (z_1', z_2', z_3', z_4'),$$

where

$$(z_1, z_2, z_3, z_4) = \frac{(z_1 - z_2)(z_3 - z_4)}{(z_1 - z_4)(z_3 - z_2)}.$$

20. Deduce from Question 19 that Möbius transformations send circles and straight lines to circles and straight lines.

21. Show that if $(z_1, z_2, z_3, z_4) = \lambda$, then the possible values of any permutations of z_1, z_2, z_3, z_4 are

$$\lambda, \frac{1}{\lambda}, 1 - \lambda, \frac{1}{1 - \lambda}, 1 - \frac{1}{\lambda}, \frac{\lambda}{\lambda - 1}.$$

22. Show that there is a subgroup, K, of S_4, the group of all permutations of the symbols (z_1, z_2, z_3, z_4), which preserves the cross-ratio (z_1, z_2, z_3, z_4), is of order 4, and is normal in S_4.

23. Summarize your findings in Questions 21 and 22 by showing that there is a homomorphism from S_4 onto S_3 with kernel K.

24. Obtain Papperitz's result [1889]: the differential equation satisfied by

$$P \left\{ \begin{matrix} a & b & c \\ \alpha & \beta & \gamma & z \\ \alpha' & \beta' & \gamma' \end{matrix} \right\}$$

is

$$\frac{d^2 w}{dz^2} + \sum \frac{1 - \alpha - \alpha'}{z - a} \frac{dw}{dz} + \sum \frac{\alpha\alpha'(a - b)(a - c)}{(z - a)} \cdot \frac{w}{(z - a)(z - b)(z - c)} = 0$$

by showing firstly that the equation is satisfied by two linearly independent determinations of P, say P_1 and P_2, so it is of the form

$$\frac{d^2 w}{dz^2} + p \frac{dw}{dz} + qw = 0,$$

where

$$p = -\frac{P_2 P_1'' - P_1 P_2''}{P_2 P_1' - P_1 P_2'}$$

and

$$q = \frac{P_2' P_1'' - P_1' P_2''}{P_2 P_1' - P_1 P_2'}.$$

(Hint: write down a suitable determinant).

Now deduce that

$$p = \frac{1 - \alpha - \alpha'}{z - a} + u_1(z)$$

near $z = a$ from the monodromy relation of $z = a$, and therefore that

$$p = \frac{1 - \alpha - \alpha'}{z - a} + \frac{1 - \beta - \beta'}{z - b} + \frac{1 - \gamma - \gamma'}{z - c} + u_2(z),$$

where $u_2(z)$ is analytic at infinity. Since p is analytic at infinity,

$$p = \frac{2}{z} + 0(z^{-2});$$

deduce that $u_2(z) = 0$, and $\alpha + \alpha' + \beta + \beta' + \gamma + \gamma' = 1$. Conduct a similar argument to find q; $q(z) = 0(z^{-4})$.

25. Which of the following differential equations are hypergeometric?

(i) Legendre's equation:

$$(1 - x^2)\frac{d^2 y}{dx^2} - 2x\frac{dy}{dx} + n(n + 1)y = 0;$$

(ii) Bessel's equation:

$$x^2 \frac{d^2 y}{dx^2} + x \frac{dy}{dx} + (x^2 - n^2)y = 0;$$

(iii) A second order linear differential equation with constant coefficients.

26. (i) Show that, when n is not an integer, two independent solutions of Legendre's equation are

$$y_1 = P_n(x) = F\left(n + 1, -n, 1, \frac{1-x}{2}\right),$$

and $y_2 = P_n(-x)$.

(ii) Find the monodromy group of the equation by finding the monodromy relations at $x = +1$ and $x = -1$.

27. Show that the monodromy matrices of a differential equation of the form $y'' + py = 0$ all have determinant 1, by considering

$$D := \begin{pmatrix} y_1' & y_1 \\ y_2' & y_2 \end{pmatrix},$$

where y_1 and y_2 are a basis of solutions. This means that the monodromy group of a second order linear equation can always be made to lie in $PSL(2; \mathbf{C})$. If the equation has order n, the same method shows that the group may be made to lie in $PSL(n; \mathbf{C})$.

28. (Riemann's calculation of the ratios $\alpha_\beta / \alpha'_\beta$, etc.)

(i) Show

$$\alpha_\gamma e^{2\pi i \gamma} P^\gamma + \alpha_{\gamma'} e^{2\pi i \gamma'} P^{\gamma'} = (\alpha_\beta e^{-2\pi i \beta} P^\beta + \alpha_{\beta'} e^{-2\pi i \beta'} P^{\beta'}) e^{-2\pi i \alpha}$$

by considering the LHS as the result of analytically continuing $P^\alpha = \alpha_\gamma P^\gamma + \alpha_{\gamma'} P^{\gamma'}$ on a positive circuit around $c = 1$ and the RHS as the result of analytically continuing $P^\alpha = \alpha_\beta P^\beta + \alpha_{\beta'} P^{\beta'}$ on a negative circuit around $a = 0$ and $b = \infty$.

(ii) Use $\alpha_\beta P^\beta + \alpha_{\beta'} P^{\beta'} = \alpha_\gamma P^\gamma + \alpha_{\gamma'} P^{\gamma'}$ to deduce

$$\alpha_\gamma \sin(\sigma - \gamma)\pi e^{\pi i \gamma} P^\gamma + \alpha_{\gamma'} \sin(\sigma - \gamma')\pi e^{\pi i \gamma'} P^{\gamma'}$$
$$= \alpha_\beta \sin(\sigma + \alpha + \beta)\pi e^{-\pi i(\alpha+\beta)} P^\beta$$
$$+ \alpha_{\beta'} \sin(\sigma + \alpha + \beta')\pi e^{-\pi i(\alpha+\beta')} P^{\beta'}.$$

(iii) Eliminate $P^{\gamma'}$ from the equation in (ii) and its counterpart in which α has been replaced by α', to obtain two equations of the form

$$C'P^{\gamma} = C''P^{\beta} + C'''P^{\beta'}$$

and deduce

$$\frac{\alpha_{\gamma}}{\alpha'_{\gamma}} = \frac{\alpha_{\beta}\sin(\alpha+\beta+\gamma')\pi e^{-\alpha\pi i}}{\alpha'_{\beta}\sin(\alpha'+\beta+\gamma')\pi e^{-\alpha'\pi i}} = \frac{\alpha_{\beta'}\sin(\alpha+\beta'+\gamma')\pi e^{-\alpha\pi i}}{\alpha'_{\beta'}\sin(\alpha'+\beta'+\gamma')\pi e^{-\alpha'\pi i}}.$$

(iv) Set $\sigma = \gamma$ and deduce

$$\frac{\alpha_{\gamma'}}{\alpha'_{\gamma'}} = \frac{\alpha_{\beta}\sin(\alpha+\beta+\gamma)\pi e^{-\alpha\pi i}}{\alpha'_{\beta}\sin(\alpha'+\beta+\gamma)\pi e^{-\alpha'\pi i}} = \frac{\alpha_{\beta'}\sin(\alpha+\beta'+\gamma)\pi e^{-\alpha\pi i}}{\alpha'_{\beta'}\sin(\alpha'+\beta'+\gamma)\pi e^{-\alpha'\pi i}}.$$

These equations are consistent because $\alpha + \alpha' + \cdots + \gamma' = 1$, and $\sin s\pi = \sin(1-s)\pi$, as Riemann remarked. Riemann also showed that

$$\alpha_{\beta} = \frac{\sin(\alpha+\beta'+\gamma')\pi}{\sin(\beta'-\beta)\pi} \qquad\qquad \alpha_{\beta'} = -\frac{\sin(\alpha+\beta+\gamma)\pi}{\sin(\beta'-\beta)\pi}$$

$$\alpha'_{\beta} = \frac{\sin(\alpha'+\beta'+\gamma)\pi}{\sin(\beta'-\beta)\pi} \qquad\qquad \alpha'_{\beta'} = -\frac{\sin(\alpha'+\beta+\gamma')\pi}{\sin(\beta'-\beta)\pi}$$

$$\alpha_{\gamma} = \frac{\sin(\alpha+\beta'+\gamma')\pi e^{(\alpha'+\gamma)\pi i}}{\sin(\gamma'-\gamma)\pi} \qquad \alpha_{\gamma'} = \frac{-\sin(\alpha+\beta+\gamma)\pi e^{(\alpha'+\gamma')\pi i}}{\sin(\gamma'-\gamma)\pi}$$

$$\alpha'_{\gamma} = \frac{\sin(\alpha'+\beta+\gamma')\pi e^{(\alpha+\gamma)\pi i}}{\sin(\gamma'-\gamma)\pi} \qquad \alpha'_{\gamma'} = -\frac{\sin(\alpha'+\beta'+\gamma)\pi e^{(\alpha+\gamma')\pi i}}{\sin(\gamma'-\gamma)\pi}$$

29. The uniqueness of a P-function with given data.

$$P = P\begin{pmatrix} a & b & c \\ \alpha & \beta & \gamma & z \\ \alpha' & \beta' & \gamma' \end{pmatrix},$$

and let P_1 be another P-function with the same data. To show $\dfrac{P_1}{P}$ is constant, show that

(i) $\quad P^{\alpha}P_1^{\alpha'} - P^{\alpha'}P_1^{\alpha} = (\det B')(P^{\beta}P_1^{\beta'} - P^{\beta'}P_1^{\beta})$
$\qquad\qquad\qquad\qquad = (\det C')(P^{\gamma}P_1^{\gamma'} - P^{\gamma'}P_1^{\gamma}).$

(ii) $\quad (P^{\alpha}P_1^{\alpha'} - P^{\alpha'}P_1^{\alpha})z^{-\alpha-\alpha'}$ is single-valued and finite at $z = 0$, as is

$$(P^{\beta}P_1^{\beta'} - P^{\beta'}P_1^{\beta})z^{\beta+\beta'} \quad \text{at } z = \infty,$$

and

$$(P^{\gamma}P_1^{\gamma'} - P^{\gamma'}P_1^{\gamma})(1-z)^{-\gamma-\gamma'} \quad \text{at } z = 1.$$

Deduce that $(P^\alpha P_1^{\alpha'} - P^{\alpha'} P_1^\alpha) z^{-\alpha-\alpha'} (1-z)^{-\gamma-\gamma'}$ is continuous, single-valued, and behaves at infinity like

$$(P^\beta P_1^{\beta'} - P^{\beta'} P_1^\beta) z^{\alpha+\alpha'+\gamma+\gamma'} = (P^\beta P_1^{\beta'} - P^{\beta'} P_1^\beta) z^{\beta+\beta'-1},$$

whence

$$(P^\alpha P_1^{\alpha'} - P^{\alpha'} P_1^\alpha) z^{-\alpha-\alpha'} (1-z)^{-\gamma-\gamma'}$$

is bounded everywhere and must be a constant; indeed must be zero, its value at $z = \infty$. Deduce that

$$\frac{P_1^{\alpha'}}{P^\alpha} = \frac{P_1^\alpha}{P^\alpha} \frac{P_1^\beta}{P^\beta} = \frac{P_1^\beta}{P^{\beta'}} = \frac{\alpha_\beta P_1^\beta + \alpha_{\beta'} P_1^{\beta'}}{\alpha_\beta P^\beta + \alpha_{\beta'} P^{\beta'}} = \frac{P_1^\alpha}{P^\alpha},$$

and

$$\frac{P_1^\gamma}{P^\gamma} = \frac{P_1^{\gamma'}}{P^{\gamma'}} = \frac{P_1^\alpha}{P^\alpha}.$$

It remains to ensure that P^α and $P^{\alpha'}$ do not simultaneously vanish for some $z \neq 0, 1, \infty$. This follows from an analogous consideration of

$$\left(P^\alpha \frac{dP^{\alpha'}}{dz} - P^{\alpha'} \frac{dP^\alpha}{dz} \right) z^{-\alpha-\alpha'+1} (1-z)^{-\gamma-\gamma'+1},$$

which is a non-zero constant (provided $\alpha \neq \alpha'$, as has been assumed).

Chapter II

Lazarus Fuchs

Introduction

In the years 1865, 1866, and 1868, Lazarus Fuchs published three papers, each entitled "Zur Theorie der Linearen Differentialgleichungen mit veränderlichen Coefficienten" ("On the theory of linear differential equations with variable coefficients"). These will be surveyed in this chapter. In them he characterised the class of linear differential equations in a complex variable x, all of whose solutions have only finite poles and possibly logarithmic branch points. So, near any point x_0 in the domain of the coefficients, the solutions become finite and singled-valued upon multiplication by a suitable power of $(x - x_0)$ unless it involves a logarithmic term. This class came to be called the Fuchsian class, and equations in it, equations of the Fuchsian type. As will be seen, it contains many interesting equations, including the hypergeometric. In the course of this work, Fuchs created much of the elementary theory of linear differential equations in the complex domain: the analysis of singular points; the nature of a basis of n linearly independent solutions to an equation of degree n when there are repeated roots of the indicial equation; explicit forms for the solution according to the method of undetermined coefficients, He investigated the behaviour of the solutions in the neighbourhood of a singular point, much as Riemann had done, by considering their monodromy relations — the effect of analytically continuing the solutions around the point — and, like Riemann, he did not explicitly regard the transformations so obtained as forming a group. One problem which he raised but did not solve was that of characterising those differential equations all of whose solutions are algebraic. It became very important, and is discussed in the next chapter.

Fuchs

Fuchs' career may be said to have begun with these papers. Born in Moschin near Posen in 1833, he went to Berlin University in 1854 and started his graduate studies under Kummer, writing his dissertation on the lines of curvature of a surface (1858), and then papers on complex roots of unity in algebraic number theory.[1] Just as Fuchs was the first to present his thesis to the new Berlin University under Kummer's direction, so his friend Koenigsberger was the first to present one under Weierstrass. But where Kummer's lectures were always clear and attractive, Weierstrass at first was a disorganised lecturer, and Fuchs does not seem to have come under his influence until Weierstrass lectured on Abelian functions in 1863. Weierstrass' approach was based on his theory of differential equations, and that subject became Fuchs' principal field of interest, Fuchs in turn becoming its chief exponent at Berlin.

When Fuchs presented his *Habilitationsschrift* in 1865, Kummer was the principal referee and Weierstrass the second, but by 1870 Fuchs spoke to himself as "a pupil of Weierstrass" (letter to Casorati, quoted in Neuenschwander [1978b, p. 46]).

In 1866 Fuchs succeeded Arndt at Berlin, before going to Greifswald in 1868. He returned to Berlin in 1884 as Professor Ordinarius, occupying the chair vacated by Kummer. His friend Koenigsberger described him as irresolute and anxious, a temporizer and one easily persuaded by others, but humorous and unselfish. He played a weak role in the battle between Weierstrass and Kronecker that polarized Berlin in the 1880s, thereby disappointing Weierstrass, and was no match for Frobenius in the 1890s, but several of his students, notably Schlesinger, spoke warmly of him.

2.1 Fuchs' theory of linear differential equations

The study of complex functions and their singularities was a matter of active research in Berlin in the 1860s.[2] The general method in use was the analytic continuation of power series expansions. If the function could be expanded as a series of the form $\sum_{n=0}^{\infty} a_n (x - a)^n$, convergent on some disc with centre $x = a$, then a was a non-singular point. If a term of the form $(x - a)^\alpha$ appeared in the expansion then a was a branch point, and it was a logarithmic branch point if $\log(x - a)$ appeared. No distinction was made between single-valued, many-valued, and even infinitely, many-valued functions. All were referred to as functions and distinguished as single-valued or many-valued as appropriate (and that usage will be followed here). If a function was analytic in a neighbourhood of $x = a$ Fuchs and others called it finite and continuous; the term "regular" introduced by Thomé (see below, p. 59) will sometimes be used here.

The points at which a function became infinite were called its points of discontinuity, but the nature of these points was obscure. Several writers, for example

Briot and Bouquet, equivocated over whether the function was defined at that point (and took the value) or was not defined at all. However, Weierstrass had developed the theory of power series expansion[3] of the form $\sum_{n=-k}^{\infty} a_n(x-a)^n$ in 1841, although he did not publish it until 1894, and on the basis of this theory, he made a clear distinction between functions admitting a finite "Laurent" expansion, which are infinite at $x = a$, and those of the second kind, which are not defined at $x = a$. The behaviour of a function near such singular points depends dramatically on the kind of expansion they admit. In the former case, its value tends so infinity but in the second case the Casorati–Weierstrass theorem asserts that it comes arbitrarily close to any pre-assigned value. Neuenschwander [1978a, 162] has considered the history of this theorem and finds that "Weierstrass ... had the Casorati–Weierstrass theorem already in the year 1863", the year he lectured on Abelian functions. So Fuchs picked the largest class of linear differential equations whose solutions can be defined at every point of the domain of the coefficients, including their singular points, logarithmic cases aside.

Of Fuchs' three papers considered here, the [1866] one is the most important. Its first three sections are identical with those of the [1865], but thereafter it diverges to give a much more thorough analysis of the case where the monodromy matrices have repeated eigenvalues and logarithmic terms enter the solution. It was also published in a more accessible journal, *Crelle's*, rather than the *Jahresbericht* of the Berlin *Gewerbeschule*, and it was more widely read. [1868], regarded by Fuchs as the concluding paper in the series, goes into more detail in special cases, extends the method to inhomogeneous equations of the appropriate kind, and discusses accidental singular points, the unusual case when a singularity of the coefficients does not give rise to a singularity in the solution. Therefore the two earlier papers will be considered first.

When Fuchs published his first paper on the theory of linear differential equations in 1865, he quite correctly stated that the task facing mathematicians was not so much to solve a given differential equation by means of quadratures, but rather to develop solutions defined at all points of the plane. Since analysis, he said, teaches how to determine a function if its behaviour in a neighbourhood of its points of discontinuity and its multiple-valuedness can be ascertained, the essential task is to discover the position and nature of these points. This approach had been carried out, he observed, by Briot and Bouquet in their paper [1856b] on the solution of differential equations which are of the form $f\left(y, \dfrac{dy}{dx}\right) = 0$, where $f\left(y, \dfrac{dy}{dx}\right)$ is a polynomial expression in y and $\dfrac{dy}{dx}$, and by Riemann in his 1857 lectures on Abelian functions. Fuchs proposed to investigate homogeneous linear ordinary differential equations. The conventions of his 1865 and 1866 papers will be followed fairly closely if such an equation is written as

$$\frac{d^n y}{dx^n} + p_1 \frac{d^{n-1}y}{dx^{n-1}} + \cdots + p_{n-1}\frac{dy}{dx} + p_n y = 0, \tag{18}$$

where $p_1, \dots, p_n$ are single-valued meromorphic functions of x in the entire complex x-plane, or on some simply-connected region $T \subset C$. Fuchs called the points in T where one or more of the p_i were discontinuous *singular* points, a term he attributed to Weierstrass, and he assumed each p_i had only finitely many singular points.[4]

Solutions near a non-singular point

Fuchs first considered a solution of (18) in a neighbourhood of a non-singular point x_0, and showed that it is a single-valued, finite, continuous function which is prescribed once values for y, $\dfrac{dy}{dx}, \dots, \dfrac{d^{n-1}y}{dx^{n-1}}$ are given at $x = x_0$. His method was to replace (18) by a second differential equation known to have a solution in a neighbourhood of x_0 and so related to (18) that its solution guarantees the existence to a solution to (18). The question of convergence was thus handled by Cauchy's method of majorants, much in the manner of Briot and Bouquet.

The solutions obtained in this way to (18) are power series of the form $\sum a_r (x - x_0)^r$, so it follows at once that the singular points of the solution can only be located among[5] the singular points of the coefficients $p_1(x), \dots, p_n(x)$. They are therefore fixed, and determined by the equation. This is in sharp contrast with the case of non-linear differential equations, where the singular points may depend on arbitrary constants involved in the solutions. (Fuchs later classified those non-linear equations whose branch points are fixed, see his [1884a] and also Ince [1926, Chs. 13,14] and Matsuda [1980, 11–18.])

Solutions near a singular point

To study the solutions in the neighbourhood of a singular point, Fuchs proceeded as Riemann had. With the singular points removed, T becomes a multiply connected region, T', and if circles are drawn enclosing each singular point and the circles joined by cuts to the boundary of T, a simply connected region T'' is marked out. Inside T'', a single-valued, continuous, and finite solution y of (18) can be found by the methods just described, which can be represented by an infinitely many leaved surface spread over T', the leaves joining along the cuts in a fashion to be determined by the method of monodromy matrices.

The special case of the second-order equation

In order to make Fuchs' argument more immediately comprehensible, I shall specialise to the case of a second-order differential equation

$$\frac{d^2y}{dx^2} + p_1 \frac{dy}{dx} + p_2 y = 0. \tag{19}$$

Fuchs worked throughout with an equation of degree n, and I shall indicate his more general results in Appendix 3.

Fuchs said that the two solutions y_1 and y_2 to (19) form a fundamental system (i.e., a basis of solutions) if their Wronskian

$$D := \begin{vmatrix} \dfrac{dy_1}{dx} & y_1 \\ \dfrac{dy_2}{dx} & y_2 \end{vmatrix}$$

never vanishes in T'. He proved that the general solution to (19) can always be written in the form $\eta = c_1 y_1 + c_2 y_2$, where c_1 and c_2 are constants. He also proved theorems about the freedom to choose the elements of a fundamental system, nowadays familiar from the theory of finite dimensional vector spaces but here developed ad hoc, as was often the case in the contemporary work of Weierstrass and Kronecker.[6]

Fuchs next showed that the prescribed conditions on p_1, p_2 imply that at least one solution of the equation would be single-valued upon multiplication by a suitable power of $x - a_1$, where a_1 is a singular point. Let y_1 and y_2 be a fundamental system near a_1, and y_1 and y_2 the result of analytically continuing the solutions y_1 and y_2 once round a_1. There are then equations of the form

$$y_1 = \alpha_{11} y_1 + \alpha_{12} y_2$$
$$y_2 = \alpha_{21} y_1 + \alpha_{22} y_2,$$

where the α's are constants, and $\alpha_{11}\alpha_{22} - \alpha_{12}\alpha_{21} \neq 0$.

Fuchs showed that the roots[7] of what he called the fundamental equation
$$\begin{vmatrix} \alpha_{11} - w & \alpha_{12} \\ \alpha_{21} & \alpha_{22} - w \end{vmatrix} = 0 \quad (20)$$ are independent of the choice of fundamental system, and neither can be zero. Let w_1 be one of them. Then y_1 can be chosen so that $y_1 = w_1 y_1$, and if $w_1 = e^{2\pi i r_1}$, then $y_1(x - a_1)^{-r_1}$ has the required property of being single-valued near a_1. The case in which (20) has repeated roots was first discussed by Fuchs in his [1865], then more fully in the later papers, and will be described (Appendix 3) when it will be seen that the general solution involves a logarithmic term. Indeed, if the transformations are

$$\tilde{y}_1 = w y_1$$
$$\tilde{y}_2 = y_1 + w y_2$$

(which is the general case if $w_1 = w_2 = w$), then the solutions must be of the form

$$y_1 = (x - a_1)^r \theta_1$$
$$y_2 = (x - a_1)^r \theta_2 + (x - a_1)^r \theta_1 \log(x - a_1),$$

where θ_1 and θ_2 are single-valued and holomorphic near a_1 and $w = e^{2\pi i r}$. So the general solutions contain $(x - a_1)^r \theta$, where θ is a polynomial in $\log(x - a_1)$ with holomorphic coefficients.

Equations of the Fuchsian class

The natural question was then: what conditions must be placed on p_1 and p_2 so that all the solutions are regular, i.e., have only finite poles at the singular points $a_1, \ldots, a_\rho, a_{\rho+1} = \infty$, and are at worst logarithmic upon multiplication by a suitable power of $(x - a_i)$ or $\frac{1}{x}$? From now on, Fuchs took ∞ to be a singular point, and T to be $\hat{\mathbf{C}} : \mathbf{C} \cup \{\infty\}$, so the coefficient functions must be rational functions, i.e., quotients of polynomials in x. His solution to this question [1865, §4, 1866, §4] produced a class of differential equations readily characterised by the purely algebraic restrictions upon the coefficients p_i; this class has become known as the Fuchsian class, and such equations will henceforth be referred to as "equations of the Fuchsian class." Fuchs himself always preferred more modest circumlocutions. (The phrase Fuchsian equation refers to a much more general linear differential equation; see Chapter VI.)

To answer this question, he took a fundamental system y_1^i, y_2^i, such that near a_i, $y_k^i = (x - a_i)^{r_{ik}} \theta_k^i (k = 1, 2)$ for some single-valued function θ_k^i. If I denote[8] the pair (y_1^i, y_2^i) by y^i, then the $y^i (i > 1)$ are related to y^1 by equations $y^1 = B_i y^i$, B_i matrices of constants, and if the analytic continuation of y^1 once around[9] a_1 produces $\tilde{y}^1$, then $\tilde{y}^1 = R_1 y^1$. Likewise a circuit of a_i produces $\tilde{y}^i = R_i y^i$ and the analytic continuation of y^1 around a_i produces $B_i R_i B_i^{-1} y^1$. Since a circuit of the finite singular points $a_1, \ldots, a_\rho$ is simultaneously a circuit of $a_{\rho+1}$ in the opposite sense

$$R_1 (B_2 R_2 B_2^{-1}) \cdots (B_{\rho+1} R_{\rho+1} B_{\rho+1}^{-1}) = I, \text{ the identity matrix}$$

so $\det R_1 \cdots \det R_{\rho+1} = 1$. But $\det R_i = w_{i1} w_{i2} = e^{2\pi i (r_{i1} + r_{i2})}$ so $\sum_{i=1}^{\rho+1} (r_{i1} + r_{i2}) =$

k is an integer.

Fuchs' notation, and his use of the monodromy matrices, follows Riemann's here. He even wrote his matrices in brackets, $(R)_i$, (B), and so forth. It is likely that Kummer encouraged Fuchs to follow Riemann's ideas; Kummer had already commented most favourably on the thesis of Riemann's student Prym.[10] On the other hand, his study of the eigenvalues of these matrices is in the thorough-going Berlin spirit.

The integer k was made precise by a technically complicated study of the analytic continuation of a fundamental system of solutions. The underlying idea is that one knows what happens near $x = \infty$ in two ways (from the monodromy and by transforming the equation), so one can obtain an equation for the sum $\sum r_{ij}$, and hence for k. The result is $k = \rho - 1$, and the same argument also showed that $\lim_{x \to \infty} x p_1$ is finite. More work of this kind enabled Fuchs to find p_2, and therefore to show that the most general equation of the Fuchsian class (and the second order) has the form

$$\frac{d^2 y}{dx^2} + \frac{F_{\rho-1}(x)}{\psi} \frac{dy}{dx} + \frac{F_{2(\rho-1)}(x)}{\psi^2} y = 0, \tag{20}$$

where $\psi = (x - a_1) \dots (x - a_p)$ and $F_s(x)$ is a polynomial in x of degree at most s. This is Fuchs main theorem, but as it applies to second-order equations.

Fuchs then looked again at solutions valid near the singular point $x = a_i$, which are necessarily of the form $u = (x - a)^{r_{ij}} \sum_{k=0}^{\infty} c_k (x - a_i)^k$, and showed that the r_{ij} satisfy $r(r - 1) - r P_{i1}(a_i) - P_{i2}(a_i) = 0$, where $P_{im} = (x - a_i)^m F_m(\rho - 1)\psi^{-m}$. This last equation was later called the determining fundamental equation (*determinirende Fundamentalgleichung*) by Fuchs [1868, 367 = 1904, 220] and subsequently the *indicial equation* by Cayley ([1883, 5]). The r's are related to the w's obtained earlier by the equations $w = e^{2\pi i r}$. Fuchs here showed how to obtain the solutions valid near a singular point a_i directly, without considering the monodromy relations, but he did not discuss the case when two of the r's differ by an integer, and a w is therefore repeated. He only discussed that situation in terms of w's and so of analytic continuation, never in terms of r's and the formal derivation of the solutions.

Having shown that any differential equation all of whose solutions are of the form required has necessarily the form of (2.1.4), Fuchs next had to show that, conversely, any equation of that form had solutions of that type. But this was a routine exercise in computing with power series and using the method of majorants to establish their convergence, so I omit the details.

The *n*th order equation

Since the description of the second order equation was deliberately constructed to follow Fuchs' analysis of the nth order equation, it is possible simply to state his more general results. He found that if the fundamental equation for any singular point a had no repeated roots, then the solutions near that point were all of the form $(x - a)^{\alpha} \phi(x - a)$, where ϕ is a holomorphic function, what he called a finite, single-valued, continuous function. But if the fundamental equation had a k-fold repeated root, the general solution was of the form of a polynomial in $\log(x - a)$ of degree $k - 1$ with holomorphic coefficients. The determining fundamental equation and the method of obtaining solutions was described quite generally. The general form of an nth order equation of the Fuchsian class was shown to be

$$\frac{d^n y}{dx^n} + \frac{F_{\rho-1}(x)}{\psi} \frac{d^{n-1} y}{dx^{n-1}} + \frac{F_{2(\rho-1)}(x)}{\psi^2} \frac{d^{n-2} y}{dx^{n-2}} + \dots + \frac{F_{n(\rho-1)}(x) y}{\psi^n} = 0, \quad (21)$$

where $\psi = (x - a_1) \cdots (x - a_\rho)$; the finite singular points are $a_1, a_2, \dots, a_\rho$.

It is very interesting to note that exactly this equation, in the form it takes when ∞ is not a singular point, was given by Riemann in a lecture on 20 February 1857, when he stated that no solution of it becomes infinitely great to an infinitely great order (*Werke* XXI). It was not published, however, until 1876, a decade after Fuchs' independent rediscovery of it.

Inhomogeneous equations

Fuchs also showed [1868 §5] how his methods could deal with the inhomogeneous equation

$$p_0 \frac{d^n y}{dx^n} + p_1 \frac{d^{n-1} y}{dx^{n-1}} + \cdots + p_n y + q = 0, \tag{22}$$

which he abbreviated to $Y + q = 0$.

He showed that every solution of $Y + q = 0$ and $Y = 0$ is a solution of the equation

$$q \frac{dY}{dx} - Y \frac{dq}{dx} = 0, \tag{23}$$

and that conversely every solution of (23) is either a solution of $Y = 0$ or of the inhomogeneous equation $Y + q = 0$. Furthermore, if the equation $Y = 0$ is of the Fuchsian class, then the equation $Y + q = 0$ will have all its solutions regular, provided $\frac{d \log q}{dx}$ becomes finite and single-valued on being multiplied by a finite power of $(x - a)$, whenever a is a singular point of (22). The converse is also true, and the function q must be of the form $C(x - a_1)^{\mu_1} \ldots (x - a_\rho)^{\mu_\rho}$, where $a_1, \ldots, a_\rho$ are the finite singular points of the coefficients $p_1, \ldots, p_n$, and $C, \mu_1, \ldots, \mu_\rho$ are constants. The conditions on q are simply derived from insisting that (23) be an equation of the Fuchsian class.

Corollaries of Fuchs' work

Fuchs drew several conclusions from his work. One concerned the problem with characterising those differential equations whose solutions were algebraic, and another, the special role played by the hypergeometric equation.

In [1865, §7] and [1866, §6] he observed that the class of differential equations, all of whose solutions are algebraic, is contained within the Fuchsian class, since Puiseux had showed [1850, 1851] that in a neighbourhood of a singular point, a, an algebraic function admits an expansion of the form

$$\eta = \sum_0^{\mu-1} c_k (x - a)^{-r+k/\mu}$$

where μ and σ are integers. This implies that

$$\eta = (x - a)^{-r} \cdot \phi + (x - a)^{-r+1/\mu} \cdot \phi_1 + \cdots + (x - a)^{-r+\frac{\mu-1}{\mu}} \phi_{\mu-1}.$$

where the ϕ's are single-valued, continuous, and finite functions near a, and each term $(x - a)^{-r+j/\mu}$ satisfies the conditions required to make η a solution to differential equation of the Fuchsian class. It would, Fuchs said, be an interesting problem to seek a precise characterisation of the equations which have only algebraic solutions, but one which he was then unable to solve. He was able to solve the much simpler special case: when the solutions are always rational functions. The necessary and sufficient conditions are that the roots of the indicial equation are all real integers, for then the solutions are single-valued in the entire complex plane.

The question of when a differential equation has only algebraic solutions was taken up by many mathematicians in the 1870s and is the subject of the next chapter. It is a question with a long history. Liouville had made a thorough, if ultimately inconclusive, study of it in the 1830s and 1840s, which has recently been thoroughly analysed in Lützen's analysis of Liouville's early work [Lützen, 1984], where the even earlier work of Abel and later 19th century and some 20th century works are also discussed.

Fuchs also observed ([1865 §6, 1866 §6]) that the hypergeometric equation is of the Fuchsian class. It appears in the form of (20) as

$$\frac{d^2y}{dx^2} + \frac{f_0 + f_1 x}{(x - a_1)(x - a_2)}\frac{dy}{dx} + \frac{g_0 + g_1 x + g_2 x^2}{(x - a_1)^2(x - a_2)^2}y = 0. \tag{24}$$

If $a_1 = 0$, $a_2 = 1$, the indicial equations at 0, 1, and ∞ are

$$r(r - 1) - f_0 r + g_0 = 0$$
$$r(r - 1) + (f_0 + f_1)r + g_0 + g_1 + g_2 = 0 \text{ and}$$
$$r(r - 1) + (2 - f_1)r + g_2 = 0 \quad \text{respectively.}$$

If the roots are denoted by α and α': 0, γ'; and β, β', respectively, where $\alpha + \alpha' + \beta + \beta' + \gamma' = 1$ in conformity with the determination of k in this case, the differential equation takes Riemann's form

$$(1 - x)\frac{d^2y}{d\log x^2} - [\alpha + \alpha' + (\beta + \beta')x]\frac{dy}{d\log x} + (\alpha\alpha' - \beta\beta')y = 0,$$

where it is assumed that none of $\alpha - \alpha'$, $\beta - \beta'$, γ' are integers.

Fuchs was also able to give a significant characterisation of the hypergeometric equation. In the general differential equation of the Fuchsian type, each function $F_{k(\rho-1)}$ contains $k\rho - k + 1$ constants, so there are $\frac{1}{2}n(n + 1)\rho - \frac{1}{2}n(n - 1)$. There are n exponents at each of the $\rho + 1$ singular points (including ∞). So there are $n(\rho + 1)$ in all, of which only $n(\rho + 1) - 1$ are arbitrary by virtue of Fuchs' equation for k. Accordingly, if a set of given exponents is to determine the equation, then

$$\frac{n(n + 1)\rho}{2} - \frac{n(n - 1)}{2} = n(\rho + 1) - 1. \tag{25}$$

so $\rho = 1 + \frac{2}{n}$ and either[11] $n = 1$, $\rho = 3$, or $n = 2$, $\rho = 2$. Accordingly, if the first-order equations are excluded, the class of equations of the Fuchsian type, for which the exponents at the singular points determine the coefficients of the equation, contains precisely the hyper-geometric equation. This explains why Riemann's methods worked so well for the Gaussian equation, for it is precisely in that case that his initial data (the exponents) characterise not only the solution functions but also the equation itself. Contrariwise, in almost all other cases, the number of exponents is too few to characterise the equation, and the excess numbers, called the accessory parameters, have since proved to be quite intractable.

2.2 Generalisations of the hypergeometric equation

After 1868, Fuchs turned his attention to finding applications and consequences of his theory, rather than to finding ways of extending it. Among the problems then arousing the greatest interest in mathematics was the investigation of hyperelliptic functions. Weierstrass had published two important papers on Abelian functions and the inversion of hyperelliptic integrals in Crelle's *Journal für Mathematik* for 1854 and 1856 (Weierstrass [1854, 1856]) which were instrumental in securing him an invitation to become associate professor at the University of Berlin. Liouville hailed the first papers as "one of those works that marks an epoch in science" (quoted in Bierman [1976, 221]), and Weierstrass' influential lecture cycle also dealt largely with elliptic and Abelian functions. Riemann's [1857c] had solved the inversion problem for an arbitrary integral of an algebraic function by means of the theory of θ-functions in several variables, and this work was also much discussed in Berlin (see Chapter V).

Fuchs first proposed, in his [1870a], to study the periods of hyperelliptic integrals from the standpoint of his new theory, reserving the full treatment of Abelian integrals for a subsequent opportunity [1871 a, b]. These later works require a knowledge of the theory of θ-functions, and will not be discussed. It will be clearest to begin with the example that dominates the [1870] paper, the hyperelliptic integral

$$\int \frac{dx}{[(x - k_1) \cdots (x - k_{n-1})(x - u)]^{\frac{1}{2}}} = f(z). \tag{26}$$

Fuchs introduced this example in §7 to illustrate the more general integral $\int \dfrac{dx}{s}$, where $s^2 = \phi(x, u)$ and $\phi(x, u)$ is algebraic of degree n in x, "partly" as he said "to elucidate the proceeding and partly for later use." It certainly presents a simpler piece of analysis, and one not without applications. When $n = 3$, his equation for the periods has become famous as the Picard–Fuchs equation. Its significance in algebraic geometry is well described in Clemens [1980].

Fuchs' strategy for finding the periods of the integral (26) was straightforward. Suppose u is fixed for the moment. Then the periods are found by cutting the x-plane along a curve, called the principal cut, joining u to k_1, k_1 to k_2, …,k_{n-2} to k_{n-1}, and (if n is odd) k_{n-1} to ∞. If a, b are any two consecutive points of the sequence $(k, k_1, \ldots, k_{n-1}, \infty)$, the period corresponding to the cut, joining a to b is found by evaluating $f(x)$ around a closed curve γ enclosing that part of the principal cut but no other points in $\{u, k_1, \ldots, k_{n-1}, \infty\}$. Precisely,

$$\eta = \frac{1}{2} \int_\gamma y \, dx, \quad y = [(x_1 - k_1) \cdots (x - k_{n-1})(x - u)]^{-\frac{1}{2}}.$$

The first question is, how to arrange all this when u is allowed to vary? Fuchs showed that the periods were continuous, single-valued functions of u away from the branch points. Continuity is evident; to show they are single-valued Fuchs considered

$$s^2 = [(x - k_1) \cdots (x - k_n)(x - u)].$$

For each u, s is defined as a two-leaved Riemann surface T_u over the complex x-plane. The branch points are at u, at $k_1, k_2, \ldots, k_{n-1}$ and possibly ∞, all of which are independent of u.

Accordingly, the principal cut is independent of u once it has reached k_1. It can furthermore be made to lie in the upper leaf of T_u for all u, since, as u makes a circuit σ around some k_λ, the factor $x - u$ of s becomes $e^{2\pi i}(x - u)$ if x is enclosed within σ while all the others remain unchanged, so s becomes $-s$. On the other hand, if x is not within σ, then $x - u$ does not change and s remains unaltered. The circuit σ has transformed T_u into T_u' (the lower leaf) and outside σ upper leaves are joined to upper leaves because σ did not change. But σ could be taken arbitrarily small, whence the desired result.

Of course, the periods are not single-valued functions of u, but Fuchs could now write down (§7) a sequence of equations connecting the periods before and after a circuit is made by u around any k_λ. This gave him the monodromy relations at each singular point of the differential equation he sought, connecting the periods. To obtain the differential equation, he observed that it must be of order $n - 1$ since there are $n - 1$ linearly independent periods. It therefore has the form

$$\beta_{n-1}\frac{d^{n-1}\eta}{dx^{n-1}} + \cdots + \beta_0\eta = 0, \tag{27}$$

whose solutions, $\eta_1 \ldots, \eta_{n-1}$, are the periods along the parts of the principal cut. After some calculation he was able to show that $\dfrac{\beta_{(n-1)-i}}{\beta_{n-1}}$ is necessarily of the form $\dfrac{F_{i(n-2)}(u)}{\psi^i(u)}$ where $F_{i(n-2)}$ is a polynomial of degree at most $i(n - 2)$ and $\psi(u) = (u - k_1) \ldots (u - k_{n-1})$. So the singular points are precisely $k_1, \ldots, k_{n-1}$ (and ∞ if and only if n is odd); none are accidental. The differential equation is of Fuchsian type and all of its solutions are regular.[12]

When $\phi(x, u) = A[(x - k_1) \ldots (x - k_{n-1})](x - u) = \psi(x)(x - u)$, the calculations to find $\beta_{n-1-i}/\beta_{n-1}$ are not too difficult. Fuchs did not work directly with the monodromy relations he had obtained in §7, but preferred the expressions involving partial derivatives of y with respect to u. The result is clear, even if the method chosen pass through some murky country; Fuchs found (§19, equation 21) $\beta_{n-i} = \rho_0 \tau_{n-1}\psi^{(i-1)}(u)$ where ρ_0 is a constant,

$$\tau_{n-i} = \frac{1 - 2n + 2}{2\sigma_{n-i}(i - 1)!},$$

and

$$\sigma_k = \frac{1.3\ldots(2k - 1)}{2^k}.$$

The monodromy relations give directly (§13) that the roots of the indicial equation are rational numbers. This result remains true if the general integral

$$\int \frac{f(x)}{s} \, dx, \quad s^2 = \phi(x, u), \quad \phi \text{ or order } n \text{ in } x,$$

is considered, but the rest of the analysis is more complicated. Some of the periods may now depend linearly on others; so, if there are exactly p linearly independent ones, the differential equation has order only p. It remains of the Fuchsian class, but the β's are much harder to find. There are n finite singular points (∞ is singular if and only if n is odd) which are actual, and there may also be accidental singularities, depending on $f(x)$. Fortunately the examples Fuchs gave are of the simpler type (26).

For the elliptic integral $n = 3$ and $\psi(x) = x(x - 1)$, $\beta_2 = -2u(u - 1)\rho_0$, $\beta_1 = -2(2u - 1)\rho_0$, $\beta_0 = -\frac{1}{2}\rho_0$, and the differential equation is

$$2u(u - 1)\frac{d^2\eta}{du^2} + 2(2u - 1)\frac{d\eta}{du} + \frac{\eta}{2} = 0. \tag{28}$$

This takes Legendre's form on substituting $u = \dfrac{1}{k^2}$, $\eta = k\zeta$:

$$k(1 - k^2)\frac{d^2\zeta}{dk^2} + (1 - 3k^2)\frac{d\zeta}{dk} - k\zeta = 0, \tag{29}$$

equivalent, as he said, to the system of simultaneous differential equations deduced by Koenigsberger in the first volume of the *Mathematische Annalen*.

Fuchs concluded the paper with a detailed study (§21) of the case of the elliptic integral, which turned out to be important for his later work. He knew the singular points of (28) were at $u = 0$, 1, and ∞, and he knew the monodromy relations at each point; taking them in order he found: at $u = 0$ the indicial equation is $r^2 = 0$, so one solution is

$$v_{01} = 1 + \sum_k \left(\frac{1.3 \ldots 2k - 1)}{2.4 \ldots 2k}\right)^2 u^k,$$

and substituting $\zeta = v_{01} \displaystyle\int \zeta_0 \, du$ into (28) he found $\zeta_0 = -\dfrac{C}{u} + CG_0(u)$, where

$$G_0(u) = \frac{1 + v_{01}^2(u - 1)}{v_{01}^2 u(u - 1)},$$

and C is an arbitrary constant.

Accordingly, a second solution of (28) valid near $u = 0$ is $v_{02} = H_0(u)v_{01} - v_{01} \log u$, where

$$H_0(u) = 4 \log 2 + \int_0^u G_0(u) \, du.$$

But the monodromy relations between η_1 (the period along $(u, 0)$) and η_2 (the period along $(0, 1)$) relate η_1 and η_2 to v_{01} and v_{02}. The monodromy relations for η_1 and η_2 are $\tilde{\eta}_1 = \eta_1$ and $\tilde{\eta}_2 = 2\eta_1 + \eta_2$. The relations for v_{01} and v_{02} are immediate from their explicit representations: $\tilde{v}_{01} = v_{01}$ and $\tilde{v}_{02} = -2\pi i v_{01} + v_{02}$. So the relationships $\eta_1 = c_{11}v_{01} + c_{12}v_{02}$, $\eta_2 = c_{21}v_{01} + c_{22}v_{02}$ simplify, because $c_{12} = 0$, $c_{11} = -\pi i c_{22}$, to become: $\eta_1 = c_{11}v_{01}$, $\eta_2 = c_{21}v_{01} - \dfrac{1}{\pi i}c_{11}v_{02}$. The constants c_{11}, c_{21} can be found from

$$\eta_1 = \int_u^0 \frac{dx}{[x(x-1)(x-u)]^{\frac{1}{2}}}, \qquad \eta_2 = \int_0^1 \frac{dx}{[x(x-1)(x-u)]^{\frac{1}{2}}},$$

on letting $u \to 0$. It turns out that $\eta_1 = -\pi v_{01}$, $\eta_2 = \pi v_{01} - i v_{02}$. For $\eta_1(0) = -\pi = c_{11}v_{01}(0) = c_{11}.1$, and, as $u \to 0$

$$\eta_1(u) \simeq \int_0^1 \frac{1 - (1-x)^{\frac{1}{2}}dx}{[x(x-1)(x-u)]^{\frac{1}{2}}} + \int_0^1 \frac{dx}{[x(u-x)]^{\frac{1}{2}}}$$
$$= -2i \log 2 \quad + -2i \log((u-1))^{\frac{1}{2}} + i) + i \log u.$$

$H_0(0) = 4\log 2$, so $v_{02}(u) \simeq 4\log 2 - \log u$, and $\eta_2(u) = c_{21}v_{01}(u) - i v_{02}(u)$ implies that

$$i \log u - 4i \log 2 + \pi = c_{21} - i(4\log 2 - i \log u),$$

so $c_{21} = \pi$. An exactly similar calculation near $u = 1$ produces solutions v_{11}, v_{12} to (28) related to η_1 and η_2 by

$$\eta_1 = -\pi i v_{11} + v_{12}$$
$$\eta_2 = -2\pi i v_{11} - v_{12}.$$

Finally, near $u = \infty$, where the indicial equation is $(r - \frac{1}{2})^2 = 0$,

$$v_{\infty 1} = \left(\frac{1}{u}\right)^{\frac{1}{2}} \left[1 = \sum_{K=1}^{\infty} \left(\frac{1.3 \ldots (2K-1)}{2.4 \ldots 2K}\right) \left(\frac{1}{u}\right)^K\right]$$

and

$$v_{\infty 2} = v_{\infty 1} - 4\log 2 + \int_0^t \left(\frac{t + v_{\infty 1}^2 (t-1)}{v_{\infty 1}^2 (1-t)t}\right) dt - v_{\infty 1} \log u$$
$$\eta_1 = -\pi v_{\infty 1} - i v_{\infty 2}$$
$$\eta_2 = -\pi v_{\infty 1}.$$

The substitution $u = \dfrac{1}{k^2}$, $x = z^2$ transforms the integral

$$\int \frac{dx}{\sqrt{[x(x-1)(x-u)]}} \quad \text{into } 2k \int \frac{dx}{\sqrt{[(1-z^2)(1-k^2x^2)]}}.$$

So, if

$$K = \int_0^1 \frac{dz}{\sqrt{[(1-z^2)(1-k^2z^2)]}} \text{ and } K'i = \int_1^{1/k} \frac{dz}{\sqrt{[(1-z^2)(1-k^2z^2)]}},$$

then

$$K = -\frac{1}{2k}\eta_2 = \frac{\pi}{2k}v_{\infty 1}$$

and

$$K'i = \frac{1}{2k}(\eta_1 - \eta_2) = -\frac{i}{2k}v_{\infty 2},$$

which give the well-known power series for K and K'.

Fuchs' techniques were thus able to give him a new derivation of the solutions to Legendre's equation, and in principle they were capable of handling the analogues of Legendre's equation for the periods of hyperelliptic integrals.[13] Legendre had also considered what he called elliptic integrals of the second kind, for which the periods are

$$J = \int_0^1 \frac{k^2y^2dy}{[(1-y^2)(1-k^2y^2)]^{\frac{1}{2}}}, \text{ and } J'i = \int_1^{1/k} \frac{k^2y^2dy}{[(1-y^2)(1-k^2y^2)]^{\frac{1}{2}}}.$$

J and J' satisfy the differential equation

$$(1-k^2)\frac{d^2y}{dk^2} + \frac{1-k^2}{k}\frac{dy}{dx} + y = 0, \text{ Legendre [1828, I, 62]}.$$

They are connected to K and K' by Legendre's relation $KJ' - K'J = \dfrac{\pi}{2}$, Legendre [1825, I, 61].

As I shall discuss in Chapter IV, the J's behave in some ways surprisingly differently to the K's, and when Fuchs was led to investigate this at Hermite's request a decade later, the consequences for the theory of modular functions were to be significant.

2.3   Conclusion

In these papers Fuchs succeeded in characterising those linear ordinary differential equations, none of whose solutions have an essential singular point anywhere in the extended complex plane. In so doing he developed the theory of their singularities in terms of their monodromy matrices and their eigenvalues — the roots of the associated fundamental equation. He showed how these roots were connected to the exponents of the branch points, which themselves satisfy the indicial or determinantal fundamental equation. He gave a thorough analysis of the situation when one or more eigenvalues are repeated and logarithmic terms enter the solution. He indicated the special position occupied by the hypergeometric equation among equations of the Fuchsian class, and raised the problem of isolating those equations which have algebraic solutions. He followed Riemann in using monodromy matrices to study the analytic continuation of the solutions around the singular points, but did not rely on Dirichlet's principle to obtain the solutions globally. Rather he preferred to study the global forms using the analytic continuation of power series in the Weierstrassian style. His study of hyperelliptic integrals is his most "Riemannian" paper, later ones are couched more and more in the theory of power series. This shift of emphasis reflects a general tendency during the 1870s to avoid Dirichlet's principle, and to seek to obtain Riemann's results in other ways (see Chapter V p. 148 and Appendix 1). This tendency was most marked in Berlin.

This achievement raised three areas for future work. One, with which I shall be little concerned, is the problem of going outside the Fuchsian class to study equations whose solutions have essential singularities. This problem was taken up by Thomé in the 1870s, and later by Poincaré. Although power series solutions to such equations can be found formally, they frequently do not converge in the neighbourhood of singular points of the coefficients which are not of Fuchs' type. Instead they provide asymptotic solutions to the equation, as was first realized by Poincaré [1886]. There is a thorough discussion of this topic in Schlissel [1976/77].

The second area was the use of the methods of Riemann and Fuchs to explore equations of the Fuchsian class. This included using the methods to reformulate and add to the knowledge of familiar equations, such as Legendre's, and attempting to discuss new equations. By and large Fuchs took the former path, quite successfully as will be seen in Chapter IV, but attempts to delineate new equations with precision were less successful. The pioneer in this direction was Thomae.[14] The difficulty here is precisely Fuchs' discovery that the hyper geometric equation is the only equation of order greater than 1 for which the exponents at the branch points uniquely determine the coefficients.

The third area was the attempt to formulate a theory of differential equations whose coefficients were not rational functions, but elliptic functions. This class includes, for example, Lamé's equation and was studied accordingly by several writers in the 1870s. It forms an intriguing stepping stone on the way to the general theory of differential equations with algebraic coefficients, and will be described in Chapter VI.

Fuchs turned his attention to non-linear equations and movable singularities in the 1880s and 1890s. For a brief account of that work and its consequences, see Gray [1984a].

2.4   The new methods of Frobenius and others

In this section some developments are considered which lie to one side of the main story: Frobenius' simplification of Fuchs' work, his introduction of the idea of irreducibility of an equation, Thomé's study of equations with irregular singular points, and the introduction of Jordan's theory of canonical forms for matrices. These ideas will not be referred to again, and they are included purely for the sake of completeness.

Fuchs' arguments were, one might say, solidly contemporary adaptations of general methods for studying algebraic differential equations. Consequently they are long and more difficult than they need be. The man who first proposed the simpler methods which have since become customary in treatments of linear ordinary differential equations was Georg Frobenius, in 1873. Frobenius, then 24, had not yet presented his *Habilitationsschrift*, and even before he did, he was nominated by Weierstrass for a newly created Professorship Extraordinarius the next year. He held that post for only half a year before going to the Polytechnic in Zurich, one of a series of distinguished professors the Swiss recruited from Berlin; see K. R. Biermann [1973b, 96].

Frobenius' main aim in his paper [1873a] was, as he said, to find simpler methods for obtaining Fuchs' results. He found that this could be done by working directly with power series. He considered the differential equation (where x and y are complex and $y^{(n)} = \dfrac{d^n y}{dx^n}$)

$$x^n p(x) y^{(n)} + \cdots + p_n(x) y = 0,$$

which he denoted $P(y) = 0$, in a neighbourhood of $x = 0$. The solutions were to be bounded near 0 when multiplied by a suitable power of x. He assumed, as he may without loss of generality, that $p(x) = 1$ and the other p_i have power series expansions near $x = 0$.

For y, he substituted the power series $g(x, \rho) = \displaystyle\sum_{j=0}^{\infty} g_j x^{\rho+j}$, containing a parameter ρ, and observed that $P(x^\rho) = x^\rho f(x, \rho)$, where $f(x, \rho)$ is a polynomial in ρ:

$$
\begin{aligned}
f(x, \rho) &= \rho(\rho - 1) \ldots (\rho - n + 1) p(x) + \rho(\rho - 1) \ldots + (\rho - n + 2) p_1(x) \\
&\quad + \cdots + p_n(x) \\
&= \sum_j f_j(\rho) x^j,
\end{aligned}
$$

here each $f_j(\rho)$ is a polynomial function of ρ. Accordingly, he obtained a recurrence relation for the g_j's:

$$
\begin{aligned}
&g_0 f_0(\rho) = 0 \\
&g_1 f_0(\rho + 1) + g_0 f_1(\rho) = 0, \\
&\cdots, \\
&g_j f_0(\rho + j) + g_{j-1} f_1(\rho + j - 1) + \cdots + g_0 f_j(\rho) = 0.
\end{aligned}
$$

Since it is assumed that $g_0 \neq 0$, the equation $g_0(\rho) = 0$ determines ρ. Frobenius observed this, but preferred to let ρ remain a variable and determine the g_j as functions of ρ instead. Each g_j is a rational function,

$$g_j(\rho) = g_0(\rho) h_j(\rho) / f_0(\rho + 1) \ldots f_0(\rho + j),$$

where $h_j(\rho)$ can be determined explicitly as a function of $f_j(\rho + k)$'s. If ρ is restricted to suitably small neighbourhoods of the roots of $f_0(\rho) = 0$, then the denominators of the g_j's vanish only at the roots of $f_0(\rho) = 0$, and $g_0(\rho)$ can be chosen so that the functions $g_j(\rho)$ are bounded. So, if the series $g(x, \rho) = \sum g_j x^{\rho+1}$ converges, it represents a solution of the differential equation. Frobenius next established the necessary convergence, by means of the ratio test and a simple majorizing argument, and showed explicitly that within that domain, the convergence is uniform. This established that the power series can be differentiated term by term, and concluded the proof that the series represents a solution of the differential equation.

In the next two sections of his paper Frobenius considered the nature of the solutions which are obtained once ρ is a root of $f_0(\rho) = 0$. He simplified Fuchs' treatment of repeated roots (when polynomials in $\log x$ enter the solution) by exploiting the parameter ρ. Since $P(g(x, \rho)) \equiv f_0(\rho)g(\rho)x$, if ρ_k is a k-fold root of $f(\rho) = 0$, then differentiating both sides of this identity k times with respect to ρ and setting $\rho = \rho_k$ yields $P\left(\dfrac{d^k}{d\rho^k}g(x, \rho_k)\right) = 0$. This implies that $\dfrac{d^k}{d\rho^k}g(x, \rho_k) = g^{(k)}(x, \rho_k)$ is a solution of the differential equation $P(y) = 0$, and since $g(x, \rho) = x^\rho \sum g_j(\rho)x^j$, the solution obtained is

$$g(k)(x, \rho_k) = x^{\rho_k} \sum (g_j^{(k)}(\rho_k) + k g_j^{(k-1)}(\rho_k) \log x$$
$$+ \frac{k(k-1)}{2!} g_j^{(k-1)}(\rho_k)(\log x)^2 + \cdots + g_j(\rho_k)(\log x)^k x^j .$$

Consequently the solution to the differential equation has no logarithmic terms if the equation $f_0(\rho) = 0$ has no repeated roots. Frobenius concluded the paper with other methods for obtaining the coefficients $g_j(\rho)$.

Frobenius confined his first study of linear differential equations to rederiving the results of Fuchs. Almost none of the results were new: the indicial equation is to be found in Fuchs' [1865] and [1866], as are the associated power series solutions. The concern about uniform convergence is also present in the earlier work. What is new is the method, involving the parameter ρ, for proving the convergence of the solutions obtained via the indicial equation. Frobenius was quite scrupulous in acknowledging Fuchs' work, but as his simpler methods drove out those of Fuchs, the comparison became blurred until he is sometimes remembered more for the results than the methods (see e.g., Birkhoff [1973, 282], Hille [1976, 344], or Piaggio [1962, 109]).

Frobenius' new results were largely connected with the idea of the irreducibility of a differential equation, which he introduced in his next paper [1873b], published simultaneously with the one just described. By analogy with polynomial equations, he said that a differential equation was irreducible if it had no solutions in common with a differential equation of lower order, or one of the same order but lower degree. This implies, for instance, that a differential equation, all of whose solutions are algebraic, is reducible, for its solutions are the roots of a polynomial equation, which is a differential equation of order zero. More substantially, Frobenius showed

(§5) that if the hypergeometric equation is reducible, then the hypergeometric series entering in to one of its solutions must be a polynomial. Since this result displays nearly all the features of Frobenius' theory of irreducibility, and since that theory is strikingly similar to the theory of irreducible algebraic equations, the exposition may reasonably be confined to this example.

The hypergeometric equation $x(1-x)\dfrac{dy^2}{dx} + (\gamma - (\alpha + \beta + 1)x)\dfrac{dy}{dx} - \alpha\beta y = 0$ is reducible if one of its solutions is the solution of a first order equation. Frobenius took as a basis of solutions two which are defined on a neighbourhood of the singular point $x = 1$:

$$w = F(\alpha, \beta, \alpha + \beta - \gamma + 1, 1 - x),$$
$$w' = (1-x)^{\gamma - \alpha - \beta} F(\gamma - \alpha, \gamma - \beta, \gamma - \alpha - \beta + 1, 1 - x)$$

because, as he said, each has part of its domain of convergence in common with the series solutions valid near $x = 0$ and $x = \infty$. Indeed, if u and u' are a basis of solutions near $x = 0$ and v and v' are a basis of solutions near $x = \infty$, then Kummer's relations imply that

$$w = au + a'u', w = cy + c'v'$$
$$w' = bu + b'u', w' = dy + d'v'$$

for precise constants $a, a', \ldots, d'$. Accordingly, if the equation is reducible, then either one coefficient vanishes in each expression for w or in each expression for w', or an analytic continuation of w (or w') would yield two linearly independent branches, which must satisfy a differential equation of order two. Frobenius considered each possibility in turn. For example,

$$a = \frac{\Pi(\alpha + \beta - \gamma)\Pi(-\gamma)}{\Pi(\alpha - \gamma)\Pi(\beta - \gamma)}, \quad c' = \frac{\Pi(\alpha + \beta - \gamma)\Pi(\alpha - \beta - 1)}{\Pi(\alpha - \gamma)\Pi(\alpha - 1)}.$$

The function Π never vanishes, but it is infinite at the negative integers, so a and c' vanish if $\gamma - \alpha = \nu + 1$, for some non-negative integer ν. But then in the solution $u' = x^{1-\gamma} F(\alpha - \gamma + 1, \beta - \gamma + 1, 2 - \gamma, x)$ the F term is a polynomial of degree ν. The other cases proceeded similarly.

To put this conclusion in a clearer light, as he said, he rederived it starting from the observation that the hypergeometric equation is completely determined by its exponents at its singular points. Now, if analytically continuing two solutions round its singular points only multiplies them by constants, then the exponent differences must all be integers without the solution containing a logarithmic term. The general first order equation of the Fuchsian type has the form $\dfrac{dy}{dx} - \left(\dfrac{\alpha_1}{x - a_1} + \dfrac{\alpha_2}{x - a_2} + \cdots + \dfrac{\alpha_\mu}{x - a_\mu}\right)y = 0$ with solution $y = C(x - a_1)^{\alpha}(x - a_2)^{\alpha_2} \ldots (x - a_\mu)^{\alpha_\mu}$. Its singular points are $a_1, a_2, \ldots, a_\mu$ and ∞ at which the exponents are $\alpha_1, \alpha_2, \ldots, \alpha_\mu$ and β respectively, and Fuchs' condition in this case is $a_1 + \alpha + \cdots + \alpha_\mu + \beta = 0$. Frobenius had shown earlier

in this paper (§4, Theorem I) that the singular points of a differential equation of
lower order are either the singular points of an equation of highest order with which
it has solutions in common, or they are accidental singular points of the lower-order
equation for which the roots of the corresponding indicial equation are less than the
order of the higher order equation. So here, $\alpha_3 = \alpha_4 = \cdots = \alpha_\mu = 1$. Furthermore, the common solution of the two differential equations may be taken to have
its singular points at 0, 1 and ∞ with exponents α, γ, β, so it is

$$y = Cx^\alpha (1-x)^\gamma (x-a_1)(x-a_2)\dots(x-a_\mu).$$

In this way Frobenius obtained the necessary conditions $\alpha + \beta + \gamma + \nu = 0$,
$\alpha' + \beta' + \gamma' = \nu + 1$. These conditions are also sufficient, as can be seen from
consideration of the function

$$P \begin{pmatrix} 0 & \infty & 1 & \\ 0 & -\mu & 0 & x \\ \alpha' - \alpha & \beta' - \beta - \nu & \gamma' - \gamma & \end{pmatrix}$$

for which the corresponding hypergeometric series is a polynomial.

Frobenius used his theory of irreducibility to clarify three aspects of the theory
of linear differential equations: the behaviour of the solutions under analytic continuation; the nature of accidental singular points; the occurrence of solutions with
essential singularities when the equation is not of the Fuchsian type.

Typical of his results under the first heading in this ([1873b, 3, Theorem IV]):
if a differential equation is reducible, there is an equation of lower order with which
it has all its solutions in common. Indeed, if an equation of order γ has all of its
solutions in common with the equation $Q(y) = 0$ of order $\mu(\gamma > \mu)$, then each of
its solutions satisfy an equation of the form $Q(y) = w$ in which w is a solution of a
certain equation of order $\gamma - \mu$.

His analysis of the singular points has already been described. To consider
his treatment of equations not of the Fuchsian type, it is necessary to look also
at the work of his contemporary at Berlin, L.W. Thomé. Thomé was the first to
consider how Fuchs' theory might be extended to such equations, and starting in
1872 he published a series of increasingly long papers on this topic in the *Journal für
Mathematik*.[15] In the second of these [1873, 266] he introduced the convenient word
"regular" to describe solutions of the kind Fuchs had sought (*reguläre Integrale*); in
this terminology he sought properties of the coefficients of a differential equation
which indicated how many linearly independent regular integrals it would have. His
first main theorem in this direction [1872, §5, Theorem 2] applied to the equation
$\dfrac{d^m y}{dx^m} + p_1 \dfrac{d^{m-1}}{dx^{m-1}} + \cdots + p_m y = 0$, whose coefficients $p_i = p_i(x)$ were single-
valued in some neighbourhood of $x = a$ and may be written $\dfrac{f(x)}{(x-a)^{\pi_i}}$ in lowest
terms. He showed that, if in the sequence $\pi_1 + m - 1, \pi_2 + m - 2, \dots, \pi_m$, the
greatest value $g > 0$, then not all solutions of the equation are regular. Furthermore,
if $\pi_h + m - h$ is the first to equal g, then most $m - h$ solutions are regular. As he

observed, in Fuchs' study, $g \leq m$. In [1873] Thomé sought to characterise those
equations for which precisely $m - 1$ linearly independent solutions are regular, and
obtained an answer which I shall mention below. In subsequent papers Thomé
[1874, 1876] looked at the question of reducing differential equations to those of
lower order, and this work caught the attention of Frobenius.

Frobenius [1875b] considered the differential equation

$$A(y) = p_m \frac{d^m y}{dx^m} + p_{m-1} \frac{d^{m-1} y}{dx^{m-1}} \cdots + p_0 y = 0$$

and its adjoint

$$\hat{A}(y) = (-1)^m \frac{d^m}{dx^m}(p_m y) + \cdots - \frac{d}{dx}(p_1 y) + p_0 = 0,$$

an expression introduced independently by Fuchs, Thomé, and himself earlier (but
known already to Lagrange). He established that, if $B(y) = 0$ denotes another
differential equation, then the equation $A(B(y)) = 0$ has no fewer regular solutions
than $B = 0$ and no more than $A = 0$. (Solutions of $A(B(y)) = 0$ are either
solutions of $B(y) = 0$, or if $A(u) = 0$, of $B(y) = u$.) Now, if the equation $C = 0$
has a regular solution, say $x^\rho v$, where $v(x) = \sum_0^\infty a_n x^n$, then y satisfies the first
order differential equation $B_1(y) = 0$, where

$$B_1(y) := (\rho v + xv')y - xvy' = 0.$$

Frobenius redefined reducibility here so that, in the new sense of the word, $C(y) =
0$ is reducible. His definition applied only to equations whose coefficients near
$x = 0$ had the character of rational functions, and said that such a linear differential
equation was reducible if it had a solution in common with a linear differential
equation of lower order having the same property near $x = 0$. To avoid confusion
I shall refer to this property as local reducibility. In this terminology Frobenius
showed that an m-th order equation of the Fuchsian type is locally reducible to

$$B_m(B_{m-1}(\ldots(B_1(y))\ldots)) = 0,$$

where each B_i is of the first order and has a regular solutions, and that the converse
also holds. So in particular the regular solutions of a differential equation $C(y) =
0$ all satisfy a differential equation $B(y) = 0$ of the kind Frobenius considered,
and if $C = A(B(y)) = 0$, then $A(y) = 0$ has no regular solutions. Turning to
the adjoint equation $\hat{A}(y) = 0$ of $A(y) = 0$, Frobenius showed that if $A(y) =
0$ has only regular solutions, then $\hat{A}(y) = 0$ likewise has only regular solutions.
This followed from the result just cited and the reciprocity theorem due to him
and Thomé independently: that $\widehat{A(B(y))} = \hat{B}(\hat{A}(y))$. It was connected with the
number of regular solutions by the observation that the indicial equation at $x = 0$
has as many roots (counted according to multiplicity) as the differential equation

has regular solutions (§6 Theorem 3). So Frobenius obtained Thomé's result that $A(y) = 0$ of order γ has exactly β regular integrals, if and only if $\widehat{A}(y) = 0$ is reducible (in the earlier sense) to an equation of order $\gamma - \beta$, whose indicial equation is a constant. In particular, this implies Thomé's theorem mentioned above, that $A(y) = 0$ has all but one of its solutions regular if and only if its indicial equation is of order $\gamma - 1$, and the adjoint equation $\check{A}(y)$ has a solution of the form

$$\exp\left\{\left(\frac{c_n}{x^n} + \cdots + \frac{c_1}{x}\right)\right\} \sum_{\nu}^{\infty} a_\nu x^{\rho+\nu}.$$

These results conclude our discussion of Frobenius' work on differential equations, which is complete except for his short paper on the existence of algebraic solutions which will be described in Chapter III. It remains to look briefly at one further simplification of Fuchs' theory, that of Jordan and Hamburger, and one addition to it, that of Jules Tannery.

The form of the solution to a differential equation in the neighbourhood of a singular point $x = 0$ was discussed by Fuchs who, as has been seen, presented it in this form when the monodromy matrix had a k times repeated eigenvalue w and $e^{2\pi i r} = w$:

$$u_1 = x^r \phi_{11}$$
$$u_2 = x^r \phi_{21} + x^r \phi_{22} \log x$$
$$\cdots$$
$$u_k = x^r \phi_{k1} + x^r \phi_{k2} \log x + \cdots + x^r \phi_{kk} (\log x)^{k-1} \quad \text{[1866, 136, eq. 11]}.$$

The ϕ_{ij} are functions of x, single-valued near $x = 0$, and each is a linear combination of $\phi_{11}, \phi_{21}, \ldots, \phi_{k1}$. So, in the particular case when $k = 2$, the solutions in Fuchs' form are

$$u_1 = x^r \phi_{11}$$
$$u_2 = x^r \phi_{21} + x^r \phi_{22} \log x.$$

These arise, as Fuchs had shown, when the analytic continuation of solutions u_1 and u_2 around the singular point produces

$$\bar{u}_1 = w u_1$$
$$\bar{u}_2 = w_{21} u_1 + w u_2 \quad \text{[1866, 135 eq. 10 abbreviated]}.$$

It is easy to see, by writing $x e^{2\pi i}$ for x, that a suitable basis of solutions in this case is $u_1 = x^r \phi$, $u_2 = \dfrac{x^r \phi \log x}{2\pi i w}$ for which $w_{21} = 1$. In other words, the monodromy matrix with respect to the basis has the form $\begin{pmatrix} w & 0 \\ 1 & w \end{pmatrix}$, and is in Jordan canonical form.

The Jordan canonical form of a matrix had been introduced by Jordan in his *Traité des substitutions et des équations algébriques* [1870, Section II, Chapter 2] to simplify the discussion of linear substitutions. That same year Yvon Villarceau drew the attention of the Paris Académie des Sciences to a gap in the general theory of linear differential equations with constant coefficients: the method used to resolve even a second order equation into two first order ones breaks down when the characteristic equation has equal roots. Villarceau wished to resolve $\dot{Y} = AY$, where Y was a vector and A a matrix of constants, into $\dot{y}_1 = \lambda_1 y_1$, $\dot{y}_2 = \lambda_2 y_2$, but of course this is impossible in the case just mentioned.

Jordan observed [1871] that the gap is readily filled by his theory of canonical forms for matrices, which he proceeded to do, so simplifying Fuchs' form of the solution in the case of an equation with constant coefficients (which is, however, not an equation of the Fuchsian type). The introduction of Jordan canonical form into the Fuchsian theory of differential equations was carried out by Hamburger [1873]. He concluded (p. 121) that, if w is a k-fold repeated eigenvalue of a monodromy matrix, then a basis of solutions $y_1, \ldots, y_k$ can be found for which the transforms under analytic continuation are

$$\tilde{y}_1 = wy_1$$
$$\tilde{y}_2 = y_1 + wy_2$$
$$\cdots$$
$$\tilde{y}_m = y_{m-1} + wy_m .$$

Jordan's theory of canonical forms had been preceeded, in 1858 and 1868, by Weierstrass' theory of elementary divisors which he developed in connection with his study of the problem of simultaneously diagonalizing two bilinear or quadratic forms. Hamburger also sketched the correspondence between Jordan's analysis, which he had presented in quite an elementary form, and that of Weierstrass, which makes considerable use of the theory of determinants. However, the theory of matrices, especially in Germany, continued to rely on Weierstrassian ideas at least until the turn of the century, as Hawkins [1977] has established. The reader is referred to Hawkins's paper for a full discussion of Weierstrass' theory and its influence.

In 1875 Tannery published a paper on linear differential equations which clearly resembled those of Fuchs. As observed above, it corrected a mistake of Fuchs' [1866], and it also contained a pleasing converse to Fuchs' main theorem. Tannery showed [1875, 130] that if $y_1, y_2, \ldots, y_m$ were functions of x continuous except at points $a_1, \ldots, a_\rho$ around which each become, on analytic continuation, a linear combination of $y_1, y_2, \ldots, y_m$, then they were solutions of a linear differential equation of order m with single-valued coefficients:

$$\frac{d^m y}{dx^m} = p_1 \frac{d^{m-1} y}{dx^{m-1}} + \cdots + p_m y, \text{ where } p_i = D_i/D'$$

$$D = \begin{vmatrix} y_1 & \dfrac{dy_1}{dx} & \cdots & \dfrac{d^{m-1}y_1}{dx^{m-1}} \\ y_2 & \dfrac{dy_2}{dx} & \cdots & \dfrac{d^{m-1}y_2}{dx^{m-1}} \\ \cdot & \cdot & \cdots & \cdot \\ y_m & \dfrac{dy_m}{dx} & \cdots & \dfrac{d^{m-1}y_m}{dx^{m-1}} \end{vmatrix}$$

and D is obtained from D by replacing the ith column by the transpose of $\left(\dfrac{d^m y_1}{dx^m}, \dfrac{d^m y_2}{dx^m}, \ldots, \dfrac{d^m y_m}{dx^m} \right)$. So, in particular, every algebraic function of order m satisfies such a differential equation. The proof is straightforward, and I omit it.

Exercises

1. (a) Show that a necessary condition for the second order linear differential equation

    $$\frac{d^2 w}{dz^2} + p(z)\frac{dw}{dz} + q(z)w = 0$$

    to have z_0 as a regular singular point is that $(z - z_0)p(z)$ and $(z - z_0)^2 q(z)$ are analytic near z_0.

    (b) Show, by writing $\tilde{z} = \frac{1}{z}$, that the equation becomes

    $$\tilde{z}^4 \frac{d^2 w}{d\tilde{z}^2} + \tilde{z}^2(2\tilde{z} - p(\tilde{z}^{-1}))\frac{dw}{d\tilde{z}} + q(\tilde{z}^{-1})w = 0,$$

    and deduce that a necessary condition for $z = \infty$ to be a regular singular points is that $zp(z)$ and $z^2 q(z)$ be analytic at infinity.

    (c) Show that these conditions are also sufficient.

2. Show that the point $z = 0$ can be a singular point of the coefficients of a differential equation although the solutions are analytic there (such points are called accidental singular points) by carrying through the following argument. Let the equation be

    $$w'' + \frac{p(z)}{z}w' + \frac{q(z)}{z^2}w = 0, \quad \text{where } w' = \frac{dw}{dz}, \text{ etc.}$$

    and suppose $w_1(z)$ and $w_2(z)$ are independent solutions analytic near $z = 0$. Form

    $$\Delta = \begin{vmatrix} w_1' & w_1 \\ w_2' & w_2 \end{vmatrix}, \quad \Delta_1 = \begin{vmatrix} w_1'' & w_1 \\ w_2'' & w_2 \end{vmatrix}, \quad \Delta_2 = \begin{vmatrix} w_1' & w_1'' \\ w_2' & w_2'' \end{vmatrix}.$$

    (a) Show that $z^{-1}p(z) = -\Delta_1/\Delta$ and $z^{-2}q(z) = -\Delta_2/\Delta$.

(b) Show that $\Delta(0) = 0$.

(c) Show that

$$\frac{d}{dz}\log\Delta = -z^{-1}p(z)$$

$$= -z^{-1}p(0) + \frac{dG}{dz},$$

where G is analytic near $z = 0$, and deduce that

$$\Delta = cz^{-p(0)}e^{G(z)}.$$

(d) Deduce that $p(0)$ must be a negative integer, and that the roots of the indicial equation must be negative integers.

(e) For the singularity to be accidental, no logarithmic terms may appear in the solution. Show that if they are excluded, then the roots r_1 and r_2 of the indicial equation are distinct.

(f) Deduce from the fact that Δ has a pole of order $r_1 + r_2 - 1$ at 0 (why?), that $z = 0$ will be a regular if $r_1 = 1$ and $r_2 = 0$ and not otherwise. When $z = 0$ is an accidental singular point, $p(0) = 1 - (r_1 + r_2) < 0$.

3. Show that the following equations all have accidental singular points at the origin:

(a) $z^2\dfrac{dw}{dz} - (4z - z^2)\dfrac{dw}{dz} + (4 - z)w = 0$.

(b) $z^2\dfrac{d^2w}{dz^2} - 2(4 - \lambda z^2)\dfrac{dw}{dz} + (6 + \mu z^2)w = 0$
(for all values of the constants λ and μ.)

4. Show that the following third order equation, due to Hurwitz (see Chapter V, n. 13), has an accidental singular point at $z = 1$:

$$z^2(z-1)^2\frac{d^3w}{dz^3} + (7z-4)z(z-1)\frac{d^2w}{dz^2}$$

$$+ \left(\frac{72}{7}z(z-1) - \frac{20}{9}(z-1) + \frac{3z}{4}\right)\frac{dw}{dz}$$

$$+ \left(\frac{72.11}{73}(z-1) + \frac{5}{8} + \frac{2}{23}\right)w = 0.$$

5. Show that none of the following equations are of the Fuchsian type.

(a) $z^2\dfrac{d^2w}{dz^2} + z\dfrac{dw}{dz} + (z^2 - a^2)w = 0$ (Bessel's equation)

(b) $z\dfrac{d^2w}{dz^2} + (a_0 + b_0z)\dfrac{dw}{dz} + (a_1 + b_1z)w = 0$ (Laplace's equation)

(c) $\dfrac{d^2w}{dz^2} + (c - z^2/4)w = 0$ (the Hermite–Weber equation)

(d) $\dfrac{d^2w}{dz^2} + (a + b\cos 2z)w = 0$ (Mathieu's equation)

Question 5 shows that several of the better known differential equations of mathematical physics are not of the Fuchsian type, and so the study of equations of the Fuchsian type could not be directly connected to physical problems. However, Klein and Bôcher in 1894 showed that it was possible to obtain physically significant equations from those of Fuchsian type by a process of amalgamating singularities in one particular equation. This process is well described in Whittaker and Watson, 203–204, as follows.

6. Show that the general equation of the Fuchsian type with 5 singular points can be written as

$$
\frac{d^2w}{dz^2} + \left(\sum_1^4 \frac{1 - \alpha_r - \beta_r)}{z - a_r} \right) \frac{dw}{dz}
$$
$$
+ \left(\sum_1^4 \frac{\alpha_r \beta_r}{(z - a_r)^2} + \frac{Az^2 + 2Bz + C}{(z - a_1)\ldots(z - a_4)} \right) w = 0
$$

where the exponents at a_r and α_r and β_r, and at ∞ are μ_1 and μ_2, provided A is chosen so that μ_1 and μ_2 are the roots of $\mu^2 + \mu \left(\sum_1^4 (\alpha_r + \beta_r) - 3 \right) + \sum \alpha_r \beta_r + A = 0$, and B and C are constants.

What in particular does this equation look like when $\beta_r - \alpha_r = \frac{1}{2}$, $\mu_2 - \mu_1 = \frac{1}{2}$? (The equation in this form is called the generalised Lamé equation.)

7. Let $a_1 \to a_2$ in the generalised Lamé equation, and show that the new exponents at a_2 are α and β, where

$$
\alpha + \beta = 2(\alpha_1 + \beta_2)
$$

$$
\alpha\beta = \alpha_1 \left(\alpha_1 + \frac{1}{2} \right) + \alpha_2 \left(\alpha_2 + \frac{1}{2} \right) + \frac{Aa_1^2 + 2Ba_1 + C}{(a_1 - a_3)(a_1 - a_4)}.
$$

The new exponents depend on B and C, so they may be chosen arbitrarily by making a suitable choice for B and C. This process, called amalgamation, has produced an equation of this Fuchsian type in which three singular points have exponent difference $\frac{1}{2}$ and one (a_2) has arbitrary exponent difference.

8. Show that amalgamating a_3 with a_2 can produce an equation not of the Fuchsian type.

9. Show that equations can be produced in this way having ℓ regular singular points with exponent difference $\frac{1}{2}$, m with arbitrary exponent difference, and n irregular singular points, where

$$\ell = 3, m = 1, n = 0, \quad \text{(Lamé)}$$
$$\ell = 2, m = 0, n = 1, \quad \text{(Mathieu)}$$
$$\ell = 1, m = 2, n = 0, \quad \text{(Legendre)}$$
$$\ell = 0, m = 1, n = 1, \quad \text{(Bessel)}$$
$$\ell = 1, m = 0, n = 1, \quad \text{(Hermite–Weber)}$$
$$\ell = 0, m = 0, n = 1, \quad \text{(Stokes).}$$

10. Show that $\dfrac{dy}{dx} = -y \log^2 y$ has the solution $y = \exp((x - c)^{-1})$ which has an essential singular point at $x = c$ which is independent of the equation.

11. Show that the substitution $y = u e^{-\frac{1}{2} \int p \, dx}$ reduces

$$\frac{d^2 y}{dx^2} + p \frac{dy}{dx} + qy = 0 \text{ to } \frac{d^2 u}{dx^2} + Pu = 0$$

where $P = -\frac{1}{4} p^2 - \frac{1}{2} \frac{dp}{dx} + q$ (p and q being rational functions of x).

Show that if f and g are two linearly independent solutions of $\dfrac{d^2 u}{dx^2} + Pu = 0$ and $\eta = \dfrac{f}{g}$, then $\dfrac{\eta'''}{\eta'} - \dfrac{3}{2} \left(\dfrac{\eta''}{\eta'} \right)^2 = 2P$.

12. Show that a general equation of the Fuchsian type and of the second order, with m finite singularities $a_1, a_2, \ldots, a_m$ and a singular point at ∞, which has exponents $\alpha_1, \beta_1, \alpha_2, \beta_2, \ldots, \alpha_m, \beta_m$, and α, β at those points (where $\alpha_1 + \beta_1 + \cdots + \alpha_m + \beta_m + \alpha + \beta = m - 1$) is of the form

$$y'' + \left(\sum_1^n \frac{A_i}{z - a_i} \right) y' + \sum \left(\frac{B_i}{(z - a_i)^2} + \frac{C_i}{z - a_i} \right) y = 0$$

where $\displaystyle\sum_1^m C_i = 0$.

Show moreover that $A_i = 1 - \alpha_i - \beta_i$, and $B_i = \alpha_i \beta_i$.
Deduce that the equation can be written as

$$y'' + \left(\sum_1^m \frac{1 - \alpha_i - \beta_i}{z - a_i} \right) y' + \left(F_{m-2}(z) + \sum_1^m \frac{\alpha_i \beta_i \psi'(a_i)}{z - a_i} \right) \frac{y}{\psi} = 0 \quad (*)$$

where F_{m-2} is a polynomial of degree $m - 2$ with leading coefficient $\alpha\beta$, and $\psi = (x - a_1) \ldots (x - a_m)$.

Show that the substitutions $w = (z - a_1)^{\alpha_1}(z - a_2)^{\alpha_2} \ldots (z - a_m)^{\alpha_m} y$
produces an equation whose exponents are 0 and $\beta_i - \alpha_i$, and (at ∞)
$\alpha + \sum \alpha_i$, $\beta + \sum \alpha_i$. Write down the form of this equation, substituting
$\lambda_i = \beta_i - \alpha_i$, $\sigma = \alpha + \sum \alpha_i$, $\tau = \beta + \sum \alpha_i$. Show that the substitution
of Question 11 here takes the form $u = (z - a)^{\frac{1}{2}(1 - \lambda_m)}$, and converts $(*)$
into an equation of the form

$$u'' + \frac{u}{4\psi}[F_{m-2}(z) + \sum \frac{(1 - \lambda_i)^2 \psi'(a_i)}{z - a_i}] = 0,$$

where $F_{m-2}(z)$ is a polynomial of degree $m - 2$ with leading term $(1 - (\sigma - \tau)^2)z^{m-2}$. Show that the exponents are now $\frac{1}{2}(1 - \lambda_i)$, $\frac{1}{2}(1 + \lambda_i)$,
and $\frac{1}{2}(-1 + \sigma - \tau)$, $\frac{1}{2}(-1 - \sigma + \tau)$.

13. Verify Fuchs' observation that, when $m > 2$, the position and nature
    of the singular points does not determine the differential equation com-
    pletely.

14. Show that the general equation of the Fuchsian class with three finite
    singular points (Heun's equation) is of the form

$$y'' + \left(\frac{\lambda_0}{z} + \frac{\lambda_1}{z - 1} + \frac{\lambda_2}{z - a}\right) y' + \frac{\sigma z(z - q)}{z(z - 1)(z - a)} y = 0$$

where q is an arbitrary (accessory) parameter.

Chapter III

Algebraic Solutions to a Differential Equation

Introduction

This chapter considers how Fuchs' problem: when are all solutions to a linear ordinary differential equation algebraic? was approached, and solved, in the 1870s and 1880s. First, Schwarz solved the problem for the hypergeometric equation. Then Fuchs solved it for the general second-order equation by reducing it to a problem in invariant theory and solving that problem by *ad hoc* means. Gordan later solved the invariant theory problem directly. But Fuchs' solution was imperfect, and Klein simplified and corrected it by a mixture of geometric and group-theoretic techniques which established the central role played by the regular solids already highlighted by Schwarz. Simultaneously, Jordan showed how the problem could be solved by purely group-theoretic means, by reducing it to a search for all finite monodromy groups of 2×2 matrices with complex entries and determinant 1. He was also able to solve it for 3rd and 4th order equations, thus providing the first successful treatment of the higher order cases, and to prove a general finiteness theorem for the nth order case (Jordan's finiteness theorem). Later Fuchs and Halphen were able to treat some of these cases invariant-theoretically.

The problem occupied the attention of many leading mathematicians in this period, and provided an interesting test of the relative powers of the older methods of invariant theory and the new group-theoretic ones, which favoured the new techniques. It also led to Schwarz's discovery of a new class of transcendental functions associated with the hypergeometric equation which, although not appreciated at the time, were to be of vital importance in the theory of automorphic functions (discussed in Chapter VI).

3.1   Schwarz

On 22nd August 1871, at a meeting of the mathematical section of the Swiss *Natur-forschenden Gesellschaft*, H.A. Schwarz announced the solution to the problem: "when is the Gaussian hypergeometric series $F(\alpha, \beta, \gamma, x)$ an algebraic function of its fourth element?" His paper on this question appeared in the *Journal für Mathematik* for 1872 (Schwarz [1872]) and his arguments have been popular ever since. Schwarz wrote the equation for complex x and y as

$$\frac{d^2y}{dx^2} + \frac{\gamma - (\alpha + \beta + 1)x}{x(1-x)}\frac{dy}{dx} - \frac{\alpha\beta}{x(1-x)}y = 0 \qquad (30)$$

and considered two different cases: where the equation has only one algebraic solution, and where it has two linearly independent algebraic solutions. These cases must be treated separately, and Schwarz first considered the simpler case where one particular integral is algebraic; but either this algebraic function or its logarithmic derivative is a rational function, so the quotient of two branches of the function is constant, and the existence of a second algebraic solution cannot be inferred. From the work of Fuchs, he said, it is clear that the general solution of (3.1.1) has one of the following forms as a convergent power series:

$$x^a(1 + c_1x + c_2x^2 + \cdots),$$

$$\left(\frac{1}{x}\right)^b(1 + \frac{c_1}{x} + \frac{c_2}{x^2} + \cdots), \text{ or}$$

$$(1-x)^c(1 + c_1(1-x) + c_2(1-x)^2 + \cdots)$$

$$a = 0 \text{ or } 1 - \gamma$$
$$b = \alpha \text{ or } \beta$$
$$c = 0 \text{ or } \gamma - \alpha - \beta.$$

If (30) has a particular integral, y_1, which is algebraic and whose logarithmic derivative is a rational function of x, then it must be of the form

$$y_1 = x^a(1-x)^c g(x),$$

where a and c are rational numbers and g is a polynomial function of x of degree n, say. For simplicity, one may assume $b = \alpha$, when there are four sub-cases to consider, according as $a = 0$ or $1 - \gamma$ and $c = 0$ or $\gamma - \alpha - \beta$. Schwarz showed that each of them is possible and that the corresponding $F(\alpha, \beta, \gamma, x)$ is algebraic.

The rest of the paper was devoted to the case of two linearly independent algebraic solutions. In this case every solution is algebraic, and the quotient of any two solutions is also algebraic. The equation

$$\frac{d^2y}{dx^2} + p\frac{dy}{dx} + qy = 0 \qquad (31)$$

may be supposed to have two linearly independent solutions y_1 and y_2. They satisfy

$$y_2 \frac{dy_1}{dx} - y_1 \frac{dy_2}{dx} = Ce^{-\int p\,dx},$$

a result which Schwarz took from Abel [1827], a paper in which Abel derived differential equations, notably Legendre's, for functions defined by definite integrals. In this case

$$y_2 \frac{dy_1}{dx} - y_1 \frac{dy_2}{dx} = Cx^{-\gamma}(1-x)^{\gamma-\alpha-\beta-1}$$

so γ and $\alpha + \beta$ must be rational numbers. Indeed, said Schwarz, one sees by consulting entries 9 and 10 in Kummer's table of 24 solutions that α and β must themselves be rational, and for the rest of the paper he therefore assumed α, β and γ were rational numbers.

Schwarz proposed considering the quotient of y_1 and y_2, and the related quotients $s = \dfrac{C_1 y_1 + C_2 y_2}{C_3 y_1 + C_4 y_2}$ where $C_1, \ldots, C_4$ are constants. These quotients all satisfy a differential equation obtained by eliminating the three ratios $C_1 : C_2 : C_3 : C_4$ by successive differentiation:

$$\psi(s,x) = \frac{2\frac{ds}{dx} \cdot \frac{d^3 s}{dx^3} - 3\left(\frac{d^2 s}{dx^2}\right)^2}{2\left(\frac{ds}{dx}\right)^2} = 2p - \frac{1}{2}p^2 - \frac{dp}{dx} = F(x). \tag{32}$$

Following the usage later established by Cayley ([1883, 31]) $\psi(s,x)$ will be called the Schwarzian of s with respect to x or, less briefly, the Schwarzian derivative.[1] Schwarz regarded the equation $\psi(s,x) = F(x)$ as a special case of Kummer's equation in [1834] discussed above.

In the present case

$$\psi(x) = 2p - \frac{1}{2}p^2 - \frac{dp}{dx} = \frac{1 - (1-\gamma)^2}{2x^2} + \frac{1 - (\gamma - \alpha - \beta)^2}{2(1-x)^2}$$
$$- \frac{(1-\gamma)^2 - (\alpha - \beta)^2 + (\gamma - \alpha - \beta)^2 - 1}{2x(1-x)}$$

or, on setting

$$(1-\gamma)^2 = \lambda^2, \ (\alpha - \beta)^2 = \mu^2, \ (\gamma - \alpha - \beta) = \nu^2,$$

(32) becomes

$$\psi(s,x) = \frac{1 - \lambda^2}{2x^2} + \frac{1 - \nu^2}{2(1-x)^2} - \frac{\lambda^2 - \mu^2 + \nu^2 - 1}{2x(1-x)}. \tag{33}$$

λ, μ, ν will be taken to the positive roots of λ^2, μ^2, ν^2, respectively. The advantage of (32) over (31) is that, as Heine pointed out to Schwarz, if y_1/y_2 and $e^{-\int p\,dx}$

are both algebraic, then y_1 and y_2 are algebraic, for $y_2^2 \dfrac{d}{dx}\left(\dfrac{y_1}{y_2}\right) = Ce^{-\int p\,dx}$. So Schwarz needed only to consider the algebraic nature of one function, y_1/y_2, not two.[2]

The power series solutions of (30) involve x or $1 - x$, so the effect of replacing x by $z = \dfrac{c_1 x + c_2}{c_3 x + c_4}$ is therefore to be considered:

$$\psi(s, x) = \left(\frac{dz}{dx}\right)^2 \psi(s, z).$$

The effect of replacing x by $1 - x$, or $\frac{1}{x}$ or compositions thereof on the solutions to (32) is then readily seen to be, if $s(\lambda, \mu, \nu, x)$ is one solution:

$$
\begin{aligned}
s(\lambda, \mu, \nu, z) &= s(\lambda, \mu, \nu, x) && \text{if } z = x, \\
&= s(\nu, \mu, \lambda, x) && \text{if } z = 1 - x, \\
&= s(\mu, \lambda, \nu, x) && \text{if } z = 1/x, \\
&= s(\nu, \lambda, \mu, x) && \text{if } z = 1/(1 - x), \\
&= s(\lambda, \nu, \mu, x) && \text{if } z = x/1 - x, \\
&= s(\mu, \nu, \lambda, x) && \text{if } z = \frac{x - 1}{x},
\end{aligned}
$$

which agrees with Riemann's theorem concerning his P-function, so $s(\lambda, \mu, \nu, x)$ is a quotient of two linear independent branches of the P-function $P(\lambda, \mu, \nu, x)$.

To solve (32), Schwarz first considered the solutions near a point $x_0 \neq 0$, 1, or ∞. Standard power-series methods, together with Kummer's solutions to (33) enabled him to establish this theorem:

Theorem 3.1   *The map $s(\lambda, \mu, \nu, x)$ from the complex x-plane to the complex s-plane maps each simply-connected region X not containing 0, 1 or ∞ onto a simply connected region S containing ∞ once or several times in its interior but having no branch point in its interior.*

If in particular x_0 is real and neither 0, 1, nor ∞ then, since λ^2, μ^2, and ν^2 are real, s is real when x is and S is marked out by circular arcs.

Next, he considered the solutions to (33) in the neighbourhood of the singular points $x = 0$, 1, ∞. Now he could show

Theorem 3.2   *The upper-half x-plane E is mapped conformally by s, a particular integral of (32), onto a simply connected domain S, having no winding point in its interior, which is, in general, a circular arc triangle. The angles at the vertices corresponding to 0, 1, ∞ are $\lambda\pi$, $\mu\pi$ and $\nu\pi$ respectively.*

How are these circular-arc triangles connected? To avoid irksome special cases, Schwarz first assumed that none of the λ, μ, ν are integers and, to avoid overlapping the triangles unnecessarily, he reduced λ, μ and ν mod 2. It then turns out by the reflection principle that each domain S is a circular arc triangle for which

$$\lambda + \mu + \nu > 1 \qquad \text{and}$$

$$-\lambda + \mu + \nu < 1, \quad \lambda - \mu + \nu < 1, \quad \lambda + \mu - \nu < 1.$$

In precisely this case, S can be taken to be bounded by great circles. An adjoining triangle S_1 comes from a second copy of the half plane E, say E_1, and corresponding points in E and E_1 are mapped onto reciprocal points in S and S_1, that is, to points which are images under the Möbius transformation of inversion in the common side. For reasons of symmetry, each triangular region S is the image of its neighbour, and the question reduces to finding all circular-arc triangles which, when so transformed, give only a finite covering of the sphere. This is equivalent to finding such triangles that only occupy a finite number of positions upon successive reflections in their sides, and in this case s is an algebraic function of $x \in E$, since, said Schwarz, the Riemann surfaces of x and s are then closed surfaces with finitely many leaves. Schwarz observed that this problem had already been discussed to some extent by Riemann, in a paper published just after his death ([Riemann 1867]) where, in §12, Riemann considered the case when $\dfrac{du}{d \log \eta}$ is an algebraic function of η, and in §18 alluded to the conformal representations of regular solids onto the sphere.

Indeed, Schwarz noted, the solution of his (Schwarz's) problem is precisely that the triangles must either fit together to form a regular double pyramid (angles $\frac{\pi}{2}$, $\frac{\pi}{2}$, $\nu\pi$, $\nu = \frac{1}{n}$) or a regular solid. This gave him a table of 15 cases (up to an ordering of λ, μ, ν) in which s was algebraic, given in Table 3.1, and in all other cases (when $\lambda + \mu + \nu \leq 1$ or $\lambda + \mu + \nu > 1$ but λ, μ, or ν not as tabled) s was transcendental. In each case it is possible to write down the associated Gaussian hypergeometric series and exhibit it directly as an algebraic function. Schwarz gave as an explicit example the case $\lambda = \frac{1}{3} = \mu$, $\nu = \frac{1}{2}$, for which the regular solid is a tetrahedron, divided by its symmetry planes into 24 triangles with angles $\frac{\pi}{3}$, $\frac{\pi}{3}$, $\frac{\pi}{2}$ and, in slightly less detail, the cases of the octahedron, icosahedron, and dodecahedron. His analysis was extended by Brioschi [1877a, b] and completed by Klein [1877], see Section 3.3.

It remained for Schwarz to consider the special cases where some of λ, μ, ν are integers. If $\lambda = 0$, the function s is necessarily transcendental. When $\lambda = m \neq 0$, he showed that s can only be algebraic if $B_m = 0$ and $x = 0$ is an accidental singularity. This, it turned out, would happen if and only if one of $|\lambda| - |\mu + \nu|$ or $|\lambda| - |\mu - \nu|$ was an odd positive integer. Further conditions are necessary for s to be algebraic, namely, that all of λ, μ, ν are non-zero integers, their sum is odd, and the sum of the absolute value of any two exceeds the absolute value of the third.

Schwarz did more than solve the problem of finding algebraic solutions to the hypergeometric equation. His thorough treatment of the tetrahedral case revealed elegant connections with elliptic function theory, as one might expect, but even more important for the direction of future work was his investigation of the simplest transcendental cases when $\lambda + \mu + \nu < 1$. This occupied §5 of his paper.

Here he showed that a circular-arc triangle with angles $\lambda\pi$, $\mu\pi$, $\nu\pi$ can be formed with sides perpendicular to a fixed boundary circle, and that successive reciprocation can then fill out the interior of this fixed circle with copies of the original triangle. A picture of the case $\lambda = \frac{1}{5}$, $\mu = \frac{1}{4}$, $\nu = \frac{1}{2}$ was supplied. The function s is necessarily transcendental, but it is single-valued whenever $\frac{1}{\lambda}$, $\frac{1}{\mu}$, $\frac{1}{\nu}$ are integers.

No.	λ''	μ''	ν''	area π	
I.	$\frac{1}{2}$	$\frac{1}{2}$	ν	ν	dihedron
II.	$\frac{1}{2}$	$\frac{1}{3}$	$\frac{1}{3}$	$\frac{1}{6} = A$	tetrahedron
III.	$\frac{2}{3}$	$\frac{1}{3}$	$\frac{1}{3}$	$\frac{1}{3} = 2A$	
IV.	$\frac{1}{2}$	$\frac{1}{3}$	$\frac{1}{4}$	$\frac{1}{12} = B$	cube or
V.	$\frac{2}{3}$	$\frac{1}{4}$	$\frac{1}{4}$	$\frac{1}{6} = 2B$	octahedron
VI.	$\frac{1}{2}$	$\frac{1}{3}$	$\frac{1}{5}$	$\frac{1}{30} = C$	dodecahedron
VII.	$\frac{2}{5}$	$\frac{1}{3}$	$\frac{1}{3}$	$\frac{1}{15} = 2C$	or
VIII.	$\frac{2}{3}$	$\frac{1}{5}$	$\frac{1}{5}$	$\frac{1}{15} = 2C$	icosahedron
IX.	$\frac{1}{2}$	$\frac{2}{5}$	$\frac{1}{5}$	$\frac{1}{10} = 3C$	
X.	$\frac{3}{5}$	$\frac{1}{3}$	$\frac{1}{5}$	$\frac{2}{15} = 4C$	
XI.	$\frac{2}{5}$	$\frac{2}{5}$	$\frac{2}{5}$	$\frac{1}{5} = 6C$	
XII.	$\frac{2}{3}$	$\frac{1}{3}$	$\frac{1}{5}$	$\frac{1}{5} = 6C$	
XIII.	$\frac{4}{5}$	$\frac{1}{5}$	$\frac{1}{5}$	$\frac{1}{5} = 6C$	
XIV.	$\frac{1}{2}$	$\frac{2}{5}$	$\frac{1}{3}$	$\frac{7}{30} = 7C$	
XV.	$\frac{3}{5}$	$\frac{2}{5}$	$\frac{1}{3}$	$\frac{1}{3} = 10C$	

Table 3.1

In such a case the fixed circle is a natural boundary of s, and s cannot be analytically continued onto the boundary. This phenomenon had been noticed earlier by Weierstrass in 1863, indeed, Schwarz went on, Kronecker had pointed out that the θ-series $\sqrt{\frac{2K}{\pi}}$

$$\sqrt{\frac{2K}{\pi}} = 1 + 2q + 2q^4 + 2q^9 + \cdots$$

gives an example where q cannot be taken on or outside the unit circle, nor can the function be analytically continued past $|q| = 1$ any other way.[3] Other examples pertained to the case $\lambda + \mu + \nu = 0$, and had been discussed by Weierstrass in 1866. Thus, if in the usual notation:

$$K = \int_0^1 \frac{dx}{\sqrt{[(1 - x^2)(1 - k^2 x^2)]}}, \quad K' = \int_0^1 \frac{dx}{\sqrt{[(1 - x^2)(1 - k'^2 x^2)]}}$$

where $k^2 + k'^2$ then

$$s = \frac{a'K + b'K'}{aK + bK}$$

is a function of k^2 when a, b, a', b' are given, and s is bounded but k^2 unbounded.

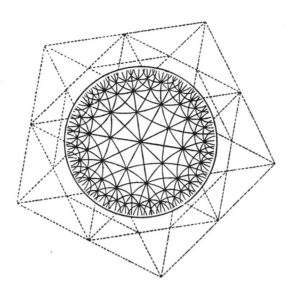

Figure 3.1

Yet another Weierstrassian example given by Schwarz pertained to the case $\lambda + \mu + \nu = 1$, when the triangles tessellate the plane and s is a single-valued function provided, up to order, either $\lambda = \frac{1}{3} = \mu = \nu$, or $\lambda = \frac{1}{2}, \mu = \nu = \frac{1}{4}$, or $\lambda = \frac{1}{2}, \mu = \frac{1}{3}, \nu = \frac{1}{6}$. In these cases x is a single-valued function of s as s runs over the plane.

Schwarz's elegant solution of the all-solutions algebraic problem for the case of the hypergeometric equation indicated the important role to be played by geometrical reasoning in analysing such problems. But it was an open question for his contemporaries as to how to proceed with the attractive problem of dealing either with the second order equation when more than three singular points are present, or with equations of higher order. The challenges were to be met both with the then-traditional methods of invariant-theory, and, more directly, with the then-novel methods of group theory. I shall suggest that the relative success of the new group-theoretic methods, coupled with their triumphant extension into Schwarz's transcendental case ($\lambda + \mu + \nu < 1$), did much to advance group theory at the expense of invariant theory.

3.2 Generalizations

Schwarz's solution to the algebraic-solution problem for the hypergeometric equation had been simple, because the hypergeometric equation is unique amongst second order equations of the Fuchsian class in that the exponents at the singular points uniquely determine the nature of the solutions, as Fuchs had shown. So the question of extending his results was a difficult one. It was to prove much easier to treat the second order case with n singular points and several attempts were made at this problem. These may be divided into two kinds, and briefly characterised as invariant-theoretic and group-theoretic.

In the early 1870s invariant theory was a central domain of mathematics. It is the study of forms (homogeneous polynomials of some degree in various indeterminates) enjoying special properties under linear changes of the indeterminates. Expressions in the coefficients of a form, which were unchanged by all such transformations were called absolute invariants, or else, if they altered by a constant which depended on the coefficients of the form but was independent of the linear transformation, they were called relative invariants. Analogous expressions involving the indeterminates were called covariants. A computational procedure was developed for producing new covariants from old, and the central question in any given problem in invariant theory was always to find a complete basis of covariants in terms of which all other covariants could be expressed as sums and products. Gordan, the leading invariant theorist of his day, had established the existence of a finite basis for binary forms in his important [1868].

The well-developed theory of binary forms, however, stood in marked contrast to the difficulties encountered in extending the study of invariants to forms in three or more variables, and so it is not surprising to find that Fuchs and Gordan, who sought to use invariant theory to solve the algebraic-solution problem, first considered the second-order differential equation.

Group theory on the other hand was a new subject in the 1870s although implicitly group-theoretic ideas had been around for some time (see Wussing [1969]). Crucial books and papers stressing the importance of the group idea were those of Jordan: *Traité des substitutions et des équations algébriques* (1870), Sylow [1872], and, if only for what it tells us of Klein himself, Klein's *Erlanger Programm* [1872] (see Klein [1998]). Klein drew his inspiration from Jordan, whose *Traité* he was later to describe as a book with seven seals (Klein [1922]), and he was at pains in the *Erlanger Programm* to stress the value of relating group theory to the invariants as a classificatory principle. (He had gone to Paris in 1870 with his friend Sophus Lie to learn the new ideas, although his studies were interrupted by the Franco–Prussian war.)

Jordan's *Traité* contains a range of applications of group theory to mathematics, and so it was natural for him and Klein to take up the algebraic-solutions problem in differential equations from that point of view, but the *Traité* is even more remarkable for its reformulation of the ideas of Galois theory.[4] Lagrange's work on the theory of equations emphasized polynomials called resolvents with certain invariance properties under permutations of the roots. Replacing the indeterminate, say x, in these polynomials by a homogenous coordinate $x_1 : x_2$ produces a binary form, and indeed invariant theory derives in part from this kind of study. Jordan's *Traité* is the first book to place the underlying groups of permutations in the foreground and to diminish the importance of specific polynomials. Jordan took the burden of Galois' ideas to lie in a permutation-theoretic form of the theory of groups and their normal subgroups which Galois had begun, and he developed this theory of such groups without regard for the hierarchies of invariant forms which correspond to the subgroups of a given Galois group. German mathematicians, on the other hand, notably Kronecker and Dedekind, sought to explore this correspondence between groups and families of invariant functions, as is described in Chapter IV. This is also the approach Klein envisaged for the algebraic-solutions problems, as is particularly clear in his treatment of the transcendental functions discovered by Schwarz. Klein was also extremely interested in the resolvents of equations and their geometric representation. This particular emphasis, present in his work and Gordan's, but absent from Jordan's, led Gordan to call their study jocularly "hyper-Galois" theory, which Klein thought somewhat unfair to Galois (Klein [1922, 261: 1967, 90]). So one may say the schools of thought brought to bear on the algebraic-solutions problem derived from the traditional approach to the theory of polynomials, as it had developed into invariant-theory; the modern group-theoretical approach, pioneered by Jordan, of Galois theory; and a geometric blending of the two, preferred by Klein.

The first person to follow up Schwarz's paper was Lazarus Fuchs, returning to the problem he had first stated in 1865. His methods were traditional, and, as will be seen, by no means completely successful. Later work by Gordan resolved the matter fully in invariant-theoretic terms, but by then Klein and Jordan had also solved the problem group-theoretically, and Jordan had gone beyond it to a study of the third order case. It seems that on this question the new methods surpassed the old.

Fuchs' solutions

Fuchs published five accounts of his work: a summary [1875], a complete account [1875], a complete account [1876a], a short note for French readers in the form of a letter to Hermite [1876b], and more definitively, [1878]. I shall proceed to give a summary of the main paper, [1876a].

If a second order differential equation with rational coefficients

$$\frac{d^2u}{dz^2} + p\frac{du}{dz} + qu = 0 \tag{34}$$

has an algebraic solution u_1, then u_1 is a root of some irreducible polynomial equation

$$A_m u^m + A_{m-1} u^{m-1} + \cdots + A_0 = 0 \tag{35}$$

where the A_i, $i = 0, 1, \ldots, m - 1$, are rational functions of z. Any other root of (35) is also a solution to (34), because (35) is irreducible, so the question arises: how are the roots of (35) related? Essentially two cases can arise, which are the cases considered by Schwarz. Either there are two roots, say u_1 and u_2, that do not have a quotient which is a rational function of z, in which case u_1 and u_2 can be taken as a basis of solutions to (34), or there are not. This latter case is simpler and will be dispensed with at once. If u_k is another root of (35) and $u_k = ju_1$, say, then one sees at once on substituting ju_1 into (35), that j must be a constant and indeed a primitive root of unity.[5] Now this is true for every root of (35), so it reduces to

$$A_m u^m + A_0 = 0$$

and u_1 is a root of a rational function. Accordingly, in this case every solution to (34) is a root of a rational function. But, said Fuchs, this case can be detected in advance by purely algebraic means which are adequate in fact to deal with the nth order differential equation, and need be discussed no further.[6] There remains the more interesting case when (34) has two independent algebraic solutions, and hence all its solutions are algebraic.

Fuchs reduced (34) in this case to

$$\frac{d^2 y}{dz^2} + Py = 0 \tag{36}$$

by means of the substitution $u = \mu y$, $\mu = e^{-\frac{1}{2}\int p\, dz}$, where

$$P = q - \frac{1}{4}p^2 - \frac{1}{2}\frac{dp}{dz}.$$

Since the general theory of equations of the Fuchsian class implies that

$$p(z) = \sum_{i=1}^{\rho} \frac{\alpha_i}{a - z_i}, \quad q(z) = \sum_{i=1}^{\rho} \frac{\beta_i + (z - a_i)\gamma_i}{(z - a_i)^2}, \tag{37}$$

where $a_1, \ldots, a_\rho$ are the singular points of (34) and the α_i and β_i, and γ_i, $i = 1, \ldots \rho$, are constants such that $\sum \gamma_i = 0$,

$$\mu = (z - a_1)^{-\frac{1}{2}\alpha_1} \ldots (z - a_\rho)^{-\frac{1}{2}\alpha_\rho}.$$

So all the solutions to (36) are algebraic if all the solutions to (34) are, and conversely all the solutions to (34) are algebraic if all solutions to (34) are and, in addition, $\alpha_1, \ldots, \alpha_0$ are rational (i.e., ρ is the logarithm of a rational function). Henceforth Fuchs worked with (36).

The general solution to (36) has the form

$$\alpha y + \beta \phi(y) \tag{38}$$

where α, β are constants, y is an algebraic solution to (36), and $\phi(y)$ a rational function of y and z, say

$$\phi(y) = c_0 + c_1 y + \cdots + c_m y^m, \tag{39}$$

where $c_0, \ldots, c_m$ are rational functions of z. Taking y and $\phi(y)$ as a basis of solutions to (36), Fuchs next considered the monodromy of the differential equation, and obtained various constraints upon the nature of the solutions that can arise, which will be discussed below.

If y is any algebraic solution to (36), then it satisfies an irreducible equation of degree m in y, and Fuchs showed that the number m does not depend on the choice of the solution of the differential equation [1876a, §8]. Suppose the equation for y is

$$A_m y^m + A_{m-1} y^{m-1} + \cdots + A_0 = 0, \qquad (40)$$

and $y_1, \ldots, y_{m-1}$ are its other roots. These roots may be divided up into equivalence classes, where one is equivalent to another if and only if their quotient is constant. A family of pair-wise inequivalent roots was said by Fuchs to form a *reduced root system* $y, y_1, \ldots, y_{n-1}$, and $y_i, y_i j, \ldots, y_i j^{\ell_i - 1}$ to be the roots corresponding to y_i, j being a primitive ℓ_ith root of unity. Fuchs called the least common multiple of the ℓ's as y_i runs through a reduced root system the *index* of the equation, and he showed [§9 Theorem 2] that

$$\ell n = m, \qquad (41)$$

since (40) is irreducible. For that reason, too, the analytic continuation of any two distinct roots of (40) along a closed path produces distinct roots [§9 Theorem 3].

Fuchs was now ready to translate his problem into invariant-theoretic terms. He let $\eta, \eta_1, \ldots, \eta_{n-1}$ be a reduced root system of index ℓ, and y_1 and y_2 any basis of solutions of the differential equation (37). Then $\eta_i = A_{i1} y_1 + A_{i2} y_2$, and a form of the nth degree is constructed:

$$f(y_1, y_2) = (A_{01} y_1 + A_{02} y_2) \ldots (A_{n-1,1} y_1 + A_{n-1,2} y_2). \qquad (42)$$

Any circuit in z changes f by at most a multiple of a primitive ℓth root of unity, so f is a root of a rational function of z, with highest exponent ℓ. Conversely, any form, which is a root of a rational function of z and has $\eta = A_{01} y_1 + A_{02} y_2$ as a factor also has factors $\eta_1 = A_{i1} y_1 + A_{i2} y_2$, where $\eta, \eta_1, \ldots, \eta_{n-1}$ are a reduced root system for the irreducible equation which η satisfies. Fuchs [§10] called any form which only has as factors the members of a reduced root system and in which every factor has degree one, a ground form (*Primform*). Accordingly, every ground form is the root of a rational function, and if two ground forms have a common factor, they are identical up to a common factor. Conversely, any form which is the root of a rational function can be written as a product of ground forms and contains each member of a reduced root system equally often. Fuchs' purpose in introducing these forms was to replace the solutions to the original differential equation, which are algebraic functions, with related quantities which are simpler, being roots of rational functions.

Although m depends only on the differential equation, the index ℓ, and the order of the reduced root system n, may depend upon the choice of solution y_1. Fuchs therefore chose y_1 so that n was as small as possible, say N, and let the corresponding index be L, when $m = NL$. N is then the least degree of a form which is a root of a rational function, and any such form is a ground form. Fuchs' approach was to calculate the possible values of N, using the machinery of invariant theory whenever possible. As it stands, the problem would be solved if one could characterise these binary forms of degree N all of whose covariants of lower degree vanish identically, but this Fuchs confessed himself unable to do, in a letter to Hermite, January 1876 (= Fuchs [1876b]). He therefore resorted to less direct methods.

The Hessian covariant of a form $\Phi(y_1, y_2)$ is the determinant

$$\begin{vmatrix} \dfrac{\partial^2 \Phi}{\partial y_1^2} & \dfrac{\partial^2 \Phi}{\partial y_1 \partial y_2} \\[3mm] \dfrac{\partial^2 \Phi}{\partial y_2 \partial y_1} & \dfrac{\partial^2 \Phi}{\partial y_2^2} \end{vmatrix} .$$

If Φ is a ground form of degree N, its Hessian, which Fuchs denoted Ψ, is again a root of a rational function, and is of degree $2N - 4$. It is therefore a ground form, since it cannot be a product of ground forms. It cannot vanish identically unless it is a power of a linear function, in which case $N = 1$, so Fuchs henceforth assumed $N > 1$.

The case $N = 2$ was disposed of simply enough [§13]. Fuchs was able to show that in this case the solutions can be written in the form

$$y_1 = \psi(z)^{-\frac{1}{4}}[p + q(z)^{\frac{1}{2}}]^{\alpha}$$

$$y_2 = \psi(z)^{-\frac{1}{4}}[p + q(z)^{\frac{1}{2}}]^{-\alpha} \tag{43}$$

for p, q rational functions of z and α a rational number. They are algebraic if and only if $\int \psi(z)^{\frac{1}{2}} dz$ is the logarithm of an algebraic function.

The remaining possible values of N are associated with the more complicated forms of monodromy relations, and required more *ad hoc* methods. After four paragraphs and seven pages, Fuchs finally produced the following table of possibilities:

N	$N > 2$ L	$L > 2$ $\phi(y_1, y_2)$
4	3 or 6	$a_0 y_1^4 + a_3 y_1 y_2^3$
6	4	$a_1 y_1^5 y_2 + a_3 y_1^3 y_2^3 + a_5 y_1 y_2^5$
6	8	$a_1 y_1^5 y_2 + a_5 y_1 y_2^5$
8	3 or 6	$a_1 y_1^7 y_2 + a_4 y_1^4 y_2^4 + a_7 y_1 y_2^7$
10	8	$a_1 y_1^9 y_2 + a_5 y_1^5 y_2^5 + a_9 y_1 y_2^9$
12	5 or 10	$a_1 y_1^{11} y_2 + a_6 y_1^6 y_2^6 a_{11} y_1 y_2^{11}$

together with the cases already discussed: $N = 4$, $L = 2$, $N = 4$, $L = 1$; $N = 2$; and $N = 1$ (§19).

The most interesting fact in the table is the upper bound on $N \leq 12$. It is testimony to the intricacies of Fuchs' method that it contains incorrect cases ($N = 6$ and $L = 4$; $N = 8$, $L = 3$ or 6; and $N = 10$) which are not possible values, as Klein was quick to notice (Klein [1876, = 1922, 305]). Klein also commented on these cases in *Fortschritte*, VII, 1877, 172–3.

Fuchs summarized his finding in two theorems [§20 Theorems 1 and 2]: if a differential equation has algebraic solutions, then either the general solution or one of the tabulated forms in y_1 and y_2 (a basis of arbitrary solutions) is a root of a rational function. Conversely, if a fundamental system of solutions can be made to yield a form (of degree > 2 and not a power of a form of degree 2) which is a root of a rational function, then the original differential equation has algebraic solutions.

It can hardly be said that this conclusion, although very interesting, amounts to a characterisation of the differential equations having algebraic solutions. It does not yield for instance, a simple test which can be applied to a given equation. Fuchs spent the last fifteen pages trying to improve matters, as follows.

Setting $v = y''$ turns the original equation (36) into

$$\frac{d^{\mu+1}v}{dz^{\mu+1}} + p_{\mu-1}\frac{d^{\mu-1}v}{dz^{\mu-1}} + \cdots + p_0 v = 0 \qquad (44)$$

with the same singular points as the original equation, and Fuchs showed that the necessary condition for (36) to have algebraic solutions is that for some value of $\mu = 1, 2, 4, 6, 8, 10, 12$ (44) is satisfied by the root of a rational function. If this is the case for $u = 2$, then conversely, if (44) has algebraic solutions so does (36). The case $u = 2$ reduces to the case $N = 2$.

He made various attempts to avoid the passage to (44) but without much success. Two theorems appear that are worthy of notice (24). They concern the solutions with exponent $r = p/q$ in neighbourhood of a singular point $z = a$. A circuit of that singular point sends y to yj^P, j a primitive qth root of 1, and $q|m$, the degree of the irreducible equation satisfied by y. So, if ν is the number of members of the reduced root system containing y, $\nu \geq N$, so $\nu \leq m/q$, i.e., $\nu \leq NL/q$ from which it follows that $q \leq L$. If $N > 2$, then from the table above, $q \leq 10$. Fuchs stated the theorem [24]: if any denominator, in lowest terms, of the exponents is greater than 10, then (36) has no algebraic solution unless it, or (43) with $\mu = 2$ is satisfied by a root of a rational function. He went on to prove that if (44) is satisfied by a root of a rational function and the denominators of the exponents are not 1, 2 or 3 then (36) is satisfied either by no algebraic functions or by roots of rational functions.

He compared this result with the result of Schwarz on the hypergeometric equation where unless two of λ and μ and ν equal 2, no value of them in Schwarz's table exceeds 5. Fuchs noted in the case $(2, 2, n)$ that if all the denominators of the exponents, q_i say, are equal to 2, and the necessary subsidiary condition for the avoidance of logarithmic terms is satisfied, then (36) is satisfied by a square root of a rational function.

3.3   Klein and Gordan

Klein first heard of Schwarz's work in the late autumn of 1874 or spring of 1875, after Gordan came to join him at Erlangen. Klein wrote, in the introduction to the *Vorlesungen über das Ikosaeder* [1884] "I had at that time already commenced the study of the Icosahedron for myself (without then knowing of Professor Schwarz's

earlier works, ...) but I considered my whole manner of attacking the question
rather in the light of preliminary training". He always spoke warmly of the semester
he spent with Gordan, his last semester at Erlangen before going to Munich, and it
seems to have been a rewarding clash of styles and approaches to mathematics for
both men. Klein, only 25 in 1874, had already published over thirty papers on line
geometry, on surfaces of the third and fourth degrees, on non-Euclidean geometry,
and on the connection between Riemann surfaces and algebraic curves. He was an
enthusiast for *anschauliche* geometry, stressing the importance of a visual and tan-
gible presentation of mathematical ideas, and preferring the conceptual framework
of group theory to the more algebraic and computational study of explicit invariants. Gordan, then 37, had spent a year in Göttingen with Riemann, who, however,
had been very ill. Together with Clebsch, Gordan had attempted to put Riemann's
function-theoretic ideas on a sound algebraic footing in their *Theorie der Abelschen
Funktionen* [1866], but Gordan soon turned to what was to be his great love: the
formal side of the theory of invariants. In his [1868] he proved the important theorem that for any binary form there is always a finite basis for its rational invariants
and covariants. He became the acknowledged master of the theory of invariants,
greatly preferring the mechanisms of the algebra to the more suggestive domain of
geometry, quarrying deeply where Klein chose to soar aloft, seeking detailed and
difficult results where Klein sought to unify the disparate parts of mathematics.

    Their work from 1875 to 1877 reveals many reciprocal influences and contrasts.
Klein wrote several papers on invariants, chiefly on the icosahedral invariants which
he was beginning to connect with the unsolvability of the quintic equation. In 1875
his work received a special impulse, as he later said (Klein [1922, 257]) from Fuchs'
work on the algebraic solutions problem. It seemed to Klein that, with his understanding of the role of the regular solids, he could complete Fuchs' treatment and,
indeed, simplify it. His approach, as he was later to stress to Poincaré, was much
closer to that of Schwarz than that of Fuchs, and he seems to have had little respect for Fuchs' achievements, which he found ungeometric (a letter from Klein to
Poincaré, 19 June 1881 = Klein [1923, 592]). Klein placed the regular solids and
their groups at the centre of his study of the algebraic-solution problem, and when
he produced the appropriate forms it was done "*without any complicated calculation only with the ideas of invariant theory*". Klein [1875, = 1922, 276] (italics in
original). Gordan for his part found Klein's geometric considerations "very abstract
and not bound up of necessity with the question in any way", and proposed to show
how the finite groups of linear transformations in one variable could be found algebraically (Gordan [1877a, 23]). Klein replaced Fuchs' indirect resolution of his
form problem with an ingenious reduction of the whole question to Schwarz's five
cases. Gordan accepted Fuchs' terms and solved the form-problem directly. Klein's
method will now be discussed.

Klein's solution

Klein began his [1875/76] with the problem of finding all finite groups of motions of the sphere or, equivalently, of finding all finite groups of linear transformations of $x + iy$. This connection between metric geometry and invariant theory was one he had been pleased to make in the Erlanger Programm (see particularly §6). The root of the connection is that, in the case of the icosahedron, there is a group of order 60, the symmetry group of proper motions of the solid. A typical point of the sphere can be moved to a total of 60 different points under the action of this group, but certain points have smaller orbits. There is an orbit of order 12 corresponding to the vertices of the icosahedron, an orbit of order 20 corresponding to the midpoints of the faces, and an orbit of order 30 corresponding to the midpoints of the edges. An orbit of each of these four kinds can be specified as the zeros of a form of the appropriate degree. Klein denoted the forms of order 12, 20 and 30 by f, H and T, respectively, and calculated, each explicitly. But between three forms of order 60 there must be a linear relationship (since a form is known up to multiplication by a constant once its zeros are given), and in this case the relationship is $T^2 + H^3 - 1728f^5 = 0$. Furthermore, as one would expect, H and T are known when f is known (the vertices specify the icosahedron completely) so it is enough to determine f and to explain its relationship to H and T. Accordingly the invariant theory can be easily developed, once it has been established what the possible finite groups are, and Klein accomplished this task in §2 of the paper.

Klein regarded the general motion of the sphere as screw-like or loxodromic about an axis. If a finite group is to be constructed each element in it must have finite order, and so be of the form $z' = \epsilon z$, ϵ a rational root of unity. Furthermore, the axes of any two motions must meet inside the sphere, and indeed all the axes must meet at the same point. But this reduces the finite group to one of the five classical examples: the cyclic, dihedral, tetrahedral, octahedral, and icosahedral groups. As Klein remarked in §3, this argument simultaneously determines all the finite groups of non-Euclidean motions.

The groups having been found, the forms in each case can be written down and related to one another by what Klein called a general principle (§5). This asserted that, if $\Pi = 0$ and $\Pi' = 0$ are the equations of two sets of points which arise from two given points by the action of a finite group and they are of the proper degree (say, the order of the group), then $\kappa \Pi + \kappa' \Pi' = 0$ represents any other such system of points, for some choice of parameter κ/κ'. Otherwise put, if Π and Π' are two G-invariant polynomials of the same degree, any third G-invariant polynomial of that degree is linearly dependent on them. This is true, Klein observed, because there is only a single infinity of orbits of the same degree, parameterised, one might add, by the sphere.[7]

To give the forms explicitly in the case of the octahedron, Klein took the canonical form obtained by Schwarz and wrote it in homogeneous co-ordinates as

$$f = x_1 x_2 (x_1^4 - x_2^4) \quad (\S6).$$

The general procedures of invariant theory then suggested the following maneuvers:

$f = a_x^6$, which is the symbolic notation for f as a binary form of degree 6; the 6th, 4th, and 2nd transvectants[8] of f with itself are: $(ab)^6 = A$, a constant;

$$(ab)^4 a_x^2 b_x^2, \text{ of degree 4; and } (ab)^2 a_x^4 b_x^4 = H, \text{ of degree 8, the Hessian of } f.$$

Any form of the sixth degree with isolated roots can be reduced to the form $x_1 x_2 \phi_4$, where ϕ_4 is of degree four and the coefficients of x_1^4 and x_2^4 are, respectively, 1 and -1. Accordingly, if its fourth transvectant (*Überschiebung*) $(ab)^4 a_x^2 b_x^2$ is made to vanish identically, it corresponds to the canonical form of the octahedron, and the converse is also true. H itself cannot vanish identically (if it did all the roots of f would coincide).

The eight roots of $H = 0$ give the centres of the eight faces of the octahedron. For the same reason the functional determinant of f and H, T, is of the twelfth degree and equated to zero locates the twelve mid-edge points of the octahedron. Together with the invariant A these forms satisfy

$$\frac{A f^4}{36} + \frac{1}{2} H^3 + T^2 = 0$$

as is easily seen in the canonical case.

Analogous reasoning dispatched the icosahedral equation, and enabled Klein to discuss the irrational covariants in each case. Klein concluded the paper with a brief reference to the Galois group of the icosahedral equation. There are, he said, 60 proper motions of the icosahedron itself, and for each motion four elements of the Galois group corresponding to the map $\epsilon \to \epsilon^\nu$, $\epsilon = e^{2\pi i/5}$, $\nu = 1, 2, 3, 4$. The Galois group, then, has 240 elements, but adjoining ϵ reduces it to a group of 60 elements. These sixty motions permute the five octahedrons inscribed in the icosahedron, which establishes an isomorphism between the Galois group and the even permutations on five elements.

Klein's next paper [1876] described the implications of this work for the algebraic-solutions problem. Suppose the equation is $y'' + py' + qy = 0$. It is algebraically integrable if and only if p is the logarithm of an algebraic (indeed, rational) function, so Klein assumed this was the case. Then y, a quotient of linearly independent solutions to the second order equation, is a solution of the Schwarzian equation

$$[n]_z = f(z) = 2q - \frac{1}{2}p^2 - p', \tag{45}$$

where $[\eta]_z$ is Klein's notation for the Schwarzian derivative, nowadays written $\{\eta, z\}$, and Klein sought to characterise the rational function $f(z)$.

The monodromy group of the original equation is known because it is finite and so a specific one of the known possibilities. To that group G, is associated a canonical G-invariant rational function $Z = Z(Y)$, which is such that the inverse function $Y = Y(Z)$ satisfies a canonical Schwarzian equation

$$[\eta]_z = R(Z). \tag{46}$$

But y is also such that its inverse function $z = \zeta(y)$ is G-invariant, and rational, and so ζ and Z are rational functions of each other. So finally, the substitution $Z = \phi(z)$, where ϕ is a rational function, converts $Y = Y(Z)$ into $Y(\phi(z)) = y(z)$ and the canonical equation (46) into the given (45); conversely the inverse rational substitution $\zeta = \psi(Z)$ converts (45) into (46). The functions ϕ, ψ, are subject to no other constraint, since any rational function of a G-invariant rational function is G-invariant, and so Klein deduced the elegant result that those second order linear homogeneous differential equations, which are algebraically integrable, are precisely those which can be obtained from the canonical ones by an arbitrary rational change of variable; so the problem is solved in principle.

Furthermore, their form is known, since if $Z = Z(Z_1)$ is any function

$$\{\eta, z_1\} = \left(\frac{dZ}{dZ_1}\right)^2 \{\eta, Z\} + \{Z, Z_1\},$$

and so the conversion of $\{\eta, Z\} = R(Z)$ by $Z = \phi(z)$ into $\{\eta, z\} = f(z)$ converts the canonical equation into $\{\eta, z\} = \left(\frac{dZ}{dz}\right)^2 R(Z) + \{Z, z\} = f(z)$. However, it must be said that the appearance of $f(z)$ in any given case is not such as to suggest immediately what substitution of ϕ should be made. Klein gave a suggestive method in his [1877 = 1922, 307–320], which is briefly discussed in Forsyth, [1900 Vol. 4, 184–187], but it must be said that neither Klein nor Forsyth is explicit about how such a calculation might actually be carried out.[9] One notes Katz' despairing remark [1976, 556] " ... but even in cases when one knows the answer ahead of time, it seems hopeless ever to carry out Forsyth's test procedure".

It may be of interest to give Klein's list of the ground forms which arise in each of the five cases as they appear in the terms of the original differential equations. Starting from the Schwarzian form

$$[\eta] = P(x),$$

where $P(x) = y_1/y_2$ is a quotient of two arbitrary particular solutions of the hypergeometric equation

$$\frac{d^2y}{dx^2} + p\frac{dy}{dx} + qy = 0,$$

he differentiated $\eta = y_1/y_2$ and wrote η' in the form

$$\frac{Ce^{-\int p\,dx}}{y_2^2},$$

C an arbitrary constant. He took for definiteness the icosahedral equation[10]

$$1728\frac{H^3(\eta)}{f^5(\eta)} = R(x),$$

and, observing that $T^2 = 12f^5 - 12^4H^3$ implies

$$T = \tilde{C}(3H'f - 5f'H),$$

deduced that $y_2^2 = C \dfrac{H^2(\eta)T(\eta)}{f^6(\eta)R'(x)} = e^{-\int p\,dx}$.

So, finally, $f(y_1, y_2) = C^6 \dfrac{R^4(1-R)^3}{R'^6} e^{-6\int p\,dx}$, and if $p = 0$ is rational. Klein listed the ground forms in the table reproduced on p.??, where C denotes an arbitrary constant.

In his paper on the Icosahedron [1877b], Klein was more explicit about the connection of the hypergeometric equation with the quintic equation. The icosahedral group consists of all rotations of an icosahedron and is a group of order 60. It acts on $\mathbf{CP}^1$ and the quotient is again $\mathbf{CP}^1$. The inverse of q, the quotient map, generally has 60 values at each point, but is branched over the points 0, 1 and ∞, where it takes only 20, 30 and 12 values, respectively. So q is a quotient of two polynomials P and Q. If Q is to vanish at just 12 points, it must be of the form f^5, where f is a polynomial of degree 12 specifying the positions of the 12 vertices of the icosahedron. It follows that $P = H^3$ where H is the Hessian of f, and $Q - P = T^2$, where T is the Jacobian of f and H. So $q = H^3/1728f^5$, and the icosahedral equation is $q(z) = u$. This equation is to be considered solved when z is known as a function of u. This, Klein argued, [1877b, §8], was known once the appropriate case of the hypergeometric equation is solved, which is the one where the monodromy group is an icosahedral group. Explicitly, this gives

$$z = F\left(\frac{11}{60}, \frac{31}{60}, \frac{6}{5}, \frac{1}{u}\right) \bigg/ \sqrt[5]{1728}\, F\left(\frac{-1}{60}, \frac{19}{60}, \frac{4}{5}, \frac{1}{u}\right).$$

So if the list of known functions includes the hypergeometric functions, the icosahedral equation is solved and with it the quintic equation.[11]

Simultaneously with Klein, the Italian mathematician Brioschi was also looking at algebraic solutions to the hypergeometric equation, and in 1877 the *Mathematische Annalen* carried two contributions from him. The first, [1877a], was in the form of a letter to Klein, which showed how the study of forms whose fourth transvectant vanishes can be connected with a second order differential equation. In his [1877b] Brioschi obtained the precise form of the Schwarzian equation for eleven of the fifteen cases in Schwarz's list. Since Schwarz had done one case (no. I), this left three (XII, XIV, XV) outstanding for which Brioschi's method was inadequate. These were tackled by Klein. Brioschi found that the equations $[t]_x = -2p$, where t was (in the order II, III, ..., X, XI XIII):

$$x, \frac{-4x}{1-x^2}, x, \frac{-(1-x)^2}{4x}, x, \frac{-4x}{1-x^2}, \frac{-(1-x)^2}{4x}, \frac{-(x-4)^3}{27x^2},$$

$$-\frac{1}{4^3}, \frac{x(x+8)^3}{(1-x)^3}, \frac{4}{27}, \frac{(x^2-x+1)}{x^2(1-x)^2}, \frac{1}{4.27}, \frac{(x^2+14x+1)^3}{x(1-x)^4}. \tag{47}$$

Brioschi's approach was more invariant-theoretic than Klein's, as befits a man of the previous generation. It was also more modest, for it accepted the solution to Fuchs' problem already proposed. The only man to offer a solution to that problem entirely in terms of invariant theory was the acknowledged master of the subject: Paul Gordan.

(I) $\quad y_1 = C_e^{-\frac{1}{2}\int p\,dx}\cdot\dfrac{R^{\frac{n+1}{2n}}}{R'},$

$\quad\quad y_2 = Ce^{-\frac{1}{2}\int p\,dx}\cdot\dfrac{R^{\frac{n-1}{2n}}}{R'}.$

(II) $\quad y_1 y_2 = Ce^{-\int p\,dx}\cdot\dfrac{R^{\frac{1}{2}}(R-1)^{\frac{1}{2}}}{R'},$

$\quad\quad y_1^n + y_2^n = 2C^{\frac{n}{2}}e^{-\frac{n}{2}\int p\,dx}\cdot\dfrac{R^{\frac{n+2}{4}}\cdot(R-1)^{n/4}}{R'^{\frac{n}{2}}},$

$\quad\quad y_1^n - y_2^n = 2C^{\frac{n}{2}}e^{-\frac{n}{2}\int p\,dx}\cdot\dfrac{R^{\frac{n}{4}}(R-1)^{\frac{n+2}{4}}}{R'^{\frac{n}{2}}}.$

(III) $\quad y_1^4 - 2\sqrt{-3}y_1^2 y_2^2 - y_2^4 = C^2 e^{-2\int p\,dx}\cdot\dfrac{R^{\frac{5}{3}}\cdot(R-1)}{R'^2},$

$\quad\quad y_1^4 + 2\sqrt{-3}y_1^2 y_2^2 - y_2^4 = C^2 e^{-2\int p\,dx}\cdot\dfrac{R^{\frac{4}{3}}\cdot(R-1)}{R'^2},$

$\quad\quad 2\sqrt[4]{-27}\cdot y_1 y_2(y_1^4 - y_2^4) = C^3 e^{-3\int p\,dx}\cdot\dfrac{R^2(R-1)^2}{R'^2}.$

(IV) $\quad \sqrt[4]{108}\cdot y_1 y_2(y_1^4 - y_2^4) = C^3 e^{-3\int p\,dx}\cdot\dfrac{R^2(R-1)^{3/2}}{R'^3},$

$\quad\quad y_1^8 + 14y_1^4 y_2^4 + y_2^8 = C^4\cdot e^{-4\int p\,dx}\cdot\dfrac{R^3(R-1)^2}{R'^4},$

$\quad\quad y_1^{12} - 33y_1^8 y_2^4 - 33y_1^4 y_2^8 + y_2^{12} = C^6\cdot e^{-6\int p\,dx}\cdot\dfrac{R^4(R-1)^{\frac{1}{2}}}{R'^6}.$

(V) $\quad \sqrt[5]{128}\,y_1 y_2(y_1^{10} + 11y_1^5 y_2^5 - y_2^{10}) = C^6 e^{-6\int p\,dx}\cdot\dfrac{R^4(R-1)^2}{R'^6},$

$\quad\quad -(y_1^{20} + y_2^{20}) + 228(y_1^{15}y_2^5 - y_1^5 y_2^{15}) - 494y_1^{10}y_2^{10}$

$\quad\quad\quad = C^{10}\cdot e^{-10\int p\,dx}\cdot\dfrac{R^7(R-1)^5}{R'^{10}},$

$\quad\quad (y_1^{30} + y_2^{30}) - 522(y_1^{25}y_2^5 - y_1^5 y_2^{25}) - 10\,005(y_1^{20}y_2^{10} + y_1^{10}y_2^{20})$

$\quad\quad\quad = C^{15}e^{-15\int p\,dx}\cdot\dfrac{R^{10}(R-1)^8}{R'^{15}}.$

From Klein [1877a], in [1922] pp. 319, 320.

Table 3.2

3.4 The Solutions of Gordan and Fuchs

Gordan considered the same geometric problem as Klein from two aspects: as a question involving finite groups, and, following Fuchs, as a question about binary forms with vanishing covariants, and dealt with them in two papers in the *Mathematische Annalen* of 1877.

In the first paper [1877a] Gordan considered transformations of the form $\eta' = \dfrac{\alpha\eta + \beta}{\gamma\eta + \delta}$, $\alpha\delta - \beta\gamma \neq 0$, essentially as rotations, which he treated in the language of invariant theory. If, he said, $\eta = \dfrac{x_1}{x_2}$ and $\eta' = \dfrac{y_1}{y_2}$, then these transformations can be thought of as the vanishing of a bilinear form. As such it has two invariants, its determinant and another which he denoted $-2\cos\phi$, where he called ϕ the argument. He showed that, if S is a transformation with argument ϕ, T has argument ψ, and ST has an argument θ, then $S^{-1}T$ has argument H, where $\cos\theta + \cos H = \cos(\theta + \psi) + \cos(\theta - \psi)$.

If the transformations form a finite group, then each transformation has finite period, and is equivalent to one of the form $\eta' = p\eta$, so each ϕ is some submultiple of π, and the transformations can be grouped into families according to the basic transformations of which they are powers. He reduced the problem to establishing the maximal periods, and then supposed S and T were each of this maximal period, and so had the same argument ϕ. Then the arguments of ST and $S^{-1}T$ would be θ and H, where $\cos\theta + \cos H = \cos 2\phi + 1$. θ and H must also be rational multiples of π, so, setting $2\phi = \phi_1$, $\pi - \theta = \phi_2$ and $\pi = H - \theta_3$, he got the equation

$$1 + \cos\phi_1 + \cos\phi_2 + \cos\phi_3 = 0,$$

which is to be solved in angles are rational multiples of ρ. He found, quoting [Kronecker 1854], that there were very few solutions, indeed just the known cases. The periods could be 2, 3, 4, or 5, giving rise to the various groups of the regular solids, whose elements he presented explicitly up to conjugacy in each case.

In his second paper [1877b] Gordan solved Fuchs' ground form problem directly, by showing, as he put it, that "the result follows immediately from the general rules which I have developed in my text *Ueber das Formensystem binärer Formen* ... ". The problem Fuchs raised was to characterise those forms, f, of least degree, all of whose covariants of lower degree vanish identically and all of whose covariants of higher order, which are powers of forms of lower order, also vanish. Gordan's approach was to consider a second form, P, of degree $\mu + 1$, with the property that its final transvectant with f, $(f, P)^{\mu+1}$, vanished. He was able to show [1877b, 147] that P enabled him to detect when covariants of f were powers of linear forms, in which case f could not be a ground form. He was able to show that the only forms of order greater than 4 satisfying Fuchs' criteria belonged to the octahedral or icosahedral system, but his argument proceeded through seven subcases and cannot be summarized here.

The work of Gordan and Klein enabled Fuchs to return to his original list of ground forms and prune it of its spurious members. In his paper [1878a] he analyzed each item in his list, indicating new properties of the genuine ones and providing valid, *ad hoc*, invariant-theoretic reasons for deleting those to which Klein had objected. For the two members that remained $N > 4$, he characterised the ground forms of each degree that can arise. For example, he showed [1878a, §7 Theorems III–VI = 1906, 128] that for $N = 6, L = 8$, every ground form of degree 24 is of the form $\nu \tilde{H}(f_6)^3 + \lambda f_6^4$, where f_6 is the unique ground form of degree 6, $\tilde{H}(f_6)$ its Hessian 10 (of degree 8), and ν and λ are constants not both zero, and such that $\dfrac{\lambda}{\nu} \neq 108$. Similarly, for $N = 12, L = 10$, every ground form of degree 60 is of the form $\sigma \tilde{H}(f_{12})^3 + \lambda f_{12}^5$, where f_{12} is the unique ground form of degree 12, and λ, ν are constants not both zero, and such that $\dfrac{\lambda}{\nu} \neq 1728$. The exceptional cases $\tilde{H}(f_6)^3 + 108 f_6^4$ and $\tilde{H}(f_{12})^3 + 1728 f_{12}^5$ are squares of rational functions.

Fuchs obtained these theorems by considering the highest degree, μ, a ground form could have, and comparing it with the lowest degree, N. If f is a form of degree $N > 4$, then $\phi := H(f)^3 - \lambda f^2 H^2(f)$ cannot vanish, for ground forms only have a common factor if they are identical, and $H(f)^3 = \partial f^2 H^2(f)$ would imply $H^2(f)$ divides $H(f)^3$ which is impossible (since the degree of $H^{(}f)^3$ is greater than the degree of $H^2(f)$). However, $\dfrac{f^2 H^2(f)}{H(f)^3}$ is not altered on substituting jf for f, where j is a root of unity, and so it is not altered by any circuit of z. So ϕ must be a root of a rational function, and is therefore a product of ground forms. However, if F is a ground form of degree μ having no factors in common with f, or $H(f)$, of $H^2(f)$, then for some suitable λ, F divides ϕ, and so $\mu \leq 6N - 12$. Consequently when $N = 6$ the maximal degree of a ground form is 24, and when $N = 12$, the maximal degree is 60. The bounds are attained, as we have seen. Furthermore, μ is related to m, the degree of the algebraic equation satisfied by a solution of the differential equation, for Fuchs showed [1878a, §2 Thm. III = *Werke* II, 120] that $\mu = m$ or $m/2$, and if L is 8 or 10, then in fact $\mu = m/2$. This result was implicit in one of his earlier tabulations of L and N, [1876, 17 = *Werke* II, 39].

Fuchs also deduced that any ground form of maximal degree μ is a rational function, and that between any three there is a linear relation with constant coefficients. But this relationship was, for him, a consequence of the Theorems III–VI just quoted, and not, as it was for Klein, a means to understanding the forms.

3.5   Jordan's solution

Camille Jordan's solution to the algebraic solutions problem, couched in terms of the finiteness of the related group of linear transformations, was put forward in a note in the *Comptes Rendus* of March 1876, [1876a]. It erred in omitting one of the possible cases, the icosahedral one, and in November of the same year he restored the missing case, [1876b], remarking:

> "a calculating error, which in no other way invalidates the principles of our reasoning, caused us to omit one of these groups . . . ",

and he acknowledged that the first solution of the group-theoretic problem was due to Klein.

In a third short paper, [1877], Jordan sketched, for the first time, the answer to the problem for a third-order differential equation, and in June 1877 he submitted his long paper [1878] on the question to the *Journal für Mathematik*. It gives not only a treatment of the nth order differential equation, but a full account of Jordan's methods. It is this paper which will now be discussed.

Jordan observed that the solutions to a given linear differential equation are all algebraic if and only if the corresponding monodromy group is finite. So, in order to enumerate the different types of mth order equation with that property, it is enough to construct the different groups of finite order which can be represented as linear groups in m variables. He found five such groups when $m = 2$, eleven when $m = 3$, and was able to show in the general case that the finite groups which can arise (and hence the solutions of the differential equation) satisfied certain additional conditions. His methods, as befits the leading group theorist of the day, were those of the newly-discovered Sylow theory.[12] Jordan first employed them in the case of the second-order differential equation, to which he devoted Chapter I of the paper.

He denoted the typical linear substitutions[13] in two variables

$$S = |u_1, u_2, \quad \alpha u_1 + \beta u_2, \gamma u_1 + \delta u_2|.$$

After a linear change of variable, S can be written either in the canonical form

$$|x, y \quad ax, by|,$$

when he said it was of the first kind, or in the form

$$|x, y \quad ax, a(y + \lambda x)|.$$

He said it was of the second kind if $\lambda = 0$, and otherwise, of the third kind. In either case the roots of the characteristic equation for S coincide. If G is to be a finite group, all of its elements S must be of finite order, so no element can have infinite order. G cannot then contain any element of the third kind.

Jordan next showed that the elements T, which commute with an element of S of the first kind, are precisely those which are also in canonical form when a basis is chosen with respect to which S is diagonalized (i.e., S and T have the same eigenspaces). Consequently, the elements T, which commute with an S of the first kind, commute with each other (indeed, form a commutative subgroup of G). The finite groups G can now be divided into two types. Those of the first type contain sets of elements of the form

$$S = |x, y \quad ax, by|$$

with respect to a fixed basis $\{x, y\}$, where a and b are roots of unity. Those of the second type are formed from the first, by adjoining to a single set of the first type, an element of the form

$$T = |x, y \quad y, kx|,$$

where k is a root of unity. Thus, every finite group H all of whose elements are conjugate to those of a group of the first type, and which contains an element of the first kind, belongs to the first or second type.

Jordan's problem was now to determine all finite groups of either type. He let G be such a group, Ω its order, and g the subgroup of elements of the second kind, with order ω, and sought a formula for Ω analogous to Schwarz's formula for the λ, μ, ν above (p. 74). He let $S \in G$ be an arbitrary element of the first kind, and defined F_S to be elements of G which commute with S. Evidently $F_S \supset g$ and F_S has $\mu\omega$ elements, say with $\mu > 1$, since $S \in F_S$, Sg. So G is the union of sets F_S, and no S appears in two different sets F_S, $F_{S'}$, as can be easily seen. If S has the form

$$S = |x, y \quad ax, by|$$

and E is the subgroup of elements, g, of G such that $g^{-1}F_S g = F_S$, then the elements of E have one of the forms

$$|x, y \quad \alpha x, \delta y|,$$

in which case they lie in F_S, or

$$|x, y \quad \beta y, \gamma x|.$$

Therefore E either has order $K\mu\omega = 2$ or μ, ω, depending on whether or not it has an element of the form $|x, y \quad \beta y, \gamma x|$. The number of sets F_S conjugate to F_S in G is then $\Omega/k\mu\omega$. Each F_S contains the ω elements of g and $(\mu - 1)\omega$ elements of the first kind. The total number of elements of the first kind in the totality of sets conjugate to F_S is therefore $\dfrac{\Omega}{k\mu\omega}(\mu - 1)\omega = \dfrac{\Omega(\mu - 1)}{k\mu}$. If this argument is repeated for each F_S, not conjugate to $F_{S'}$ until the group is exhausted, the following formula for the order of the group G is obtained

$$\Omega = \omega(1 - \frac{\mu - 1}{k\mu} - \frac{\mu' - 1}{k'\mu'} - \cdots) \geq 1. \tag{48}$$

This is the necessary formula, It can only contain two or three terms $\dfrac{\mu - 1}{k\mu}$. If it has two terms $\dfrac{\mu - 1}{k\mu}$ and $\dfrac{\mu' - 1}{k'\mu'}$, then either $k = 2$, $k' = 1$, $\mu' = 2$ and $\Omega = 2\mu\omega$, G is of the second type, or $k = 2$, $\mu = 2$, $k' = 1$, $\mu' = 3$, $\Omega = 12\omega$. If there is also a term $\dfrac{\mu'' - 1}{k''\mu''}$, then necessarily $k = k' = k'' = 2 = \mu''$ and either $\mu' = 2$ and $\Omega = 2\mu\omega$, or $\mu' = 3$ and $\mu = 3$, 4, or 5, in which cases $\Omega - 12\omega$, 24ω, or 60ω, respectively. In short, every G of the required kind and not of the second type has order $r\omega$ where r is either 12, 24, or 60.

There is an evident representation of G as 2×2 matrices, by means of the function $z = \dfrac{mx + ny}{px + qy}$, which annihilates the w elements of g. The image of G

under this representation is a group Γ of order r, and Jordan studied Γ using Sylow theory. His treatment of the case $r = 60$ is typical. Γ in this case contains 6 groups of order 5 and indeed is the homomorphic image of a group of permutations of 6 letters of order 60. But then it must also be isomorphic to the alternating group on five letters, A_5. Its generators can be written down; as permutations they are

$$A^1 = (\alpha\ \beta\ \gamma\ \delta\ \epsilon), \quad B^1 = (\beta\ \epsilon)(\gamma\ \delta), \quad C^1 = (\beta\ \delta)(\gamma\ \epsilon)$$

and as linear substitutions, they are

$$A = |x, y \quad \theta x, \theta^{-1} y|, \quad \theta^{10} = 1,$$
$$B = |x, y \quad y, -x|,$$
$$C = |x, y \quad \lambda x + \mu y, \mu x - \lambda y| \text{ where } \mu^2 + \lambda^2 + 1 = 0,$$
$$\lambda = \frac{1}{\theta^2 - \theta^{-2}}.$$

G is obtained from Γ by adjoining elements $|x, y \quad ax, ay|$ where a is a primitive ωth root of unity.

The five types of groups which Jordan found are, as one would expect, the cyclic and dihedral groups of arbitrary order, and the tetrahedral, octahedral and icosadedral groups. Corresponding to each group is a particular type of solution function to the appropriate differential equation. For instance, in the case of the icosahedral group A_5, let z be an arbitrary solution to a differential equation whose monodromy group is A_5. Then z^ω is a rational function whose discriminant is a perfect square. If u is another solution, then u^ω is a rational function of z^ω and the independent variable t.

In the third and final chapter of his [1878], Jordan sought to give a complete analysis of the third order linear differential equation by extending the methods used to discuss the second order case. There are trivial extensions of the two dimensional groups to three dimensions, whereby a finite cyclic group is added on as a direct summand and alone affects the z-variable [no. 62]. Jordan looked for non-trivial three dimensional representations as groups of matrices with determinant 1. If one such group is H, his first task was to find an equation for the order Ω of H, analogous to (48). H may well have a sub group, K, consisting of the direct sum of a two-dimensional group K^1, and a one-dimensional group. A lengthy consideration of the various cases that can arise, depending in part on the choice of K^1 [nos. 71–96], finally yielded the sought-after equality [no. 96, equation 63]

$$\Omega = \frac{\Phi}{1 - \sum},$$

where $\sum$ is a sum of terms from the following list:

$$1 - \frac{1}{120\lambda}, 1 - \frac{1}{48\lambda}, \frac{65}{96}, \frac{33}{48}, 1 - \frac{1}{24\lambda}, \frac{15}{24}, \frac{m-1}{km}, \frac{1}{2}, \frac{1}{4}, \frac{1}{8}.$$

Here $k = 1, 2, 3$ or 6, m is subject to certain restrictions, and λ is arbitrary.

Jordan's next task was to extract from this formula a list of the possible groups it could refer to. He considered the fourteen different summations that $\sum$ could be, and came up with a table of 47 associated orders for groups [no. 124], with, in each case, an indication of the kinds of subgroups that would be present. To get some control over the proliferating chaos, Jordan next observed that the groups he sought were either simple groups from the table, which could be added to the list in no. 62, or had a group in the new list as a normal subgroup – let them be added to the list – or had a group in the extended list as a normal subgroup, and so on [no. 125].

It turned out that there were very few simple groups in the table. Relatively simple considerations involving Sylow theory eliminated all but seven of them [nos. 127–149], and six of those in fact correspond to no group at all [nos. 150–186]. The outstanding case, XXXII of order 60Φ, corresponds to the icosahedral group, and an extension of it to a group of order $60.3 = 180$ [nos. 187–193].

The construction of a group whose elements normalized a simple group in the list could be carried out in three distinct ways, yielding a group of order 27.2.12 and two of its subgroups, of orders 27.2.4, 27.2.2 [no. 202]. No new group had the icosahedral group as a normal subgroup. So finally Jordan produced the six groups (in addition to the given trivial types of no. 62) listed in footnote.[14]

As has been remarked, this list is incomplete, since it lacks the simple group of order 168. When Jordan revised his *Journal für Mathematik* paper for the *Atti della Reale Accademia* in 1880, he rectified this omission, which had been brought to light by Klein.[15] It had derived from a too-hasty interpretation of his equation 63 [no 96]. Correctly interpreted, it led directly to a group of order 24.7 containing 8 cyclic groups of order 7, which necessarily must be isomorphic to the group of transformations

$$\left| t \frac{\alpha t + \beta}{\gamma t + \delta} \right|$$

$t = \infty, 0, 1, \ldots, 6 \bmod 7$ and $\alpha\delta - \beta\gamma$, a quadratic residue mod 7, i.e., the simple group G_{168}. Jordan showed this group had generators

$$A = \begin{pmatrix} \tau & 0 & 0 \\ 0 & \tau^2 & 0 \\ 0 & 0 & \tau^4 \end{pmatrix} \quad \tau^7 = 1, \text{ of order 7,}$$

$$B = \begin{pmatrix} 0 & 1 & 0 \\ 0 & 0 & 1 \\ 1 & 0 & 0 \end{pmatrix} \quad \text{or order 3, and}$$

$$C = \begin{pmatrix} a & c'' & b' \\ c'' & b' & a \\ b' & a & c'' \end{pmatrix}, \quad \text{or order 2, where}$$

$$a\tau + b'\tau^2 + c''\tau^4 = 0$$
$$a\tau^{-1} + b'\tau^{-2} + c''\tau^{-4} = 0$$

in which form it precisely matched Klein's description of it in Klein [1878/79, §4], discussed in Chapter V. Jordan was lazy about such matters and mentioned Klein but did not give the reference.[16] Jordan also showed that G_{168} was the only group which had been omitted of order 24.7.q, the case under discussion.

Jordan has also shown in his [1878] that in some sense only finitely many linear differential equations of order n have a finite monodromy group.[17] He showed that the finite subgroups of $G\ell(n; \mathbf{C})$ could be classified into types, in such a way that the situation for arbitrary n resembled that for $n = 2$, when there are five types; two infinite families and three other groups. This result, for general n, is of independent interest in the study of groups, and is known as Jordan's finiteness theorem. He found his proof of 1878 imperfect and reworked it for a second publication in 1880. This proof, which amplifies and clarifies the earlier one, will only be discussed briefly here it is [1880, $=$ *Oeuvres*, II, 177–217]. References are to his numbered paragraphs.

He took G to be a fixed finite subgroup of the linear group in n variables (7) and F to be an abelian subgroup (*faisceau*). F can be simultaneously diagonalized, and when this is done its elements may be written in the form

$$(a_{1i}, a_{2i}, \ldots, a_{ni}) \text{ for some index } i, 1 \le i \le |F|$$

(modifying his notation slightly). Jordan defined F to be irreducible with respect to an integer λ, if each ratio

$$a_{\ell i}/a_{ki} \quad 1 \le k < \ell \le |F|$$

took more than λ values as i went from 1 to $|F|$, provided $a_{\ell i} \ne a_{ki}$, (4). He said F was maximal (*complet*, 7) if it was the largest abelian subgroup in G whose elements had the same form, say, for example

$$(a_{1i}, a_{2i}, \ldots, a_{ni}), \quad a_{1i} = a_{2i}.$$

His precise result is (8).

Theorem. *There are integers λ_n and μ_n depending on n, such that G has a maximal abelian normal subgroup F irreducible with respect to λ_n and of index $\le \mu_n$ in G, and F contains every other abelian subgroup irreducible with respect to λ_n.*

His proof was by induction on n; $n = 1$ is trivial, and $\lambda_1 = \mu_1 = 1$. To prove the theorem for n when it was known for 1, 2, ..., $n - 1$, he considered two cases. He said G was decomposable (9) if $\mathbf{C}^n$ can be decomposed into proper subspaces, such that each element of G either preserves the subspaces or permutes them. In this case, Jordan showed that G contained a subgroup G' which preserved the subspaces, and the inductive hypotheses readily supplied a subgroup F of G' and integers λ_n and μ_n satisfying the theorem with respect to $G'(14)$. Jordan then showed that F also satisfied the theorem with respect to $G(15 - 20)$. If a group is not decomposable Jordan called it indecomposable; the modern word is "primitive". This paper is one of many displaying a high level of abstraction in Jordan's work, which goes far beyond the theory of permutations in which it is couched. In this paper the theory of "systems of imprimitivity" is thoroughly outlined. The preface to his [1880] is interesting for the light it sheds on Jordan's approach to mathematics. Speaking of the problem of determining the finite groups of linear transformations in two variables, he remarked that it was first solved by Klein [1875/76] and then confirmed by Fuchs [1876] and Gordan [1877a, b] using entirely different methods. Then he went on:

"In spite of the considerable interest which attaches to the work of these eminent geometers, one could want a more direct method for solving this question. The determination of the sought-after groups is in effect only a problem of substitutions, which must be capable of being treated by the sole resources of that theory without recoursing as M. Klein, to non-Euclidean geometry or, as MM Fuchs and Gordan, to the theory of forms. Besides, the new method which is to be found, in order to be entirely satisfactory, must be capable of being extended to groups in more than 2 variables."

Jordan as a mathematician believed in propaganda by deeds rather than words, but he was here asserting that the proper approach to the question originally raised by Fuchs is group theory, and proposing that the test for all methods must be their capability to deal with the higher order cases, for which group-theoretic methods were currently the only ones. By his example Jordan established group theory as a subject in its own right, and as one capable of many applications. It became increasingly regarded as the "natural" abstract structure underlying many mathematical problems, and following Jordan's example, French mathematicians came to prefer group theory to the theory of invariants. So, although Hermite had been strongly attracted to invariant theory, the next generation in France was not, and the subject developed much more strongly in Germany. On the other hand, Halphen's successful treatment of the algebraic solutions problem for differential equations of higher order did depend essentially on invariant theory, as will be seen. But it was eclipsed almost at once by the group-theoretic methods of Poincaré, so once again invariants seemed to be less powerful than the newer techniques.

3.6 Equations of higher order

Frobenius had observed [1875a] that if $P = 0$ is a homogenous linear differential equation of order λ all of whose solution are algebraic, then the solutions satisfy an irreducible algebraic equation of order $\nu \geq \lambda$, all the roots of which, $y_1, \ldots, y_\nu$ also satisfy the differential equation. Only λ of these roots will be linearly independent, so constants $c_1, \ldots, c_\nu$ can be found such that

$$y = c_1 y_1 + \cdots + c_\nu y_\nu$$

takes $\nu!$ distinct values as the ν roots are permuted in all possible ways. By theorems of Abel and Galois, $y_1, \ldots, y_\nu$ are therefore expressible as rational functions of y and x. Frobenius investigated the converse, and found that if an irreducible linear differential equation of order greater than 2 has a solution in terms of which all the other solutions can be written rationally, then all the solutions are algebraic functions. He argued as follows. If the differential equation has a solution y and all the other solutions can be expressed rationally in terms of y but y is transcendental, then under analytic continuation, y can only transform as $y \mapsto \dfrac{ay+b}{cy+d}$, where a, b, c, and d are rational functions. Straightforward monodromy considerations produce a transformation $y \mapsto ky + r$, where k is a constant, so the rational function r satisfied the differential equation. Any rational function satisfies a first order linear differential equation, and so the original equation is reducible. It might happen that r was zero, but then the original equation can only be of first or second order. The reverse of this conclusion is that, if the given differential equation is irreducible and or order greater than 2, then y, if it exists, is algebraic, and Frobenius' conclusion is established.

   Not much else was done with the differential equations of order greater than 2 all of whose solutions were algebraic, for Jordan's work indicated how technically complicated it could become. In [1822a, b] Fuchs showed that if y_1, y_2 and y_3 are a basis of solutions to a third order differential equation of the Fuchsian class which furthermore satisfy a homogenous polynomial $f(y_1, y_2, y_3) = 0$, of order $m > 2$, the equation is satisfied by the square of the solutions of a certain second-order differential equation. Such equations had been studied earlier and in a different way by Brioschi [1879]. The polynomial f is a projective embedding of a Riemann surface, and Fuchs showed that the genus, p, of the surface has a strong effect on the number, n, of reduced roots of any algebraic equation satisfied by any solution of the differential equation: $p > 1$ implied $n \leq 4$; $p = 1$ implied $n = 2, 3, 4$ or 6; and $p = 0$ reduced to the case of an algebraically integrable second order differential equation. Fuchs' method involved the pth order differential equation satisfied by the ϕ's which appear in the homogeneous form of the integrands of the first kind on the Riemann surface

$$\frac{\phi(y_1, y_2, y_3) \sum \pm c_1 y_2 dy_3}{\sum c_1 \dfrac{\partial f}{\partial y_1}}$$

(see Chapter V), and, when $p = 1$, the earlier results of Briot and Bouquet [1856a]. It would be too long an excursion to show more precisely how Fuchs obtained this result, but the simpler result that the differential equation is integrable algebraically if the curve $f(y_1, y_2, y_3)$ is algebraic, was somewhat simplified by Forsyth [1902, IV, 214–216], and can be presented. It rests on the observation that, n being greater than 2, the Hessian of f is a single-valued non-constant function of z, and in fact a rational function (since the differential equation is of the Fuchsian class). Consequently, every other covariant of f is a rational function of z (or a constant). Let $k = \psi$ be a non-constant covariant other than H. Then $f = 0$, $H = \phi$ (a rational function) and $K = \psi$ provide three algebraic equations from which y_1, y_2, and y_3 can be found, and the result is proved.

   Halphen devoted most of his paper [1884] to the relationships which exist between a linear differential equation of order q and the curve defined projectively by a basis of solution $(y_1 : \ldots : y_q)$. He was particularly interested in the differential invariants which survive the transition from the equation

$$\frac{d^q Y}{dX^q} + q P_1 \frac{d^{q-1}Y}{dX^{q-1}} + \frac{q(q-1)}{2!} P_2 \frac{d^{q-2}Y}{dX^{q-2}} + \cdots q P_{q-1} \frac{dY}{dX} + P_q Y = 0$$

to the equation

$$\frac{d^q y}{dx^q} + q p_1 \frac{d^{q-1}y}{dx^{q-1}} + \frac{q(q-1)}{2!} p_2 \frac{d^{q-2}}{dx^{q-2}} + \cdots + q p_{q-1} \frac{dy}{dx} + p_q y = 0 \quad (49)$$

under the arbitrary changes of variable $\frac{dx}{dX} = \mu(X)$, $Y = yu(x)$, where $\mu(X)$ and $u(X)$ are indeterminate functions. He defined an absolute invariant as a function ϕ of P_i and their derivatives such that

$$\phi\left(P_1, P_2, \ldots, \frac{dP_1}{dX}, \ldots\right) = \phi\left(p_1, p_2, \ldots, \frac{dp_1}{dx}, \ldots\right).$$

Such an invariant is

$$V = -p'' + 3(P_2' - 2P_1 P_1') - 2(P_3 - 3P_1 P_2 + 2P_1^3),$$

where $p_i' = \frac{dp_i}{dX}$ etc. [1884, 112] and he noted that, surprisingly this invariant does not depend on the order of the differential equation. He found a sequence of $q - 1$ invariants for equations of order more than three, which enabled him to prove quite general results about differential equations of arbitrary order $q > 3$. He observed that if the coefficients are algebraic, they have a genus, which he termed the genus of the equation. A reduction of (49) to an equation with constant coefficients, or rational coefficients (genus zero), or doubly periodic coefficients (genus 1) or to an equation of genus p being sought, Halphen could express necessary and sufficient conditions for this to be possible. The first task is possible if and only if the absolute invariants are all constants, the others if and only if the $q - 2$ relations between $q - 1$ absolute invariants are of genus p [1884, 126–130]. For a third-order equation he showed that V vanished identically if and only if the curve defined by $(Y_1 : Y_2 : Y_3)$ was a conic.

In awarding this essay the *Grand Prix* in 1881, Hermite said that it showed a talent of the highest order.

> Nothing is more interesting than to see the introduction of the algebraic notions of invariants into this research into the integral calculus, which have originated in the theory of forms, and these new combinations make the hidden elements appear on which, in its various analytic guises, the integration of a given equation depends .... They are here joined to a consideration which equally plays an essential role in these researches: that of the genus of an algebraic equation between two variables, introduced into analysis by Riemann and which is so often employed in the works of our time.

Exercises

1. Verify Schwarz's derivativation of (33) from (30).

2. Verify that if $z = \dfrac{c_1 x + c_2}{c_3 x + c_4}$, $\Psi(s, x) = \dfrac{d^2 z}{dx^2} \Psi(s, z)$.

3. Show that the only rational values of λ, μ and v satisfying Schwarz's conditions on p. 74 are those given in the accompanying table.

4. Show that if $u_k = j u_i$ (as on p. 79) then j is a primitive root of unity.

5. Verify Fuchs' Theorem 3 (p. 81).

6. Verify the claim made about the functions defined in (43) on p. 80.

7. Verify Klein's claim (p. 83) that the axes of any two motions R_1 and R_2 of the sphere (which lie in the same finite group) must intersect. [Consider the plane which perpendicularly bisects p and $R_1(p)$, where p is a fixed point of $R_2 R_1$.]

8. Let G be a finite group of 2×2 matrices, and let $\{\eta, Z\} = R$ be the canonical Schwarzian equation associated to G, with algebraic solution $z = z(Z)$ as described on p. 85. Let $\{\eta, \zeta\} = P$ also have monodromy group G, and suppose $y(\zeta)$ is an algebraic solution of it. Suppose $Z = Z(y)$ is a G-invariant rational function inverse to $z = z(Z)$. Show ζ is a rational function of Z by showing that $\zeta = \zeta(x(Z))$ is algebraic and single-valued.

9. Verify Klein's table of solutions on p. 87.

10. Verify Gordan's result (p. 88) that if S has argument ϕ, T argument ψ, and ST argument θ, then $S^{-1}T$ has argument H, where $\cos\theta + \cos H = \cos(\theta + \psi) + \cos(\phi - \psi)$. [Gordan's quantity $-2\cos\phi$ is the negative of the trace of the rotation matrix considered.]

11. Verify Kronecker's result quoted on p. 88.

12. Show that the symmetry group of the projective self-transformations of Desargues' configuration is the full permutation on five elements by finding five objects in the configuration which are permuted.

13. Find the symmetry of Pappus' configuration and relate it to Hesse's group.

    I am indebted to Robert Syddall who took this cryptic question above and unpacked it, enabling me to set the questions that follow. The elements of the automorphism group of the affine plane over the field of 3 elements, denoted $AGL(2, 3)$, may be written

    $$\begin{pmatrix} x \\ y \end{pmatrix} \mapsto \begin{pmatrix} a & b \\ c & d \end{pmatrix} \begin{pmatrix} x \\ y \end{pmatrix} + \begin{pmatrix} e \\ f \end{pmatrix},$$

where $a, b, c, d, e, f \in Z/3Z$ and $ad - bc \neq 0$. Show that this is a finite group of order 432. Hesse's group is the subgroup of index 2 in $AGL(2, 3)$ and order 216 consisting of those elements for which $ad - bc = 1$ (the direct symmetries). Show (following Syddall) that the automorphism group of the Pappus configuration is the subgroup of index 4 in $AGL(2, 3)$ and order 108 consisting of those elements whose matrix parts are lower triangular, so $ad \neq 0$. Show that this is not a subgroup of Hesse's group.

14. To construct a circular-arc triangle with angles, α, β and γ (where $\alpha + \beta + \gamma < \pi$), draw a circle centre 0 and mark off two radii OA and OB enclosing an angle of 2β, draw OC where $B\hat{O}C = \pi - (\alpha + \beta + \gamma)$, then draw OD where $C\hat{O}D = 2\gamma$. Let AB and DC meet at E. Show that the circular-arc triangle BCE has angles β, at B, γ at C, and α at E. Show how to draw a circle centre E which is orthogonal to the arc CE. How is the construction to be modified if $\alpha + \beta + \gamma > n\pi$?

Construct Schwarz's tessellation. You will find it helpful to draw all arcs in the figure out to the boundary circle, C. Verify the following simple construction for inverting an arc a in circle b, where both arcs are orthogonal to the boundary circle C. Join the extremities A_1 and A_2 or a to the centre of b and let these lines meet C at A'_1 and A'_2. Then the image of a is the arc joining A'_1 and A'_2 which is orthogonal to C.

These constructions can be interpreted as isometries for the non-Euclidean geometry of this disc, see Appendix 4. Verify what Schwarz's tessellation suggests but Poincaré was the first to prove rigorously, that the tessellation reaches arbitrarily close to the boundary circle.

Chapter IV

Modular Equations

The study of the algebraic-solutions problem for a second-order linear ordinary differential equation had brought to light the conceptual importance of considering groups of motions of the sphere, and, in particular, finite groups. Klein connected this study with that of the quintic equation, and so with the theory of transformations of elliptic functions and modular equations as considered by Hermite, Brioschi, and Kronecker around 1858. Klein's approach to the modular equations was first to obtain a better understanding of the moduli, and this led him to the study of the upper half plane under the action of the group of two-by-two matrices with integer entries and determinant one; his great achievement was the production of a unified theory of modular functions. Independently of him, Dedekind also investigated these questions from the same standpoint, in response to a paper of Fuchs. So this chapter looks first at Fuchs' study of elliptic integrals as a function of a parameter, and then at the work of Dedekind. The algebraic study of the modular equation is then discussed; the chapter concludes with Klein's unification of these ideas.

4.1 Fuchs and Hermite

Fuchs' interest in the elliptic integrals K and K', J and J' had been reawakened by a letter he received from Hermite written on 1st July 1876 (Fuchs, [1906, 113]). Hermite wrote: "You should without doubt be able to show, by means of the principles at your command, that on setting $\dfrac{K'}{K} = \omega$ and $k = f(\omega)$, k is a single-valued function of $\omega = x + yi$ for all positive x, but what I cannot work out, and it interests me very much, is how to see clearly that on setting $\dfrac{J'}{J} = x + yi$ one ceases to have a single-valued function. Your methods, I don't doubt, should immediately give the reason for the difference in nature of the functions defined by the two equations".

Fuchs replied in November 1876 and a lengthy extract was published [1877a

= Fuchs 1906, 85–114]. As before (see Ch. II.2) Fuchs studied K and K' as functions of k^2, the modulus, by means of the differential equation which they satisfy. Analytic continuation around closed circuits in the k^2-plane transform K and K' to $a_1K + b_1K'$, $a_2K + b_2K'$ where a_1, b_1, a_2, b_2 are independent of the start and finish point. Fuchs found these numbers and so was able to show that the real part of

$$H = \frac{a_2K + b_2K'}{a_1K + b_1K'} \tag{50}$$

is always positive or zero. When $q = e^{-\pi H}$ is considered as the independent variable, k^2 as a function of q is holomorphic inside the unit circle in the q-plane, but cannot be continued analytically onto or beyond the circle. On the other hand, while J and J', the integrals of a second kind, permit the definition of a function

$$Z = \frac{\alpha_1 J + \beta_1 J'}{\alpha_2 J + \beta_2 J'} \tag{51}$$

and the introduction of a new independent variable $s = e^{\pi Z}$, when $\dfrac{1 - k^2}{k^2}$ is considered as a function of s, it can be extended analytically to a holomorphic function in the whole finite s-plane.

In more detail, Fuchs argued that K and K' are functions of k^2, which, on setting $k^2 = 1/u$, yield two functions η_1 and η_2 of u, $K = \frac{1}{2}\sqrt{u} \cdot \eta_1$, $K' = \frac{1}{2}\sqrt{u} \cdot \eta_2$. These satisfy Legendre's equation

$$2u(u - 1)\frac{d^2\eta}{du^2} + 2(2u - 1)\frac{d\eta}{du} + \frac{1}{2}\eta = 0. \tag{52}$$

η_1 and η_2 can be given power series expansions in a neighbourhood of the singular points 0, 1 and ∞, of this equation, and they transform upon analytic continuation around closed circuits of each singular point according to the following monodromy matrices:

around $u = 1$: $\quad S_1 = \begin{pmatrix} 1 & -2i \\ 0 & 1 \end{pmatrix}$,

around $u = \infty$: $\quad S_\infty = \begin{pmatrix} -1 & 0 \\ 2i & -1 \end{pmatrix}$,

and, since $\quad S_0 = S_1^{-1}S_\infty^{-1}$,

around $u = 0$ $\quad S_0 = \begin{pmatrix} 3 & -2i \\ -2i & -1 \end{pmatrix}$.

The monodromy relation for any closed circuit is therefore given by some product of the form $S_0^k S_1^\ell S_0^{k_1} S_1^{\ell_1} \ldots$, and this, like all the separate powers of S_0 and S_1, has the form

$$\sigma = \begin{pmatrix} \lambda & \mu i \\ \nu i & \rho \end{pmatrix},$$

where λ, μ, ν, ρ are real integers, and $\lambda\rho + \mu\nu = 1$.

To specify all the values of $H = \eta_2/\eta_1 = K'/K$ at a given point u, he let H_0 be an arbitrary value. Then all values are necessarily of the form

$$\frac{\nu i + \rho H_0}{\lambda + \mu i H_0} = \frac{i}{\mu}\frac{1}{\lambda + \mu i H_0} - \frac{\rho i}{\mu}.$$

He specified a single branch of H by the conditions

$$H_0 = \begin{cases} -i & \text{at } u = 0 \\[4pt] 0 & \text{at } u = 1 \\[4pt] \dfrac{4\log 2}{\pi} + \dfrac{1}{\pi}\lim_{u\to\infty}\log u & \text{at } u = \infty. \end{cases}$$

Accordingly, H could take the values

$$\frac{\nu - \rho}{\lambda + \mu}i \quad \text{at } u = 0$$

$$\nu i/\lambda \quad \text{at } u = 1,$$

or, as $u \to \infty$, $H \to \rho i/\mu$, or to a number with real part equal to $+\infty$.

Near $u = 0, 1, \infty$, H is likewise a transform of H_0 in that neighbourhood, and H_0 can be written down explicitly from the solutions to (52). But now H_0 and H always have real part greater than or equal to zero, and so $|q| < 1$, as was to be shown. It remained to show that u is holomorphic inside the unit q-disc, but cannot be analytically continued beyond it. The representation of η_1 and η_2, as solutions to (52), established that u is holomorphic inside the q-disc but in establishing that the circle $|q| = 1$ is a natural boundary for u, Fuchs failed to notice the value ∞, perhaps sharing the widespread contemporary lack of awareness of "bad" point sets, as Schlesinger suggested [Fuchs, 1908, 113]. His mistake was detected by Dedekind and will be discussed more fully below (p. 113). In any case the conclusion that $|q| = 1$ is a natural boundary for u remains valid.

As for J and J' as functions of the modulus k^2, Fuchs again preferred to work with $u = 1/k^2$, and so he introduced $\zeta_1 = 2\sqrt{u}J$ and $\zeta_2 = 2\sqrt{u}J'$, which satisfy the differential equation.

$$2u(u-1)\frac{d^2\zeta}{du^2} + 2u\frac{d\zeta}{du} - \frac{1}{2}\zeta = 0.$$

Fuchs found that $Z = \zeta_1/\zeta_2$ transformed under analytic continuation around different circuits in the n-plane into expressions of the form

$$\frac{\lambda + \mu i Z_0}{\nu i + \rho Z_0},$$

and introducing $s = e^{-\pi Z}$, he found that u as a function of s was holomorphic inside the unit s-disc. But now s as a function of u is well behaved near $|s| = 1$, and so he showed that the inverse function $u = u(s)$ can be extended analytically to the whole s-plane.

Hermite replied to Fuchs' letter on November 27th 1876. He was, he said, delighted with it. Not only had it explained the difference between $\dfrac{K'}{K}$ and $\dfrac{J'}{J}$ but it had done so in a way "which I judge to be of the greatest importance for the history of elliptic functions. The truly fundamental point that the real part of H is essentially positive I had sought in vain to establish by elementary methods, in order not to be obliged to turn to the new method discovered by Riemann". He commented particularly on one respect of Fuchs' work, the elementary derivation of a famous equation of Jacobi

$$4\sqrt{k} = 2q^{1/8} \frac{(1+q^2)(1+q^4)\dots}{(1+q)(1+q^3)\dots}$$

as follows: Fuchs, [1906, 103]. "Is there not some point in observing that on setting $\chi^2 = f(H)$, it follows from your analysis that all solutions of $f(H) = f(H_0)$ are given by the formula

$$H = \frac{vi + \rho H_0}{\lambda + \mu i H_0},$$

and insisting on the extreme importance of this result for the determination of the singular moduli[1] of M. Kronecker, and on remarking that the beautiful discoveries of that illustrious geometer concerning the applications of the theory of elliptic functions to arithmetic seem to rest essentially on this proposition, of which a proof has not been given before?" It is interesting to observe that it was Hermite, and not Fuchs, who preferred to emphasize the inverse function to the quotient of the solutions of the differential equation which has the more readily comprehensible property

$$f(H_0) = f\left(\frac{vi + \rho H_0}{\lambda + \mu i H_0}\right).$$

Two profound ideas emerged during this exchange between Hermite and Fuchs:

(i) the study of the function inverse to the quotient of two independent solutions to a differential equation, which had earlier been broached by Schwarz [Schwarz, 1872], and

(ii) the invariance of such functions under a certain group of transformations, although the group concept was not yet made explicit in this context.

The transformation of modular functions

Hermite himself published a short but crucial paper [1858] on the transformation of modular functions. The paper was chiefly devoted to the solution of quintic equations by modular functions, and for that reason it is described in Section 5.4, but it contained a mysterious table of transformations.[2] He defined $q := e^{-\pi K'/K} = e^{i\pi\omega}$, denoted $k^{1/4}$ by $\phi(\omega)$ and the complementary modulus $k'^{1/4}$ by $\psi(\omega)$, so that

$$\phi^8(\omega) + \psi^8(\omega) = 1,$$
$$\phi(-1/\omega) = \psi(\omega),$$
$$\phi(\omega + 1) = e^{-i\pi/8}\frac{\phi(\omega)}{\psi(\omega)},$$
$$\psi(\omega + 1) = \frac{1}{\psi(\omega)}.$$

The values of

$$\phi\left(\frac{c+d\omega}{a+b\omega}\right) \quad \text{and} \quad \psi\left(\frac{c+d\omega}{a+b\omega}\right)$$

can be deduced easily from these formulae, where a, b, c, and d are integers and $ad - bc = 1$. Hermite found the value depended only on the residue class of

$$\begin{pmatrix} a & b \\ c & d \end{pmatrix} \bmod 2.$$

There are six such classes, and the corresponding transformations were, he said:

I $\quad \begin{pmatrix} a & b \\ c & d \end{pmatrix} \equiv \begin{pmatrix} 1 & 0 \\ 0 & 1 \end{pmatrix}$ $\quad \phi\left(\dfrac{c+d\omega}{a+b\omega}\right) = \phi(\omega)e^{\frac{i\pi}{8}(d(c+d)-1)}$

II $\quad\quad\quad \equiv \begin{pmatrix} 0 & 1 \\ 1 & 0 \end{pmatrix}$ $\quad\quad\quad = \psi(\omega)e^{\frac{i\pi}{8}(c(c-d)-1)}$

III $\quad\quad\quad \equiv \begin{pmatrix} 1 & 1 \\ 0 & 1 \end{pmatrix}$ $\quad\quad\quad = \dfrac{1}{\phi(\omega)}e^{\frac{i\pi}{8}(d(d-c)-1)}$

IV $\quad\quad\quad \equiv \begin{pmatrix} 1 & 1 \\ 1 & 0 \end{pmatrix}$ $\quad\quad\quad = \dfrac{1}{\psi(\omega)}e^{\frac{i\pi}{8}(c(c+d)-1)}$

V $\quad\quad\quad \equiv \begin{pmatrix} 1 & 0 \\ 1 & 1 \end{pmatrix}$ $\quad\quad\quad = \dfrac{\phi(\omega)}{\psi(\omega)}e^{\frac{i\pi}{8}cd}$

VI $\quad\quad\quad \equiv \begin{pmatrix} 0 & 1 \\ 1 & 1 \end{pmatrix}$ $\quad\quad\quad = \dfrac{\psi(\omega)}{\phi(\omega)}e^{-\frac{i\pi}{8}cd}.$

Many years later, in 1900 Hermite [1908, 13–21] wrote to J. Tannery to explain how he had come by such formulae. His method rested on a formula for transforming θ-functions due to Jacobi, but, he said (p. 20), "... it does not please me at all: it is long, above all, indirect; it rests entirely on the accident of a formula of Jacobi, forgotten and as lost among all the discoveries due to that genius".

By then the transformations had been derived several times. The first to do so was Schläfli [1870] who used the approach later employed by Fuchs (above) and considered the monodromy relations of K and iK' under analytic continuation around their poles at 0 and 1. The monodromy matrix at 0 is

$$\begin{pmatrix} 1 & 0 \\ 2 & 1 \end{pmatrix} = \begin{pmatrix} 0 & 1 \\ 1 & 2 \end{pmatrix} \begin{pmatrix} 0 & 1 \\ 1 & 0 \end{pmatrix},$$

and at 1 the monodromy matrix is

$$\begin{pmatrix} 1 & -2 \\ 0 & 1 \end{pmatrix} = \begin{pmatrix} 0 & 1 \\ 1 & 0 \end{pmatrix} \begin{pmatrix} 0 & 1 \\ 1 & -2 \end{pmatrix}.$$

Schläfli observed that a method due to Gauss [*Disq. Arith.* §27] enables one to write any matrix $\begin{pmatrix} a & b \\ c & d \end{pmatrix}$, in which $ad - bc = 1$, $a \equiv d \equiv 1$ (mod 4) and b and c are even, as a product

$$\begin{pmatrix} a & b \\ c & d \end{pmatrix} = \begin{pmatrix} 0 & 1 \\ 1 & a_1 \end{pmatrix} \begin{pmatrix} 0 & 1 \\ 1 & b_1 \end{pmatrix} \cdots \begin{pmatrix} 0 & 1 \\ 1 & a_n \end{pmatrix} \begin{pmatrix} 0 & 1 \\ 1 & b_n \end{pmatrix},$$

where each a_i and b_i is even. The expressions for a, b, c and d involve continued fractions. When the calculations are done,

$$\phi \left(\frac{c + dz}{a + bz} \right) = e^{i \frac{\pi}{8} \sum a_i} \phi(z) \quad \text{and} \quad \psi \left(\frac{c + dz}{a + bz} \right) = e^{i \frac{\pi}{8} - \sum b_i} \psi(z),$$

where

$$\sum a_i \equiv ac + a^2 - 1 \quad \text{(mod 16)} \quad \text{and} \quad -\sum b_i \equiv -ab + a^2 - 1 \quad \text{(mod 16)},$$

so Hermite's results are obtained. Schläfli also gave transformations for the function $\tilde{\psi}(z)$ defined by $\tilde{\psi}^{24}(z) = \phi \psi$, for which ([Jacobi, *Fund Nova* p. 89])

$$\tilde{\psi}(z) = \frac{2^{1/6} e^{i \pi z/24}}{\Pi(1 + q^n)},$$

and for K independently in the 6 cases distinguished by Hermite. Independently of Schläfli, Koenigsberger also solved the problem (Koenigsberger [1871]). He analysed the matrices

$$\begin{pmatrix} a & b \\ c & d \end{pmatrix}$$

as Schläfli had done, using continued fractions, but found the transformations

$$\phi \left(\frac{-1}{\tau} \right) = \psi(\tau), \quad \phi(\tau + 1) = e^{i \frac{\pi}{8}} \frac{\phi(\tau)}{\psi(\tau)}, \quad \text{and} \quad \psi(\tau + 1) = \frac{1}{\psi(\tau)}$$

via the expressions for ϕ and ψ as quotients of infinite products. He made no mention of monodromy.

Schläfli's and Koenigsberger's works suggest that the transformations which must be understood are $z \to z + 1$ and $z \to -1/z$. Although they did not say so

explicitly, when taken mod 2, these transformations generate the six element group
of matrices

$$\begin{pmatrix} a & b \\ c & d \end{pmatrix} \quad \text{for which } ad - bc \equiv 1 \quad (\text{mod } 2)$$

and each entry is 0 or 1, which, as a set, had been distinguished by Hermite. The
passage from integer elements to integers modulo 2 is not exactly explained in either
case, but, for example,

$$z = \frac{K(x)}{i K'(x)}$$

is an infinitely many-valued function of x for which the values for a common x are
of the form z and its transforms

$$\frac{cz + d}{az + b}, \quad \text{where } \begin{pmatrix} a & b \\ c & d \end{pmatrix} \equiv \begin{pmatrix} 1 & 0 \\ 0 & 1 \end{pmatrix} \quad (\text{mod } 2),$$

so x is invariant as a function of z under such transformations

$$z \to \frac{cz + d}{az + b}.$$

That the transformations $z \to z + 1$ and $z \to -1/z$ generate a group was com-
monplace by that time, for it occurs in the calculation of cross-ratios. However, the
connection between that group and the group

$$\left\{ \begin{pmatrix} a & b \\ c & d \end{pmatrix} \equiv \begin{pmatrix} 1 & 0 \\ 0 & 1 \end{pmatrix} \bmod 2 \right\}$$

seems not to have called for any explanation in the minds of Schläfli or Koenigsberger, nor need it have done so unless it would be worthwhile to place this reasoning in a larger context. The man who saw the need for that was Dedekind.

4.2   Dedekind

The first person to emphasize the "invariance" interpretation of this analysis of functions like the modulus, k^2, was Richard Dedekind, in his justly celebrated paper of
1877. It was published in the same issue of Borchardt's *Journal für Mathematik*
as Fuchs' paper, and like it is a letter, addressed in this case to Borchardt. Since
Dedekind's reputation does not even now reflect his true worth, it may not be out of
place to comment on his career at this point. Dedekind contributed much more to
mathematics than his constructive definition of the real numbers ("Dedekind cuts",
discovered in 1858 but only published in *Stetigkeit und irrationale Zahlen* [1872]).
The modern esteem in which this work is held is entirely justified, but Dedekind's
other achievements are generally known only to specialists, not just because of their
difficulty but, I fear, from an exaggerated attention paid by historians and popular-
izers to the foundations of mathematics. Dedekind did much more for mathematics

than just arithmetizing elementary analysis. He was a profound unifier of mathematics and one of the creators of modern algebraic number theory; the concepts of ring, module, ideal, field, and vector space are as much his contributions as anyone else's. He was a great problem solver, particularly in his favourite subject of algebraic number theory. He was a gifted expositor and dedicated editor of the work for others, chiefly his mentors Gauss, Dirichlet, and Riemann, all of whom he had known personally. The sensitive and illuminating article on him by K. R. Biermann [1972] will perhaps restore the balance and help to make his reputation more truly reflect his many achievements.[3]

Dedekind's first work, done under Gauss, had been on Eulerian integrals. In the 1860s he edited Gauss' manuscripts on number theory, Gauss [*Werke*, 1st ed., vol II, 1863] and in the same year brought out the first of four influential editions of Dirichlet's *Vorlesungen über Zahlentheorie*. His interest in algebraic number theory led him to most of his own original work in this period, but in 1876 he edited Riemann's papers with his lifelong friend Heinrich Weber. This paid a debt Dedekind felt to Riemann, and marked an intellectual return to Göttingen where had been a student and instructor in the 1850s. In 1880, in a joint paper with Weber, he laid the foundations of the arithmetic theory of algebraic functions and Riemann surfaces. Biermann [1972, 2] tells us that Dedekind was perhaps the first person ever to lecture on Galois theory, (these lectures are discussed in Section 3) but only two students were present to hear him replace the permutation group concept by that of the abstract group (see Purkert [1976] and below, p. 119). As we shall see, this immersion in the work of another man had its customary effect of deepening and broadening Dedekind's own understanding of mathematics.

Transformations of modular functions

Dedekind began his letter to Borchardt more or less where Hermite's comments ended. He observed that he had already been engaged for several years on the determination of the ideal class number of cubic fields, which was connected with Kronecker's work on singular moduli and complex multiplication. In seeking a simpler route to the "exceptionally beautiful" but difficult results of Kronecker, he found he had been led to appreciate the fundamental importance of Hermite's point (quoted above), that $k = k(\omega)$ is invariant under transformations

$$\omega_1 = \frac{\gamma + \delta\omega}{\alpha + \beta\omega}, \quad \alpha, \ \beta, \ \gamma, \ \delta$$

rational integers, satisfying $\alpha\delta - \beta\gamma = 1$ and β, γ even. He remarked that this observation enables one to derive the usual theory of elliptic functions without difficulty, but set as his main task the elaboration of the theory of modular functions independently of elliptic functions.

Almost all of the usual introductory material on elliptic modular functions appears for the first time in this paper.[4] The upper half of the complex plane, which Dedekind denoted S, is divided into equivalence classes by the action of the group of matrices of the form

$$\begin{pmatrix} \alpha & \beta \\ \gamma & \delta \end{pmatrix}, \quad \alpha\delta - \beta\gamma = 1, \ \alpha, \ \beta, \ \gamma, \ \delta$$

integers. Those equivalent to 0 are the rationals on the real line, together with the point $i\infty$; Dedekind denoted this set R and called it the boundary of S. As a complete system of representatives (of the equivalence classes), he chose the domain defined by the three conditions:

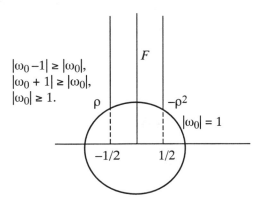

$$|\omega_0 - 1| \geq |\omega_0|,$$
$$|\omega_0 + 1| \geq |\omega_0|,$$
$$|\omega_0| \geq 1.$$

Figure 4.1

which is symmetrically situated with respect to the y-axis, and remarked that the proof of this is a complete system follows the lines of the analogous theorem for binary quadratic forms of negative determinant. The domain he called the principal field (*Hauptfeld*, denoted here by F), and he added that it is bounded by the lines $x_0 = \pm\frac{1}{2}$ and the arc $x_0^2 + y_0^2 = 1$ ($\omega_0 = x_0 + y_0 i$), each point on the boundary except i being equivalent to one other point. The effect of

$$\begin{pmatrix} \alpha & -\gamma \\ -\beta & \delta \end{pmatrix}$$

is to move the circular-arc triangle bounded by

$$\rho = \frac{-1 + i\sqrt{3}}{2}, \ -\rho^2, \ \infty,$$

to the triangle with vertices

$$\frac{-\gamma + \alpha\rho}{\delta - \beta\rho}, \ \frac{-\gamma - \alpha\rho^2}{\delta + \beta\rho^2}, \ \frac{-\alpha}{\beta}.$$

Dedekind then sought a complex-valued function v on S which took the same value at all equivalent points in such a way that, conversely, to each value of the unbounded variable v, there corresponded a unique equivalence class. To obtain such a function he followed the principles laid down by Riemann in his inaugural dissertation (§21). Each half of F can be mapped onto one half of the complex plane by the Riemann mapping principle, so that the y-axis is mapped onto the real axis, and corresponding points on the boundary are given conjugate complex values under the map. The map is extended to the whole of S by the reflection principle, and made unique by insisting that $v(\rho) = 0$, $v(i) = 1$, $v(\infty) = \infty$. The function v Dedekind denoted val(ω) and called the valency (*Valenz*); it is nowadays known as Klein's J-function. Klein studied it the next year, making due acknowledgement to Dedekind. The inverse function to the valency function Dedekind described as a covering of the complex sphere, branched over 0, 1, and ∞. The point 0 is of order 2, 1 is of order 1, and ∞ is a logarithmic branch point.

To study functions like val(ω), Dedekind introduced the differential expression

$$[v, u] := \frac{-4}{\sqrt{\left(\frac{dv}{du}\right)}} \frac{d^2}{dv^2} \left(\frac{dv}{du}\right)^{1/2},$$

which is twice the Schwarzian derivative, and therefore satisfies such identities as

$$[v, u] = \left[v, \frac{C + Du}{A + Bu}\right] \quad A, \ B, \ C, \ D \quad \text{constants}.$$

The function $v = $ val(ω) itself has the Schwarzian derivative

$$[v, \omega] = f(\omega)$$

which has a single-valued inverse $[v, \omega] = F(v)$, and $F(v)$ is finite except at $v = 1$, 0, ∞, where respectively, the products

$$(1 - v)^{-1/2} \frac{dv}{d\omega}, \quad v^{-2/3} \frac{dv}{d\omega}, \quad v^{-1} \frac{dv}{d\omega}$$

are finite and non-zero, and it is soon clear that

$$F(v) = \frac{36v^2 - 41v + 32}{36v^2(1 - v^2)}.$$

Otherwise put, $v = $ val(ω) is a solution of the third order differential equation $[v', \omega] = F(v')$ and the general solution to that question has the form

$$\text{val}\left(\frac{C + D\omega}{A + B\omega}\right).$$

Dedekind was now ready (§6) to introduce the elliptic modular functions themselves. $\left(\frac{dv}{d\omega}\right)^{1/2}$ satisfies a second order linear differential equation with respect to v, and

$$u = \text{const.} \ v^{-1/3} (1 - v)^{-1/4} \left(\frac{dv}{d\omega}\right)^{1/2}$$

satisfies the hypergeometric equation

$$v(1-v)\frac{d^2u}{dv^2} + \left(\frac{2}{3} - \frac{7v}{6}\right)\frac{du}{dv} - \frac{u}{144} = 0,$$

whose general solution in terms of the Gaussian hypergeometric series F is const. $F\left(\frac{1}{12}, \frac{1}{12}, \frac{2}{3}, v\right)$ + const. $F\left(\frac{1}{12}, \frac{1}{12}, \frac{1}{2}, 1-v\right)$, which may also be expressed as a Riemannian P-function. Dedekind found the square root of u of special interest; it is the function

$$\eta(\omega) = \text{const. } v^{-1/6}(1-v)^{-1/8}\frac{dv}{d\omega},$$

nowadays called the Dedekind η-function. It is a single-valued function on S, finite and non-zero on the interior of S, as $\omega \to \infty$, $\eta \to 0$ like $v^{-1/24}$, and thus like $1^{\omega/24} := e^{2\pi i\omega/24}$. He chose the constant so that $\eta(\omega) = 1^{\omega/24}$ at $\omega = \infty$,

$$\eta\left(\frac{\gamma + \delta\omega}{\alpha + \beta\omega}\right) = c \cdot (\alpha + \beta\omega)^{1/2}\eta(\omega)$$

where $c^{24} = 1$, in particular

$$\eta(1 + \omega) = 1^{1/24}\eta(\omega), \quad \eta\left(\frac{-1}{\omega}\right) = 1^{-1/8}\omega^{1/2}\eta(\omega)$$

where $\omega^{1/2} = 1^{1/8}$ when $\omega = 1^{1/4} = i$. Now η is completely determined, since if $f(\omega)$ is another function with these properties, then $f(\omega)/\eta(\omega)$ is an everywhere finite single-valued function of $v = \text{val}(\omega)$, i.e., a constant, which takes the value 1 at $v = \infty$, and so the constant necessarily equals 1.

Dedekind could now make clear the relationship between $\eta(\omega)$ and the modulus of an elliptic integral or its square root k. He introduced three auxiliary functions corresponding to the three transformations of order 2.

$$\eta_1(\omega) = \eta(2\omega), \quad \eta_2(\omega) = \eta\left(\frac{\omega}{2}\right), \quad \eta_3(\omega) = \eta\left(\frac{1+\omega}{2}\right),$$

for which the following identity holds:

$$\eta_1(\omega)\eta_2(\omega)\eta_3(\omega) = 1^{1/48}\eta(\omega)^3.$$

In terms of these functions it turns out that

$$k^{1/8} = 1^{1/48}\sqrt{2} \cdot \frac{\eta_1(\omega)}{\eta_3(\omega)} = \phi(\omega),$$

$$k'^{1/8} = 1^{1/48}\eta_2(\omega)/\eta_3(\omega) = \psi(\omega),$$

where ϕ and ψ are the functions introduced by Hermite [1858] (see above, p. 105), and K and K' can be defined by the equations

$$\sqrt{\frac{2K}{\pi}} = 1^{-1/24}\eta_3(\omega)^2/\eta(\omega) \quad \text{and} \quad K'i = K\omega.$$

Finally, calculating

$$\phi(1+\omega)^8 = \frac{k}{k-1} \quad \text{and} \quad \phi\left(\frac{-1}{\omega}\right) = 1-k$$

led Dedekind to the conclusion that

$$v = \text{val}(\omega) = \frac{4}{27}\frac{(k+\rho)^3(k+\rho^2)^3}{k^2(1-k)^2}.$$

The function $k = \phi(\omega)^8$ can be completely determined from this information, just as η was. It satisfies

$$[k,\omega] = \frac{(k+\rho)(k+\rho^2)}{k^2(1-k)^2},$$

which reduces to

$$\frac{d}{dk}\left(k(1-k)\frac{dK}{dk}\right) = \frac{1}{4}K,$$

which, Dedekind noted, was Fuchs' starting point, and k has therefore the same properties as the quotient of theta-functions

$$\frac{\theta_2(0,\omega)^4}{\theta_3(0,\omega)^4},$$

from which it follows that

$$\eta(\omega) = 1^{\omega/24}\prod_{n=1}^{\infty}(1-1^{\omega n}) = q^{1/12}\prod_{n=1}^{\infty}(1-q^{2n}),$$

where $q = 1^{\omega/2}$. Dedekind regretted that he could not represent η explicitly as a function of ω without invoking the theory of elliptic functions, a task later to be accomplished by Hurwitz using Eisenstein series. Dedekind had defined η as an infinite product in his earlier discussion of a fragment of Riemann's (see Riemann *Werke*, 466–478 and Dedekind's *Werke* XIII) where he also introduced the Dedekind symbol to elucidate the transformation properties of

$$\eta\left(\frac{\gamma+\delta\omega}{\alpha+\beta\omega}\right).$$

At this point Dedekind drew upon this work in editing Riemann's papers to correct the mistake by Fuchs' mentioned earlier. Riemann had considered the boundary values of elliptic modular functions [Riemann *Werke* XXVIII] and found that as $x + yi \to \frac{m}{n}$, along the vertical line $x = \frac{m}{n}$:

$$k \to \infty \quad \text{if } m \equiv n \equiv 1 \qquad \mod 2,$$
$$k \to 1 \quad \text{if } m \equiv 0 \quad n \equiv 1 \qquad \mod 2,$$
$$k \to 0 \quad \text{if } m \equiv 1 \quad n \equiv 0 \qquad \mod 2.$$

Fuchs' mistake had been, as Schlesinger observed, not to distinguish sharply between a set which continuously fills out a curve from one which is merely everywhere dense in a given curve. Fuchs' method of considering k^2 via its inverse function is, of course, less direct and informative than the Riemann–Dedekind one.

Dedekind concluded his remarkable paper with a thorough discussion of the modular equation from his point of view. For an integer matrix

$$\begin{pmatrix} A & B \\ C & D \end{pmatrix}$$

of determinant n, he set

$$v_n = \text{val } \frac{C + D\omega}{A + B\omega}$$

and asked how many different functions v_n there can be. There are, he showed, with the now customary proof,

$$\psi(n) = n \prod_{p|n} \left(1 + \frac{1}{p}\right)$$

such functions $v_1, \ldots, v_{\psi(n)}$. Then the function

$$F := \Pi_{r=1}^{\psi(n)} (\sigma - v_r) = f(\sigma, v)$$

is a single-valued function of σ and ω which is a polynomial in σ of degree $\psi(n)$ with coefficients single-valued functions of $v = \text{val}(\omega)$. The $\psi(n)$ roots of this equation are the v_n's, which are therefore algebraic functions of v. He also showed that the polynomial F is irreducible and symmetric, and conjectured correctly that its coefficients are rational integers. In conclusion, he pointed out that a study of F would illuminate the theory of singular moduli and complex multiplication, but that further developments in which the composition of quadratic forms would play an essential role would have to wait for another opportunity.[5]

Comments

One is struck by the modernity of this paper. The concept of a function, invariant under a certain group, is clearly grasped, although to be sure the matrices

$$\begin{pmatrix} \alpha & \beta \\ \gamma & \delta \end{pmatrix}$$

are not explicitly said to form a group. Even so Dedekind casually uses the fact which he stated (§6), that the matrices

$$\begin{pmatrix} \alpha & \beta \\ \gamma & \delta \end{pmatrix}$$

are all expressible as products of

$$\begin{pmatrix} 1 & 1 \\ 0 & 1 \end{pmatrix} \quad \text{and} \quad \begin{pmatrix} 0 & -1 \\ 1 & 0 \end{pmatrix}.$$

The idea that any single-valued function defined on F or, equivalently, automorphic under the matrices

$$\begin{pmatrix} \alpha & \beta \\ \gamma & \delta \end{pmatrix}$$

and single-valued on S, is a rational function of val(ω) in §3 and of $\eta(\omega)$ in §6. Many of the arguments given by Dedekind occur virtually unchanged in modern presentations of this material; all his arguments have a conceptual clarity and depth seldom found in Fuchs, for example. To what extent are they novel?

The most crucial novelty is the presentation of the theory of modular functions and the relationship between different moduli almost entirely divorced from the theory of elliptic functions. What had previously been derived by astute use of the multitude of equations concerning infinite sums and products in Jacobi's theory was here re-derived with a new lucidity and economy of ideas. Only the formula for η as an infinite product eluded Dedekind's reformulation. The theory of modular functions could now be studied independently of elliptic functions, a natural task which had been in the air for some time. In a passage quoted by Dedekind at the end of §1, Hermite had observed [1862 = 1908 II, 163n] "... no other way for reaching the modular equations has yet offered itself than that which has been given by the founders of the theory of elliptic functions", and Klein in his [1878/79] sought to emphasize the implications these new ideas might have for the study of elliptic functions.

The new theory gave a geometric interpretation of modular functions. They were now obtained by an invariance principle from their definition on a fundamental region, and could be seen to be constrained naturally to the upper half-plane. Dedekind's study of Legendre's equations and the transformations of order two

$$\left(\eta_1 := \eta(2\omega), \ \eta_2(\omega) := \eta\left(\frac{\omega}{2}\right), \ \eta_3 := \left(\frac{1+\omega}{2}\right) \right)$$

and of order N, suggested that the existence of a rich theory of modular transformations could also be obtained by extending the invariance idea, and that this should shed light on the hard-won but obscure results in the theory of moduli. Dedekind did not explicitly stress the concept of a group — that was Klein's decisive contribution — so a certain vagueness of terminology is in order.

The point of departure for Dedekind was the lattice of periods of an elliptic function. Parallelogram lattices had been studied in connection with quadratic forms by Gauss [1840 = *Werke* 1863, II, 194] and Dirichlet [1850 = *Werke* II, 1863, 194], whence Dedekind surely came to hear of them. He may well not have known of Kronecker's study, "Über bilineare Formen mit vier Variabeln" presented to the Berlin Academy in 1866 but not published until 1883 [see *Werke* II, 425–495]. In

§2 of that work Kronecker discussed the six cosets (as they would be called to-day) of $SL(2; Z)$ when the entries are reduced mod 2, and applied his results to the reduction of quadratic forms $ax^2 + 2bxy + cy^2$. The stripping-down of ellip-tic function theory to the lattice idea allowed Dedekind (and Klein) to introduce modular functions as functions on lattices. A lattice has a basis ω_1, ω_2, such that $\omega_1/\omega_2 = \tau \in H$, the upper half plane, and any other basis ω_1', ω_2' is obtained from ω_1 and ω_2 by an element of $SL(2; Z)$. The ratio of the periods, τ, is thus determined only up to the action of this group, and the inverse function (e.g., k^2) is invariant under the action. It is this realization of the importance for modular functions of a well-known result about lattices and quadratic forms that was Dedekind's starting point. He cited Dirichlet's *Zahlentheorie*, 2nd ed., §65, which discusses quadratic forms of negative determinant.[6]

The invariance of elliptic functions under transformations of the variable e.g. $\rho(\omega) = \rho(\omega + \mu_1\omega_1 + \mu_2\omega_2) = z$ was, of course, a very well-known idea. But it had always been regarded as a generalization of the periodicity of the trigono-metric functions; ω_1 and ω_2 were periods, corresponding to closed loops in the z-domain. Fuchs' presentation of the invariance of k^2 followed this approach ex-actly as did Weirstrass in his presentation of the theory of elliptic functions in the 1870s (subsequently published as *Werke*, V). Dedekind broke with it and took in-variance under a group or family of transformations as his starting point. Ultimately the two approaches are not all that different; their unification via the theory of Rie-mann surfaces was the achievement of Klein, but the group-theoretic approach is at least as natural as the topological one, and was to prove to be the way historically towards the "right" generalization of elliptic functions. In this work of Dedekind one can detect also an attempt to prefigure what a Riemann theory of functions for arbitrary closed domains might be like, and what the role of the "universal covering surface" might be. The elaboration of these ideas was due, however, to Klein and his students. To understand it, we must look first at the Galois theory of the modular equation as it had been developed in the 1850s; this will be the theme of the next two sections.

4.3 Galois theory, groups and fields

On 29 May 1832, the evening before his fatal injury in a duel, Galois wrote the now-famous letter to his friend Auguste Chevalier. It begins

"My dear friend,

I have done several new things in analysis.

Some concerning the theory of equations; others integral functions."
[*Oeuvres*, p. 25].

In the theory of equations he described how the solvability of an equation by radicals is connected to the solvability of the group of permutations of its roots. If G is such a group (Galois did not define a group, but this is the first use of the word in this sense) and H a subgroup, then one has two (coset) decompositions

$$G = H + HS + HS' + \cdots$$

$$G = H + TH + T'H + \cdots$$

The decompositions were said by Galois to be proper ("propre") if they coincide, i.e., if H is what is nowadays called a normal subgroup. If the group of the equation is successively decomposed until no further proper decomposition is possible, then Galois said of the indecomposable groups which result that:

> "If these groups each have a prime number of permutations the equations will be solvable by radicals; otherwise not". (p. 26)

Galois' proof of this theorem was found amongt his papers after his death, and published for the first time by Liouville in his *Journal de Mathématiques* in 1846. It will not be described here. In his letter, Galois went on to apply his theory of equations to the modular equations of elliptic functions, studied earlier by Abel and Jacobi.

> "One knows that the group of the equation which has its roots the sine of the amplitude of the $p^2 - 1$ divisions of a period is this:
>
> $$x_{k,\ell}, \qquad x_{ak+b\ell,ck+d\ell};$$
>
> in consequence the corresponding modular equation will have as its group
>
> $$x_{\frac{k}{\ell}}, \qquad x_{\frac{ak+b\ell}{ck+d\ell}}$$
>
> in which k/ℓ can have the $p + 1$ values
>
> $$\infty, 0, 1, \ldots, p - 1" \qquad \text{(p. 27)}.$$

He remarked that this group, let us call it G, has $(p + 1)p(p - 1)$ elements, and a normal subgroup, G' of size $\frac{1}{2}(p + 1)p(p - 1)$, consisting of the substitutions x_k, $x_{\frac{ak+b}{ck+d}}$ where $ad - bc$ is a quadratic residue mod p, but that G' has no proper decomposition $p \neq 2, 3$. Furthermore, the degree of the modular equation can sometimes be reduced. It cannot be reduced below p, because then the prime p would not appear in the size of the group of the equation, but it can be reduced from $p + 1$ to p. This can occur only if the group G has a (non-normal) subgroup H of size

$$(p + 1)\frac{p - 1}{2}.$$

For example, when $p = 7$ he paired the symbols ∞ and 0, 1 and 3, 2 and 6; 4 and 5 denoted the group of substitutions obscurely as

$$x_k, \qquad x_{a\frac{k-b}{k-c}},$$

where b and c are paired and a and c are either both residues or both non-residues mod p. He claimed this gives a subgroup of order

$$(p + 1)\frac{p - 1}{2} = 24;$$

a clearer example will be given below. A similar reduction is possible when $p = 5$ or 11, and Galois concluded,

> "Thus, for the case of $p = 5, 7, 11$, the modular equation reduces to degree p.
>
> "With complete rigour, this reduction is not possible in the higher cases". (p. 29)

The modern notation for these elements

$$\begin{pmatrix} a & b \\ c & d \end{pmatrix} a, \ b, \ c, \ d \in Z/pZ = F_p,$$

the field of p elements is quite irresistible. The action

$$k \rightarrow \frac{ak + b}{ck + d}$$

permutes the lines in the plane over this field, which have the $p + 1$ possible slopes $0, 1, \ldots, p - 1, \infty$. The substitutions of H when $p = 7$, permute the pairs of lines with slopes 0 and ∞, 1 and 3, 2 and 6, 4 and 5 so fill out an octahedral subgroup of G' of index 7. Because it is not normal in the whole group, conjugation represents the group G as acting on 7 elements.[7]

(This was not all that Galois did; let us note in passing that Galois concluded his letter with a staggering summary of the theory of Abelian integrals which remarkably foreshadows Riemann's work (p. 30).)

Responses (1) Jordan

The publication of Galois' work in Liouville's *Journal* stimulated the emerging generation of mathematicians to try to understand and apply it. This process has been described by Wussing, and may be summarized as follows (deferring details of the modular equation to the next section). There was an initial period in which several authors, notably Betti, Kronecker, Cayley and Serret, filled in holes in Galois' presentation of the idea of a group. These commentaries on Galois presented a connection between group theory and the solvability of equations by radicals, and went on to explore the solution of equations by other means, to be described below. The group idea was elaborated in terms of permutation of a finite set of objects, thus following Cauchy's presentation of the theory of permutation groups in 1844–6, and Wussing refers to the formulation as that of permutation groups. The elements are often called substitutions or operations they come with a set of objects which they permute. The crucial presentations of a permutations groups were made by Jordan in his "Commentaire sur Galois" [1869] and his *Traité des substitutions et des équations algébriques* [1870].

Jordan presented a systematic theory of permutation groups, in terms of abstract properties such as commutative, conjugacy, centralizers, transitivity, normal subgroups and quotient groups, group homomorphisms and isomorphism. It seems to me that Jordan came close to possessing the idea of an abstract group. He remarked that "One will say that a system of substitutions forms a group (or a *faisceau*) if the product of two arbitrary substitutions of the system belong to the system itself". (*Traité*, p. 22, quoted in Wussing, p. 104) and he spoke of isomorphisms (= "isomorphisme holoédrique") between groups as one-to-one correspondences between substitutions which respect products (*Traité*, p. 56, Wussing, p. 105). One might well regard the use of words like "substitution" and the permutation notation $(|x, x' \cdots ax + bx' + \cdots, a'x + b'x' + \cdots|$, for example) as well-adapted to their purpose, but not to be taken too literally. On the one hand, Jordan presented an extensive battery of technical concepts of increasing power — we have already seen his use of Sylow theory in the treatment of monodromy groups — on the other hand he analysed in the *Traité* a wide range of situations in which groups could be found permuting lines (the 27 lines on a cubic surface, the 28 bitangents to a quartic) and appearing as a symmetry group of the configuration of the nine inflection points on a cubic and of Kummer's quartic with sixteen nodal points. Jordan's capacity to articulate a powerful theory of finite groups and to recognise them "in nature" surely argues for an implicit understanding of the group idea presented for convenience only in the more familiar garb (to his audiences) of permutation groups. This is not to deny the role of permutation-theoretic ideas in Jordan's work, indicated by the emphasis on transitivity and degree (= the number of elements in the set being permuted), but rather to indicate that ideas of composition and action (as for example change of basis in linear problems) were prominent, and could be seized upon by other mathematicians.

(2) Kronecker

Nonetheless, Wussing can point to a valid distinction in the degree of abstraction between Jordan's work and the less well known treatments of Kronecker and Dedekind. Wussing shows clearly that Kronecker, who learned the new theory of solvability of polynomial equations from Hermite and others during his stay in Paris in 1853, was chiefly concerned with furthering the study of solvable equations. He sought to construct all the equations which are solvable by radicals, as for instance, the cyclotomic equations $x^p - 1 = 0$, p a prime. Those equations in particular he called "Abelian" [*Werke*, IV, 6] and it seems that this is the origin of the designation Abelian for a commutative group: the Galois group of a cyclotomic equation permutes the roots cyclically. In a later work [1870], also relating to number theory, Kronecker gave this abstract definition of a finite commutative group (quoted in [Wussing, 47]:

"Let $\theta', \theta'', \theta''' \ldots$ be a finite number of elements so constituted that from any two of them a third can be derived by means of a definite operation. By this, if the result of this operation is denoted by f, for two arbitrary elements θ', and θ'', which can be identical to one another, a θ''' shall exist which equals $(f(\theta', \theta''))$. Moreover, it will be the case that

$$f(\theta', \theta'') = f(\theta'', \theta')$$
$$f(\theta', f(\theta'', \theta''')) = f((\theta', \theta''), \theta'')$$

and also, whenever θ'' and θ''' are different from one another, $f(\theta', \theta''')$ is not identical with $f(\theta', \theta''')$.

"This assumed, the operation denoted by $f(\theta', \theta'')$ can be replaced by the multiplication of the elements $\theta'\theta''$, if one thereby introduces in place of complete equality a pure equivalence. If one makes use of the usual sign for equivalence : $\sim$, the equivalence $\theta' \cdot \theta'' \sim \theta'''$ is defined by the equation $f(\theta', \theta'') = \theta'''$."

Here the elements are abstract and not necessarily presented as permutations, but Kronecker was always concerned with using the group-idea to advance other domains of mathematics, chiefly number theory. He concentrated therefore on the study of the roots of equations, regarding them as given by some construction, and so increasingly elaborated his theory of the *"Rationalitätsbereich"*, which may be regarded as a constructive presentation of finite field extensions of certain ground fields (usually the rationals, **Q**). In this he resembles his contemporary, Dedekind, and indeed the resemblance seems to have been uncomfortable to Kronecker, who on occasion claimed priority for his theory of algebraic numbers over Dedekind's, and suggested that his ideas may have influenced Dedekind. Kronecker also delayed the publication of the important paper of Dedekind and Weber [1882] in the *Journal für Mathematik* for well over a year.[8] It seems likely that Dedekind was gradually discovering and publishing ideas which Kronecker had had earlier but had not brought forward to his (Kronecker's) own satisfaction. Kronecker often referred to ideas he had had in the 1850s when finally writing some of them down in the 1880s. In view of the probable destruction of the Kronecker *Nachlass* (Edwards [1979]), the matter of priority is unlikely ever to be settled.

(3) Dedekind

As Purkert has shown [1976], Dedekind also developed the idea of an abstract group in the context of Galois theory. He lectured on Galois theory at Göttingen in 1856–1858, although he published nothing on it until 1894, in the famous eleventh supplement to Dirichlet's lectures on number theory (4th edition). Purkert presents an updated manuscript of Dedekind, now in the Niedersächsischer Staats und Universitätsbibliotheck zu Göttingen, which he dates entirely plausibly at around 1857–1858. It goes far beyond the limited ideas about groups published by his *Werke*, II, paper LXI. In brief, the manuscript describes the following: the idea of permutations of a finite set of objects is generalised to that of a finite abstract group. For permutations, Dedekind showed

Theorem 1 *If $\theta\theta' = \phi$, $\theta'\theta'' = \psi$, then $\phi\theta'' = \theta\psi$, or, more briefly,* $(\theta\theta')\theta'' = \theta(\theta'\theta'')$.

Theorem 2 *From any two of the three equations $\phi = \theta$, $\phi' = \theta'$, $\phi\phi' = \theta\theta'$, the third always follows.*

The subsequent mathematical arguments, however, "are to be considered valid for any finite domain of elements, things, ideas, $\theta, \theta', \theta'', \ldots$, having a composition $\theta\theta'$, of θ and θ' defined in any way, so that $\theta\theta'$ is again a number of the domain and the manner of the composition corresponds to that described in the two fundamental theorems". [Purkert, 1976, 4].

The ideas of subgroup (*Divisor*) and coset decomposition of a group G are defined, and a normal subgroup (*eigentlicher Divisor*) is defined as one, K, satisfying

$$K = \theta_1^{-1}K\theta_1 = \theta_2^{-1}K\theta_2 = \cdots = \theta_h^{-1}K\theta_h$$

where the θ's are coset representatives for K (cf. Galois: "propre"). Dedekind showed that the cosets of a normal subgroup themselves formed a group, in which K played the role of the identity element. He went on to apply his theory to the study of polynomials; unhappily, it is clear that the manuscript is incomplete, and a "field-theoretic" part is missing. However, a considerable amount of the theory of polynomial equations survives, and is analysed by Purkert. One may well suppose, incidentally, that Riemann knew of Dedekind's ideas on the subject, although he never felt inclined to work in the area himself. This lends a certain irony to Klein's[9] description of his own work as a "Verschmelzung von Galois mit Riemann".

(4) Klein

To return to the theme of the development of group theory in the period up to the 1870s, there is one final source, the work of Felix Klein. Klein learned about group theory from Jordan in 1870, when he and Sophus Lie went jointly to Paris, a sojourn interrupted in Klein's case by the outbreak of the Franco–Prussian war. It came to represent the third part of his characteristic style of mathematics throughout the 1870s, the others being the invariant theory he had learned the previous year from Clebsch, and the geometric impetus which had marked him from his earliest studies with Plücker in Bonn. It is well-known that one of the earliest fruits of Klein's visit to Paris was the successful study of non-Euclidean geometry from the projective point of view, described in the two papers called "Über die sogenannte Nicht-Euklidische Geometrie" [1871, 1873], and the use of groups to classify geometries as described in the so-called Erlangen Program. As for Klein's interest in group theory *per se*, Freudenthal [1970, 226] has written with his characteristic vigour that

"... Klein's interest in groups was always restricted to those which came from well-known geometries, regular polyhedra or the non-Euclidean plane. His point of departure was never a group to which he associated a geometry – an operation of which the exploitation will be reserved for É. Cartan ... More and more, Klein's activity concerning groups recalls that of a painter of still life."

It is indeed true that Klein never considered group theory abstractly (to have done so would have been to run counter to his geometric and pedagogic inclinations) but Klein would surely have regarded his attitude to group theory as the very opposite of a painter of still life; one thinks of his disagreements with Gordan quoted in Chapter III. As to Freudenthal's passing reference to the non-Euclidean plane, I shall suggest in the final chapter how this might be otherwise expressed, the better to capture a weakness of Klein's thought.

It might be supposed that birational transformations of projective algebraic curves would be another source of group-theoretic ideas, as for instance in a study of the group of birational automorphisms of a given curve. Birational transformations had been brought to the fore by Riemann in his study of algebraic functions [1857c], but it seems that they retained their original significance of changes in the equation of a given curve for some time. The study of their group-theoretic implications will therefore be deferred until the next chapter on chronological grounds, and I shall turn from this synopsis of the history of group theory 1850–1870 to look at particular problems deriving from Galois' ideas as they were considered before Klein.

4.4 The Galois Theory of modular equations c. 1858

Betti

The first to consider Galois' work on the reduction of the degree of the modular equation was Betti, [1853].[10] He considered the equation for snw/p, where w is a period of the elliptic integral, so

$$\frac{w}{4} = \int_0^1 \frac{dx}{((1 - x^2)(1 - k^2x^2))^{1/2}},$$

and p is a prime. This equation has $p^2 - 1$ roots

$$x_{m,n} := sn\left(\frac{mw + nw}{p}\right),$$

where $\tilde{w}$ is the other period and m and n are integers $0 \leq m, n \leq p$, $(m, n) \neq (0, 0)$, and it is invariant under the substitutions

$$\left\{ \begin{matrix} x_{m,n} \\ x_{b'm+a'n,\ bm+an} \end{matrix} \right\} a, b, a', b',$$

integers between 0 and $p - 1$, in his notation. These he said formed the group G of the equation, and G was the product of a group H whose elements were

$$\left\{ \begin{array}{l} y_{q,i} \\ y_{pq,i} \end{array} \right\}$$

and K, the group of an equation of degree $p + 1$, consisting of the substitutions

$$\left\{ \begin{array}{l} y_{q,i} \\ y_{q,\frac{ai+b}{a'i+b'}} \end{array} \right\}$$

or, letting t take the values $0, 1, \ldots, p - 1, \frac{1}{0}$, K consists of the substitutions

$$\left\{ \begin{array}{l} t \\ \dfrac{at+b}{a't+b'} \end{array} \right\}$$

of which there are $p(p - 1)(p + 1)$. K is a product of a group of order 2 and one of order $\frac{1}{2}p(p - 1)(p + 1)$, consisting of the elements for which $ab' - ba'$ is a square mod p, which in turn has p subgroups of order $\frac{1}{2}(p - 1)(p + 1)$. Betti listed them explicitly when $p = 5$, defining them by stating what permutation each element was of the six objects $0, 1, \ldots, \infty$. Betti then showed that when $p = 5, 7$ or 11 it is possible to pair off the objects $0, 1, \ldots, p - 1, \infty$ as Galois had done, but that this is not possible for $p > 11$, and was led to claim, as his Theorems I and II: that the modular equation was not solvable by radicals but could be reduced from degree $p + 1$ to p when $p = 5$, defining them by stating what permutations each element was of the six objects $0, 1, \ldots, \infty$. Betti then showed that when $p = 5, 7$ or 11, and that the equation for $sn\frac{w}{p}$ decomposed into $p + 1$ factors of degree $p - 1$ which were solvable by radicals when the root of an equation of degree $p + 1$ was adjoined, but that the equation was not solvable by radicals. Its degree could come down from $p + 1$ to p when $p = 5, 7, 11$.

Hermite

Hermite published a decisive paper [1858], on the connection between modular functions and quintic equations. He began by observing that the general cubic equation can be put in the form $x^3 - 3x + 2a = 0$, $a = \sin \alpha$, when it has the three solutions

$$2\sin \alpha/3, \quad 2\sin \frac{\alpha + 2\pi}{3}, \quad 2\sin \frac{\alpha + 4\pi}{3}.$$

The general quintic equation can likewise be reduced by root extraction to

$$x^5 - x - a = 0,$$

a reduction Hermite called "the most important step ... since Abel" and attributed to Jerrard (Klein, in *Vorlesungen über das Ikosaeder*, II, 1, 2 argues convincingly that

this reduction is originally due to the Swedish mathematician E. S. Bring [1876]).[11]
In the reduction form the quintic equation is readily solvable by modular functions,
as follows. As usual, let

$$K = \int_0^{\pi/2} \frac{d\theta}{(1 - k^2 \sin^2 \theta)^{1/2}}, \quad K' = \int_0^{\pi/2} \frac{d\theta}{(1 - k'^2 \sin^2 \theta)^{1/2}}, \quad k^2 + k'^2 = 1,$$

and introduce ω by $e^{i\pi\omega} = e^{-\pi K/K'} = q$. Let λ be a modulus related to k by a
5th order transformation. Then, as Jacobi [1829] and Sohnke [1834] had shown,
$\sqrt[4]{k} = u$ and $\sqrt[4]{\lambda} = v$ are related by the modular equation

$$u^6 - v^6 + 5u^2v^2(u^2 - v^2) + 4uv(1 - u^4v^4) = 0,$$

which is to be regarded as an equation for v containing a fixed but arbitrary parameter u. If, then, $u = \phi(\omega)$ and $v = \phi(\lambda)$, the solutions of the equation are

$$-\phi(5\omega), \quad \text{and} \quad \phi\left(\frac{\omega + 16m}{5}\right), \quad m = 0, 1, 2, 3, 4,$$

which may be denoted $v_\infty, v_0, v_1, \ldots, v_4$.

Hermite then defined

$$\Phi(\omega) := \left(\phi(5\omega) + \phi\left(\frac{\omega}{5}\right)\right)\left(\phi\left(\frac{\omega + 16}{5}\right) - \phi\left(\frac{\omega + 4.16}{5}\right)\right)$$
$$\left(\phi\left(\frac{\omega + 2.16}{5}\right) - \phi\left(\frac{\omega + 3.16}{5}\right)\right)$$
$$= -(v_\infty - v_0)(v_1 - v_4)(v_2 - v_3).$$

and found that $\Phi(\omega + r.16)$, $r = 0, 1, 2, 3, 4$ were the roots of the quintic equation
for y:

$$y^5 - 2^4 \cdot 5^3 \cdot y \cdot u^4(1 - u^8)^2 - 2^6\sqrt{5^5}u^3(1 - u^8)^2(1 + u^8) = 0,$$

which can be further reduced, by means of the transformation

$$y = 2.4\sqrt{5^3}u \cdot (1 - u^8)^{1/2} \cdot t,$$

to

$$t^5 - t - \frac{2}{\sqrt[4]{5^5}} \frac{1 + u^8}{u^2(1 - u^8)^{1/2}},$$

thus exhibiting the quintic as a typical one in Bring–Jerrard form. The solution of a
given equation $y^5 - y - a = 0$ is then known once u is found such that

$$\frac{2}{\sqrt[4]{5^5}}\left(\frac{1 + u^8}{u^2(1 - u^8)}\right)^{1/2} = a,$$

which happily reduces to solving a quartic equation.

In Hermite's paper the crucial idea is to use the possibility of reducing the modular equation at the prime 5, whose solutions can be assumed to be known, to a quintic, whose solutions are thus obtained, and to show that any quintic equation can be obtained in this way. The impossibility of performing that reduction when $p > 11$ is not explained (Hermite admitted to this lacuna in the argument). Referring to Betti in a footnote, he said that Betti had published on the subject after his own first work had been done and the results announced in Jacobi's *Werke* (first edition II, 249 [2nd edition, 1969, 87–114)], while the present paper remained unpublished.

Kronecker

Other mathematicians who concerned themselves with the modular equations were Kronecker and Brioschi. Kronecker's work [1858a] on the modular equation of order 7 was presented to the Berlin *Akademie der Wissenschaften* by his friend Kummer on 22nd April 1858, shortly after Hermite's work on the quintic was presented to the Paris Acadmie. It contains a polynomial function of degree 7 in 7 variables, which takes only 30 different values under permutations of the variables, and which is Kronecker's terminology eight-fold cyclic, i.e., unaltered by 8 cyclic permutations.[12] It is clearly left invariant by a group of order $7!/30 = 168$. In general, if the seven variables are the seven roots of a seventh degree polynomial equation, then that equation has a property "which is more general than its solvability". This property Kronecker called its affect ("*Affecte*") but did not define, beyond remarking that one property of an affected equation is that all of its roots are rational functions of any three roots. He remarked that all polynomial equations of degree 7 which can be obtained by reducing the modular equation of degree 8 had this affect, and conjectured that conversely all equations of degree 7 with the affect could be solved in this way by equations derived from the theory of elliptic functions. But, he went on, "To prove this last seems indeed to be difficult; at least, I have not brought my researches, which I started for this purpose two years ago, and are concerned with the object of the present note, to conclusion". He had, he said, been able to make a direct connection between the quintic and its corresponding modular equation just as had Hermite, but he could not see that if offered any application to equations of degree 7.

On the 6th June 1858 Kronecker wrote to Hermite [1858b], enclosing a copy of his note on the equation of degree 7, remarking that it was already two years old but he had hoped to obtain more general results before publishing. However, he felt that certain only that methods such as Jerrard's would not lead to the solutions of equation of higher degree, and so he had sought instead to get a better grasp of the solution of the quintic by means of modular functions. In this case he felt that the crucial element was the existence of functions of 6 variables which are 6-fold cyclic, of which he gave examples leading to the solution of the quintic.

Brioschi

Independently of these two, Brioschi had joined the chase, also giving in his [1858] a solution of the quintic, derived from his study of the multiplier equation[13] rather than the modular equation. So one may say that the quintic equation was by then solved, but that the generalization to higher degrees remained obscure. It is likely that Brioschi also wrote to Hermite about this;[14] at all events Hermite's reply of 17 December 1858 was published in the *Annali di Mathematica* (1859, vol II = *Oeuvres*, II, 83–86). Hermite considered the reduction of the modular equation from the eighth degree to the seventh, and gave two explicit forms for the solutions of the seventh degree equation that arise. He expressed the group of the equation in this way. For x an integer modulo 7, he let $\theta(x) \equiv 2x^2 - x^5$ (mod 7) and considered the group of substitutions (generated by)

$$Z_x, \quad Z_{ax+b}, \quad Z_{a\theta(x+b)+c}$$

in his notation, where a is a quadratic residue mod 7 and b and c are arbitrary. The group has $3.7 + 3.7^2 = 168$ elements. It gave one function of 7 letters having only 30 values, and the other explicit form of the solution gave another, where in this case $\theta(x) \equiv -2x^2 - x^5$ (mod 7). He showed the two systems of substitutions were conjugate. The same issue of the *Annali di Matematica* carried Kronecker's observation, also in the form of a letter to Brioschi [*Werke* IV, 51, 52]: that the two systems of functions of 7 letters were really the same, being obtainable the one from the other by relabelling the letters.

Betti also wrote to Hermite (24 March 1859), and Hermite included the letter in one of his notes in the *Comptes Rendus* on modular equations for that year [Hermite *Oeuvres* II, 73–75]. He was led to consider the subgroup of substitutions

$$\theta(K) = \frac{aK + b}{cK + d},$$

a, b, c, d integers mod 7, consisting of

$$a\frac{K - 3b}{K - b}, \quad -a\frac{K - b}{K - 3b}, \quad aK, \quad -\frac{a}{K}$$

where a and b are residue mod 7 (a group of order 24) with analogous results for the prime 11. Again he claimed that, for number-theoretic reason, such subgroups could not be found for $p > 11$. The existence of these subgroups correspond to the reducibility of the modular equation.

So one may conclude by saying that by 1859 Betti, Brioschi, Hermite, Kronecker, at least had caught up with Galois. The modular equations at the primes 5, 7, and 11 were connected to groups of order $\frac{1}{2}p(p - 1)(p + 1) = 60, 168, 660$, respectively. The reducibility of those equations was connected to the existence of "large" subgroups of these groups, of index p in each case. Hermite had also shown that the general quintic was solvable by modular functions, Kronecker had conjectured that the general polynomial equation of degree 7 was not, and Betti had a proof that reducibility stopped at $p = 11$. Finally, Jordan in his [1868] and *Traité* (p. 348) gave a short proof of this result to Galois and Betti. These results were also published in J. A. Serret, *Cours d'Algèbre* [4th ed., 1879, 393–412], and most accessibly in Dickson, [1900, esp. p. 286].

4.5 Klein

The previous chapter (§3) discussed how Felix Klein approached the question of solving the algebraic-solutions problems from a group-theoretic point of view. The largest group which presented itself there, the group of the icosahedron, has a significance which derives from a mysterious unity between several problems.

First, there is the striking fact that the modular equation at the prime $p = 5$ can be reduced from degree six to degree five.

Second, there is Hermite's solution of the general quintic equation by means of modular functions. The occurrence of modular functions, while not unexpected, seemed to Klein to require a more profound explanation than the mere analogy with the trigonometric functions offered by Hermite. Equally, as Klein said, the studies of Kronecker and Brioschi "gave no general ground why the Jacobi resolvents of degree six are the simplest rational resolvents of the equations of the fifth degree" [1879a = 1922, 391]. (Jacobi resolvents are defined in Chapter V, n. 9.)

Third, there is the invariance of certain binary forms under the appropriate groups of linear substitutions, discussed in Chapter III.

It seemed to Klein that a deeper study of the icosahedron would not only illuminate the underlying unity of these problems in elliptic function theory, the theory of equations, and invariant theory, but also suggest generalizations to modular equations of higher degree, to ternary and higher forms and to the study of function on general Riemann surfaces. He discussed the connection between the icosahedron and the quintic equation in [1875/76] and again in [1877a]. In the first paper he proceeded from the classification of the finite groups of linear substitutions in two variables to an analysis of the covariants of the binary form corresponding to the regular solids. In the second paper he reversed the process, at Gordan's instigation,[15] deriving the theory of equations of the fifth degree from a study of the icosahedron. If μ_1 and μ_2 are projective coordinates on the Riemann sphere, then the twelve vertices of an icosahedron are specified by a binary form, f, in μ_1 and μ_2 of degree twelve. Klein took as the fundamental problem: given f and its covariants H (the Hessian of f) and T (the Jacobian of f and H) as functions of μ_1 and μ_2, find μ_1 and μ_2 as functions of f, H and T. This problem will not be pursued here, beyond noticing that an inverse is raised, and solved by Klein.

He raised the connection between the icosahedron and transformations of elliptic functions in a third paper [1878/79a] which, as he admitted, overlaps considerably with Dedekind's paper discussed above. This paper began the series of papers and books on elliptic modular functions which form Klein's greatest contribution to mathematics, and in which he and his students pioneered a geometric approach to function theory allied to a 'field theoretic' treatment of the classes of analytic functions which were brought to light. The central element in this work is the geometric role of certain Galois groups, which generalise the group of the icosahedron, and are connected to appropriate Riemann surfaces (no longer necessarily the Riemann sphere). The nature of this connection was to occupy Klein deeply for several years. Klein's analysis of the role of the icosahedron in the theory of transformations of elliptic functions and modular equations will now be presented. The nature of the generalizations he made to further problems described in the next chapter.

In Section I of his [1878/79a] he drew together certain observations on elliptic functions which, he said, were not strictly new but seemed to be little known in their totality. The elliptic integral

$$I = \int \frac{dx}{\sqrt{f(x)}}, \quad f(x) = a_0 x^4 + \cdots + a_4$$

possesses two invariants which, following Weierstrass,[16] he called

$$g_2 = a_0 a_4 - 4a_1 a_2 + 3a_2^2, \quad \text{and} \quad g_3 = \begin{vmatrix} a_0 & a_1 & a_2 \\ a_1 & a_2 & a_3 \\ a_2 & a_3 & a_4 \end{vmatrix}.$$

The discriminant of g_2 and g_3 he denoted by $\Delta := g_2^3 - 27g_3^2$. For the absolute invariant he chose g_2^3/Δ and denoted it by J, rather than by the more usual g_2^3/g_3^2. The reason for Klein's choice was that J has slightly simpler mapping properties than

$$g_2^3/g_3^2 = 27 \left(\frac{1}{J-1} \right).$$

J can be written in terms of the cross-ratio, σ, of the four roots of $f(x)$, taken in some order; Klein gave

$$J = \frac{4}{27} \left(\frac{(1 - \sigma + \sigma^2)^3}{\sigma^2(1 - \sigma)^2} \right),$$

observing that it is not altered by the substitution of any other value of the cross-ratio

$$\sigma, \quad \frac{1}{\sigma}, \quad 1 - \sigma, \quad \frac{1}{1 - \sigma}, \quad \frac{\sigma - 1}{\sigma}, \quad \frac{\sigma}{\sigma - 1}.$$

It is therefore, in the terminology of Schwarz and himself, of the double pyramid in this case having 6 faces.

Klein also gave expressions for g_2, g_3, Δ, and J in terms of the ratio, ω, of the periods ω_1 and ω_2, of the elliptic integral I, based on some of the formula in Jacobi's *Fundamenta Nova*. In particular, setting $q = e^{-\pi i \omega}$,

$$\omega_2 \sqrt[12]{\Delta} = 2\pi q^{1/6} \prod_v (1 - q^{2v})^2,$$

and in a footnote (§5 n. 8) Klein observed that $q^{1/6} \prod_v (1 - q^{2v})^2$ was the square of Dedekind's function $\eta(w)$. Furthermore

$$g_2 \left(\frac{\omega_2}{2\pi} \right)^4 = \frac{1}{12} + 20 \sum \frac{n^3 q^{2n}}{1 - q^{2n}},$$

and so as q tends to zero J behaves like

$$\frac{1}{1728q^2}.$$

To describe the mapping properties of J, Klein considered it as a function of ω, itself a function of the modulus k^2 of the integral. Indeed ω maps a half plane of k^2 onto a circular arc triangle and so the whole k^2-plane onto two adjacent triangles with all angles zero in the ω-plane as in Figure 4.2. Klein had here unknowingly rediscovered Riemann's observation on k^2 as a function of ω (Appendix 2) but not published until 1902; Klein refers only to Schwarz [1872, 241–2]. Further analytic continuation of ω extends its image by successive reflections in the sides of the triangles until the whole branching of ω as a function of k^2 is displayed, the branch points being $k^2 = 0, 1, \infty$. Now J can be described as a function of ω as follows. Each image of a k^2 half plane in the ω plane can be divided into six congruent triangles having angles $\frac{\pi}{2}$, $\frac{\pi}{3}$, 0 as shown (Figure 4.3) and, Klein said, each such small triangle is mapped by J onto a half plane. He called such a triangle elementary, and the corresponding domain, mapped by J onto the complex J-plane, an elementary quadrilateral (Figure 4.4) and observed that exactly this quadrilateral figure had been introduced by Dedekind on purely arithmetic grounds (as a fundamental domain for the group $SL(2, Z)$), although he, Klein, preferred to use the well-known results of elliptic function theory.

In terms of the effect of substitutions

$$\omega' = \frac{\alpha\omega + \beta}{\gamma\omega + \delta}$$

on the figure, Klein showed that there are elliptic substitutions, which he defined as those having fixed points. There are fixed points of period 3 at

$$\rho = \frac{-1 + \sqrt{-3}}{2}$$

and all equivalent points in H, where J is zero, and others of period two with a fixed point at i, and all other equivalent points, where $J = 1$. There are parabolic substitutions, whose fixed points are by definition $i\infty$ and all real, rational points, where J is infinite. Finally he called all other substitutions hyperbolic (they have two distinct fixed points on the real axis). It follows from Schwarz's general considerations of the mapping of triangular regions onto a half plane or plane that ω, as a function of J, satisfies the hypergeometric equation with $\alpha = \frac{1}{12} = \beta$, $\gamma = \frac{2}{3}$:

$$J(1 - J)\frac{d^2z}{dJ^2} + \left(\frac{2}{3} - \frac{7J}{6}\right)\frac{dz}{dJ} - \frac{z}{144} = 0;$$

Klein gave various forms for the solution of this equation and for the separate periods ω_1 and ω_2 as functions of J. This concluded the first section of his paper.

In the second section he investigated the polynomial equations which arise from transformations of elliptic functions. For a prime p, a pth order transformation

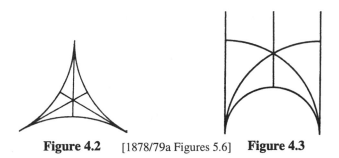

Figure 4.2   [1878/79a Figures 5.6]   **Figure 4.3**

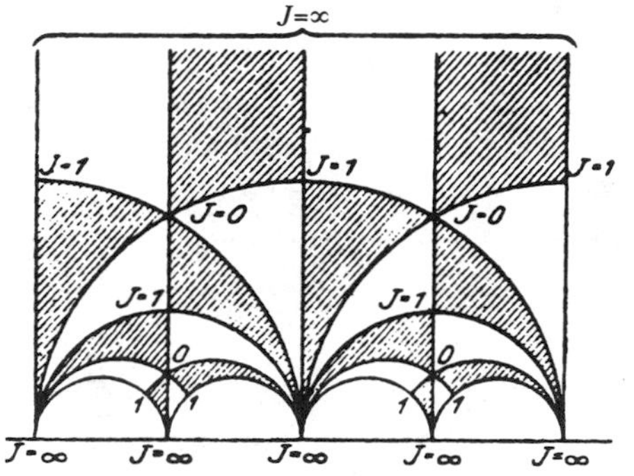

Figure 4.4 [1878/79, Figure 7]

between two elliptic functions is a transformation between a lattice of periods and one of its $p + 1$ sublattices of order p. It gives rise accordingly to a polynomial equation of degree $p + 1$ between the associated absolute invariants (J and J' in Klein's notation),

$$\phi(J, J') = 0.$$

Klein observed that these equations, first obtained by F. Müller [1867, 1872], are much simpler than the ones connecting the moduli directly, and he proposed to study them geometrically, interpreting J and J' as variables on two Riemann surfaces. He took J' so that at the value of $\omega = \omega(J)$, the corresponding $\omega' = \omega'(J)$ took the values

$$\omega' = \frac{\omega}{p}, \frac{\omega + 1}{p}, \dots, \frac{\omega + (p - 1)}{p}, \frac{-1}{p\omega}(p > 3).$$

J' is branched over $J = 0, 1$, and ∞, and Klein showed that

(i) at $J = 0$, if $p = 6m + 5$, the $p + 1$ leaves are arranged in cycles of threes, but if $p = 6m + 1$, $p - 1$ leaves are arranged in cycles of three with two leaves left isolated;

(ii) at $J = 1$, if $p = 4m + 3$, the leaves are joined in pairs, but if $p = 4m + 1$, there are two isolated leaves;

(iii) at $J = \infty$, $J' = \infty$, p leaves are joined in a cycle but one leaf is isolated.

The genus of the transformation equation ϕ was calculated from the formula

$$g = -p + \sum \frac{\sigma - 1}{2}$$

where σ is a number of leaves cyclically interchanged at a branch point and the sum is taken over all branch points. It depends crucially on the value of p mod 12, because of the branching behaviour, and Klein noted that for $p = 5, 7, 13, g = 0$; and for $p = 11, 17, 19, g = 1$; etc. He added that separate considerations showed that g was also zero when p was 2 or 3 (§8–13).

The map from J' to J, given by ϕ, associates $p + 1$ fundamental quadrilaterals in the ω-plane to each fundamental quadrilateral in the ω'-plane. The boundary of the new fundamental polygon is mapped onto itself, yielding a closed Riemann surface, by means of the substitutions

$$\omega' = \frac{\alpha \omega + \beta}{\gamma \omega + \delta}$$

which identify points in the ω-plane having the same J and J' values. In particular, ω and ω/p must be equivalent, which forces β to be divisible by p, so Klein considered those substitutions

$$\begin{pmatrix} \alpha & \beta \\ \gamma & \delta \end{pmatrix}, \quad \beta \equiv 0 \bmod p.$$

Two are parabolic

$$\begin{cases} \omega' = \omega + p \\ \omega' = \dfrac{\omega}{\omega + 1} \end{cases}$$

two are elliptic, and the rest hyperbolic. When these identifications are made, the genus g of the Riemann surface so obtainable can be calculated from Euler's formula $v + f - e = 2 - 2g$, and the results agree with the earlier calculations. This method also applied directly to the cases $p = 2, 3$, and even 4, as Klein showed.

When the genus is zero, J can be expressed as a rational function of a parameter τ. Klein discussed the case $p = 7$ in detail (§14)

$$J = \frac{\phi(\tau)}{\psi(\tau)},$$

where ϕ and ψ are polynomials of degree 8. When $J = \infty$, 7 leaves are joined in a cycle and one is isolated, so ψ factorises into a simple and a sevenfold term. Let τ

be so chosen that the sevenfold term vanishes at $\tau = \infty$, the simple term at $\tau = 0$, i.e., $\psi = c \cdot \tau$ for some constant c. Similarly ϕ has the form

$$(\tau^2 + \alpha\tau + \beta)(\tau^2 + A\tau + B)^3 ,$$

β is an arbitrary non-zero constant, Klein chose $\beta = 49$. (β cannot be zero, for then ϕ and ψ would have a common factor.) Now

$$J - 1 = \frac{\phi - \psi}{\psi}$$

should have four double zeros, arising from the nature of the branch point 1, so $\phi - \psi$ is the square of a quartic expression. This quartic must, furthermore, be a simple factor of the functional determinant

$$\begin{vmatrix} \phi & \psi \\ \dfrac{d\phi}{d\tau} & \dfrac{d\psi}{d\tau} \end{vmatrix} .$$

After a little work this leads to equations for α, A, and B, and Klein concluded that

$$\phi = (\tau^2 + 13\tau + 49)(\tau^2 + 5\tau + 1)^3$$
$$\phi - \psi = (\tau^4 + 14\tau^3 + 63\tau^2 + 70\tau - 7)^2$$
$$\psi = 1728\tau.$$

J' is likewise

$$\frac{\phi(\tau')}{\psi(\tau')}$$

for some τ', and Klein asked for the relation between τ and τ'. It must, he said, be linear,

$$\tau' = \frac{a\tau + b}{c\tau + d},$$

since τ and τ' are single-valued functions on the sphere, and of period 2, since repeating the transformation takes one from J' to J and τ' to τ. Indeed $\tau\tau' = C$ for some constant C, since $\tau = 0$ or ∞ at the two points where $J = \infty$ but these points are interchanged in the interchange of J and J'. Finally, in fact, $\tau\tau' = 49$ since, at the zeros of $\tau^2 + 13\tau + 49$, both J and J' are zero, so $\tau'^2 + 13\tau' + 49 = 0$. So $\tau\tau' = 49$.

When $p = 5$, Klein found (§15) that:

$$J : J - 1 : 1 = (\tau^2 - 10\tau + 5) : (\tau^2 - 22\tau + 125)(\tau^2 - 4\tau + 1)^2 : -1728\tau,$$

with J' being a similar function of τ, and $\tau\tau' = 125$.

The icosahedral equation

These considerations formed the basis of Klein's general approach to the question of modular equations. In Sections III and IV of the paper he turned to consider the significance of the icosahedral equation. In his terminology, the Galois resolvent of an algebraic equation is a polynomial in the roots which is altered by every elements of the group of the equation. It seemed to Klein to be a remarkable fact that all Galois resolvents containing a parameter and of genus zero could not only be determined *a priori* but indeed had also been determined, and were precisely those equations which possessed a group of linear self-transformations. The resolvents were, in short, either of the cyclic, dihedral, tetrahedral, octahedral, or icosahedral type. If the parameter is J, then the way the resolvent equation branched and the requirement that the genus be zero restricts the resolvents which can be connected to modular functions to be only of the tetrahedral, octahedral or icosahedral types. On the other hand, the Galois group of the modular (transformation) equation at the prime $n(> 2)$ is of order $\frac{1}{2}n(n^2 - 1)$. J' is branched over $J = 0, 1, \infty$ in 3s, 2s, and ns respectively, so the genus is

$$p = \frac{(n - 3)(n - 5)(n + 2)}{24},$$

which is zero only when $n = 3$ or 5. It is zero for a similar reason when $n = 2$ or 4, but for no other values.

Klein said (§4): "At $n = 3, 4, 5$ we have the same branching which the *tetrahedral, octahedral, and icosahedral equations display ... These equations are thus the simplest forms which one can give the Galois resolvents of the transformations for $n = 3, 4, 5$.* In this way the significance which above all the icosahedral equation, to which my attention in this work is particularly directed, possessed for the theory of transformations, is made as sharply recognisable as one can". (Emphasis in original).

In this spirit, Klein observed, the equation for the cross ratio $\sigma = K^2$ of the roots of $(1 - x^2)(1 - k^2 x^2)$ appears as the first of a series of equations, for

$$J = \frac{4}{27} \frac{(1 - \sigma + \sigma^2)^3}{\sigma^2(1 - \sigma)^2}$$

is obtained from the equation

$$\mu^3 + \mu^{-3} = \frac{2J - 4}{J}$$

by the linear substitution

$$\mu = \frac{\sigma + \alpha}{\sigma + \alpha^2}, \quad \alpha = e^{2\pi i/3}.$$

Since the genus of J' over J is zero, the Riemann surface for J' is a sphere branched over the complex J-sphere. The domains mapping onto a hemisphere are triangles (since there are three branch points) forming the familiar nets of regular solids.

So in Klein's interpretation the modular equation at the prime 5 is naturally connected with the quintic, since in this case the Riemann surface of J' over J is a sphere in which the faces of a naturally inscribed icosahedron are mapped onto the upper and lower half-planes.

In the fourth and final section of the paper, Klein explained the connection with the solution of the quintic equation by modular functions. He did not regard the occurrence of A_5 in the Galois group of the equations as the complete answer but sought to illuminate the question geometrically in the following way. Consider the quintic equation $x^5 + ax^4 + bx^3 + cx^2 + dy + e = 0$. The substitution $\tilde{x} = x + a/5$ reduced it to an equation of degree 5 in which the coefficient of $\tilde{x}^4$ is zero, so we need consider only equations of the form $x^5 + ax^3 + bx^2 + cx + d = 0$. Tschirnhaus was the first to consider a transformation of the form $\tilde{x} = \alpha + \beta x + \gamma x^2 + \delta x^3$. Eliminating x from these equations yields a quintic equation for x, in which the coefficient of $\tilde{x}^{5-k}$ is a homogeneous function of degree k in the indeterminates α, β, γ, and δ. In particular, the coefficients of $\tilde{x}^4$ and $\tilde{x}^3$ are, respectively, linear and quadratic in these indeterminates, and can both be made to vanish simultaneously. When this is done we obtain a quintic of the form $\tilde{x}^5 + \tilde{a}\tilde{x}^2 + \tilde{b}\tilde{x} + \tilde{c} = 0$. Let us prove these claims geometrically following Klein.

Notice first that a quintic polynomial is specified completely by its five roots x_0, x_1, x_2, x_3, x_4, whose order does not matter. These five roots x_0, x_1, x_2, x_3, x_4 define in general 120 points in the projective space $\mathbf{C}P^4$, corresponding to the 120 arrangements $[x_0; x_1; x_2; x_3; x_4]$. Consider the lines joining them to $[1, 1, 1, 1, 1]$. They meet the hyperplane $H(\sum x_i = 0)$ at 120 points. Moreover, the group S_5 acts on $\mathbf{C}P^4$ by permuting the coordinates, the point $[1, 1, 1, 1, 1]$ is fixed under the action, and the hyperplane H is mapped to itself. In the space $\mathbf{C}P^4/S^5$ the 120 points are all identified to a single point, and the 120 lines become a single line. The line is parameterized by $\lambda \in \mathbf{C}$, a typical point on it being $\lambda[1, 1, 1, 1, 1] + (1 - \lambda)[x_0, x_1, x_2, x_3, x_4]$, and for some suitable value of λ, it meets the hyperplane H. The point where the line meets H corresponds to a polynomial for which $\sum x_i = 0$, i.e., the coefficient of x^4 vanishes, so we have performed the first reduction.

To eliminate the x^3 term is to obtain a polynomial for which $\sum x_i^2$ is also zero. The locus $\sum x_i^2 = 0$ is a quadric, Q, which intersects H in a quadric hypersurface that is non-degenerate. Since H is isomorphic to $\mathbf{C}P^3$, we know that Q is doubly-ruled by two families of lines, traditionally called the A-lines and B-lines. No A-line intersects any other A-line, no B-line another B-line, but each A-line meets each B-line at a unique point (see the exercises below). To perform this elimination, suppose you can find two points in H, say $[p_1, p_2, p_3, p_4]$ and $[q_1, q_2, q_3, q_4]$, whose coordinates are rational functions of $[x_0, \ldots, x_4]$ that are invariant under the action of A_5. The line joining these points meets $Q \cap H$ in two points, and these points correspond to two quintics of the form $x^5 + ax^2 + bx + c = 0$. (We obtain two quintics, corresponding to the quadratic aspect of the Tschirnhaus transformation.)

Next, one would like to eliminate the x^2 term from the quintic. This can be done in the same way by introducing the cubic hypersurface $C: \sum x_i^3 = 0$. However, C and Q meet at $3.2 = 6$ points of H, so we are led to a sextic equation for the elimination of the x^2 term, which is not seemingly a worthwhile step to take when studying quintics. But it would be if instead the process involved only solving a cubic; Bring's discovery is that this is indeed the case. Once Bring's reduction is performed, the polynomial equation takes the form $x^5 + ax + b = 0$, which can be further reduced by setting $\tilde{x} = \rho x$ and suitably choosing ρ, to an equation of the form

$$x^5 + x + \alpha = 0, \quad \alpha \in \mathbf{C}.$$

What does it mean to have the solution to a polynomial equation? It means that given, say $x^5 + ax^4 + bx^3 + cx^2 + dx + e = 0$ (*) one has a "function" ϕ of the co-efficients a, b, c, d, e whose 5 values are the roots of the equation. Such expressions are available as algebraic functions when the polynomial is of degree 2, 3, or 4, but for no other degrees. What Hermite provided was a transcendental function of the coefficients which represented the solution to quintics. More precisely, he accepted a reduction of the problem to the study of quintics of the form $x^5 + x + \alpha = 0$ (†) and then exhibited ϕ as a function of α. The reduction was an algebraic expression which related the roots of (*) to the roots of (†). The reduced situation is that the five roots of (†) form a five-sheeted covering of the complex α-sphere, with S_5 as the corresponding monodromy group.

The task Klein set for himself was to make visible the action of S_5 as a monodromy group. To this end he stopped short of Bring's reduction, and analysed the equation $x^5 + ax^2 + bx + c$ in terms of functions ϕ of $a, b,$ and c. Now, geometrically, this polynomial is represented by a point V on $Q \cap H$. Such a point lies on an A-line and a B-line. Moreover, the action of S_5 maps Q to Q, so also $Q \cap H$ to $Q \cap H$, and such maps send lines to lines since the action is projective. Elements of A_5 permute the A-lines and permute the B-lines; elements which are odd send A-lines to B-lines and B-lines to A-lines. Now the A-lines form a family that is parameterised by any B-line and so the action on the A-lines can be thought of as an action on $\mathbf{CP}^1$, thus representing A_5 as maps

$$\frac{\lambda_1}{\lambda_2} \to \frac{a\lambda_1 + b\lambda_2}{c\lambda_1 + d\lambda_2}, \quad \frac{\lambda_1}{\lambda_2} = V \in \mathbf{CP}^1.$$

In this way Klein showed geometrically how the group A_5 enters the problem.

Klein returned to these questions, and treated them from an enlarged point of view in his famous book on the icosahedron [1884]. I have decided not to compare the treatment given there with the earlier one, but I am in the happy position of being able to refer the reader to the analyses of the book by J.-P. Serre, *Extensions Icosaédriques*, [1980] and Slodowy [1986, 1993].[17]

Reduction of the modular equation

Klein devoted a short second paper in the *Mathematische Annalen* [1879] to the question of the reduction of the modular equation. In the cases $p = 5, 7, 11$, the genus of the reduced equation is zero, so J is rational function of degree 5, 7 or 11 which Klein proceeded to calculate. Of interest here is the geometric and group-theoretic approach Klein took to these problems.

He began, following Betti [1853], with the group which I shall denote $\overline{\Gamma}(p)$ of all 2×2 integer transformations

$$\omega' = \frac{\alpha\omega + \beta}{\gamma\omega + \delta}$$

of determinant $+1$ reduced mod p ($p = 5, 7,$ or 11), which has order 60, 168, or 660 respectively, and sought subgroups H_p of index p. These he listed explicitly. The (coset) representatives for each equivalence class of elements he took as

$$\omega, \omega + 1, \ldots, \omega + (p - 1).$$

He then let y be a function of ω invariant under H_p but not G, so y takes p different values for each value of J, the absolute invariant: $y(\omega), y(\omega + 1), \ldots,$ $y(\omega + p - 1)$, and he regarded y as a p-leaved Riemann surface over the J-plane. The branching of y over J was obtained as in the previous paper, but the fundamental polygon is now made up of p elementary quadrilaterals, not $p + 1$. The explicit form of J as a rational function was only calculated for $p = 5$ and 7. For example, when $p = 5$, the branch point $J = \infty$ is a cycle of all 5 leaves, $J = 1$ two cycles of 3, $J = 0$ a cycle of 3, so the genus is 0.

Accordingly, J is a rational function of y,

$$J = \frac{\phi(y)}{\psi(y)},$$

ϕ and ψ will be a degree of 5 and from the branching behaviour $\phi(y)$ contains a term $(y - 3)^3$, $\phi(y) - 1$ a square of a quadratic term, and so indeed one is led to an explicit relationship between y and J.

Klein did not show that it is only in these cases that the modular equation is reducible by showing that it is only then that subgroups of $\overline{\Gamma}(p)$ exist of index p. Rather, he showed that in these cases it is quite easy to produce equations of degree 5 and 7 of the kind already studied by Brioschi and Hermite.

In the next chapter it will be seen that when Klein was able to extend his methods to deal with the case of higher genera, he was able to show that the modular equation of degree 8 had a certain Galois group of order 168, but when reduced to an equation of degree 7 it was only a special case of such equations. So he solved Kronecker's conjecture affirmatively: equations of degree 7 with this Galois group are solvable by means of elliptic functions.

4.6 A modern treatment of the modular equation

Let

$$M = \begin{pmatrix} \alpha & \beta \\ \gamma & \delta \end{pmatrix}$$

have integer entries and determinant $n \geq 1$, and assume that the greatest common divisor of α, β, γ, and δ is 1. Let $\mathcal{M}_n$ be the set of all such M. As usual, let $\Gamma = SL(2; \mathbf{Z})$. Γ acts on the left on $\mathcal{M}_n$, $S \in \Gamma$ sends $M \in \mathcal{M}_n$ to SM. We say that M and $\tilde{M} = SM$ are equivalent, elements of the space of equivalence classes (or orbits) $\Gamma \mathcal{M}_n$ are written ΓM. Every M in $\mathcal{M}_n$ is equivalent to a matrix of the form

$$\begin{pmatrix} \alpha & \beta \\ 0 & \delta \end{pmatrix}, \text{ and } \begin{pmatrix} \alpha & \beta \\ 0 & \delta \end{pmatrix} \text{ and } \begin{pmatrix} \alpha_1 & \beta_1 \\ 0 & \delta_1 \end{pmatrix} \text{ in } \mathcal{M}_n$$

are equivalent if

$$\begin{pmatrix} \alpha & \beta \\ 0 & \delta \end{pmatrix} = \begin{pmatrix} \pm 1 & 0 \\ 0 & \pm 1 \end{pmatrix} \begin{pmatrix} \alpha_1 & \beta_1 \\ 0 & \delta_1 \end{pmatrix}.$$

There are, accordingly

$$\psi(n) = n \prod_{p \mid n} \left(1 + \frac{1}{p} \right)$$

equivalence classes in $\mathcal{M}_n$. So, in particular, when $n = 5$ there are 6 classes:

$$\begin{pmatrix} 5 & 0 \\ 0 & 1 \end{pmatrix}, \begin{pmatrix} 5 & 1 \\ 0 & 1 \end{pmatrix}, \begin{pmatrix} 5 & 2 \\ 0 & 1 \end{pmatrix}, \begin{pmatrix} 5 & 3 \\ 0 & 1 \end{pmatrix}, \begin{pmatrix} 5 & 4 \\ 0 & 1 \end{pmatrix}, \begin{pmatrix} 1 & 0 \\ 0 & 5 \end{pmatrix}.$$

For each M in $\mathcal{M}_n$, the group Γ_M, which is defined to be $\Gamma \cap M^{-1} \Gamma M$, is now called a transformation group of order n. It is an infinite group, the word "order" refers to the transformations of functions to be introduced below. For example, when

$$M = \begin{pmatrix} 5 & 0 \\ 0 & 1 \end{pmatrix}, \quad \Gamma_M = \left\{ \begin{pmatrix} \alpha & \beta \\ \gamma & \delta \end{pmatrix} \in \Gamma : \gamma = 0 \pmod 5 \right\}.$$

The group Γ_M depends only on the equivalence class of M; $\Gamma_M = \Gamma_{\tilde{M}}$ iff M and $\tilde{M}$ are equivalent under the action of Γ on $\mathcal{M}_n$.

There is a homomorphism of Γ onto a transitive permutation group which permutes the orbits ΓM in $\Gamma \mathcal{M}_n$:

$$S \to \sigma(S), \quad \text{where } \sigma(S) \Gamma M := \Gamma M S.$$

The kernel of this map is $\Gamma^*(n) := \bigcap_{M \in \mathcal{M}_n} \Gamma_M$.

The matrices $M \in \mathcal{M}_n$ are connected with modular transformations of order n, as follows. Let f be a modular function, so $f(\gamma \tau) = f(\tau) \; \forall \gamma \in \Gamma, \forall \tau \in H$, the upper half-plane. Then the transformation $f \to f_M$, where $f_M(\tau) := f(M\tau)$, $M \in \mathcal{M}_n$, is a transformation of order n. It depends only on the equivalence class of M, since

$$f_{SM}(\tau) = f((SM)\tau) = f(S(M\tau)) = f(M\tau) = f_M(\tau).$$

Thus, if K_Γ and K_{Γ_M} denote the modular functions and the functions automorphic for Γ_M, respectively, one obtains a map $K_\Gamma \to K_{\Gamma_M}$, $f \to f_M$, and, as M runs through a set of inequivalent elements of $\mathcal{M}_n$, the functions f_M are all pairwise distinct. So one can define the polynomial

$$Q_n(x, f) := \prod_{(M)} (x - f_M),$$

where the product is taken over a complete set of inequivalent M, which is irreducible and of degree $\psi(n)$ in x. When the polynomial f is taken to be j, the Dedekind–Klein absolute invariant, the polynomial

$$p_n(x, j) := \prod_{(M)} (x - j_M)$$

is called the modular equation of degree n. It enjoys some remarkable properties: its coefficients are rational integers, it is of degree $\psi(n)$ in j, and indeed it is symmetric in x and j.

Since j_M is invariant under Γ_M, K_{Γ_M} is a Galois extension of K containing j_M and having Galois group Γ/Γ_M. So $K_{\Gamma^*(n)}$ is the splitting field of the modular equation, being the smallest field containing all the j_M, and its Galois group is $\Gamma/\Gamma^*(n)$. It is possible to describe its elements precisely; they are all the matrices

$$\begin{pmatrix} \alpha & \beta \\ \gamma & \delta \end{pmatrix} \equiv \begin{pmatrix} \alpha & 0 \\ 0 & \alpha \end{pmatrix} \bmod n \text{ in } \Gamma.$$

So, when n is prime, the index of $\Gamma^*(n)$ in Γ is $\dfrac{\mu(n)}{2}$ where $\mu(n)$ is the index in Γ of the so-called principal congruence subgroup

$$\Gamma(n) := \left\{ \begin{pmatrix} \alpha & \beta \\ \gamma & \delta \end{pmatrix} : \begin{pmatrix} \alpha & \beta \\ \gamma & \delta \end{pmatrix} \equiv \begin{pmatrix} 1 & 0 \\ 0 & 1 \end{pmatrix} \bmod n \right\}.$$

Indeed $\Gamma^*(n) = \Gamma(n)/\{\pm 1\} = PSL(2; Z/nZ)$.

Exercises

1. Check that the group generated by $\begin{pmatrix} 1 & -2i \\ 0 & i \end{pmatrix}$ and $\begin{pmatrix} -1 & 0 \\ 2i & -1 \end{pmatrix}$ consists of matrices of the form $\begin{pmatrix} a & 2bi \\ 2ci & d \end{pmatrix}$, a, b, c, d, integers.

   The next few exercises, based on Serre [1973, VII], sketch the theory of the action of the modular group $PSL(2, Z)$ on the upper half plane, H. Let $S = \begin{pmatrix} 0 & -1 \\ 1 & 0 \end{pmatrix}$ and $T = \begin{pmatrix} 1 & 1 \\ 0 & 1 \end{pmatrix}$ generate a group G. Let F be the region defined on p. ??.

2. Show that if $g = \begin{pmatrix} a & b \\ c & d \end{pmatrix} \in G'$, then $\text{Im}(gz) = \dfrac{\text{Im}(z)}{|cz + d|^2}$. Deduce that there is a $g \in G'$ for which $\text{Im}(gz)$ is a maximum. Deduce that there is an integer n such that $T^n g(z) \in F$.

3. Show that if $z \in F$ and $gz \in F$, $g \in G'$, then either $\text{Re}(z) = \pm\frac{1}{2}$ and $gz = z \pm 1$, or $|z| = 1$ and $gz = -1/z$. It is possible to suppose $\text{Im}(gz) \geq \text{Im}(z)$.

4. Show that points $z \in F$ have trivial stabilizers $S(z) = \{g \in G : gz = z\}$ unless

$$z = i, S(i) = \{I, S\},$$
$$z = \rho, S(\rho) = \{I, ST, (ST)^2\}$$
$$z = -\bar{\rho}, S(-\bar{\rho}) = \{I, TS, (TS)^2\}.$$

5. Show that $PSL(2; Z) \subseteq G'$ by considering z in the interior of F and looking at gz. Deduce that S and T generate $PSL(2; Z)$.

6. Show that $PSL(2; Z)$ has a presentation as $< S, T : S^2, (ST)^3 >$, so it is a free product of a cyclic group of order 2 with a cyclic group of order 3.

7. Show that the group

$$\left\{ \begin{pmatrix} a & b \\ c & d \end{pmatrix} : a, b, c, d \in Z, ad - bc = 1, b \equiv c = 0 \bmod 2 \right\}$$

is a free group on two generators.

8. Schläfli's method (p. 106) is best illustrated by an example.

$$\begin{pmatrix} 9 & 22 \\ 2 & 5 \end{pmatrix} = \begin{pmatrix} 0 & 1 \\ 1 & 0 \end{pmatrix} \begin{pmatrix} 0 & 1 \\ 1 & 4 \end{pmatrix} \begin{pmatrix} 0 & 1 \\ 1 & 2 \end{pmatrix} \begin{pmatrix} 0 & 1 \\ 1 & 2 \end{pmatrix}.$$

The 0, 4, 2, 2 are obtained by this procedure: in $\begin{pmatrix} a & b \\ c & d \end{pmatrix} = \begin{pmatrix} 9 & 22 \\ 2 & 5 \end{pmatrix}$

is $|d| > |c|$? Since in this case the answer is 'yes', write $\dfrac{d}{b} = \dfrac{5}{22}$ as a continued fraction with purely even entries.

$$\frac{5}{22} = 0 + \frac{1}{22/5} = 0 + \frac{1}{4 + 2/5} = 0 + \frac{1}{4 + 1/5/2}$$
$$= 0 + \frac{1}{4+} \frac{1}{2 + \frac{1}{2}}.$$

There are the 0, 4, 2, 2. Had $|d|$ been less than $|c|$, the algorithm says work out $\frac{c}{a}$, and write a 0 at the end.

(i) Consider $\begin{pmatrix} 5 & 6 \\ 4 & 5 \end{pmatrix}$

$$\frac{6}{5} = 0 + \frac{1}{2} + \frac{1}{-2} + \frac{1}{2} + \frac{1}{-2} + \frac{1}{2}$$

and indeed $\begin{pmatrix} 5 & 6 \\ 4 & 5 \end{pmatrix} =$

$$\begin{pmatrix} 0 & 1 \\ 1 & 0 \end{pmatrix}\begin{pmatrix} 0 & 1 \\ 1 & 2 \end{pmatrix}\begin{pmatrix} 0 & 1 \\ 1 & -2 \end{pmatrix}\begin{pmatrix} 0 & 1 \\ 1 & 2 \end{pmatrix}\begin{pmatrix} 0 & 1 \\ 1 & -2 \end{pmatrix}\begin{pmatrix} 0 & 1 \\ 1 & 2 \end{pmatrix}$$

(ii) $\begin{pmatrix} 9 & 2 \\ 22 & 5 \end{pmatrix}$ requires us to write $\dfrac{22}{9} = 2 + \dfrac{1}{2} + \dfrac{1}{4}$, so the numbers are 2,

2, 4, 0 and $\begin{pmatrix} 9 & 2 \\ 22 & 5 \end{pmatrix} = \begin{pmatrix} 0 & 1 \\ 1 & 2 \end{pmatrix}\begin{pmatrix} 0 & 1 \\ 1 & 2 \end{pmatrix}\begin{pmatrix} 0 & 1 \\ 1 & 4 \end{pmatrix}\begin{pmatrix} 0 & 1 \\ 1 & 0 \end{pmatrix}.$

Find e.g., $\begin{pmatrix} 9 & 10 \\ 8 & 9 \end{pmatrix}$ as a Schläfli product.

Since $\begin{pmatrix} 0 & 1 \\ 1 & 0 \end{pmatrix}\begin{pmatrix} 0 & 1 \\ 1 & 2 \end{pmatrix} = \begin{pmatrix} 1 & 2 \\ 0 & 1 \end{pmatrix}$ and $\begin{pmatrix} 0 & 1 \\ 1 & 2 \end{pmatrix}\begin{pmatrix} 0 & 1 \\ 1 & 0 \end{pmatrix} = $

$\begin{pmatrix} 1 & 0 \\ 2 & 1 \end{pmatrix}$ Schläfli's method permits one to write every matrix

$\begin{pmatrix} a & b \\ c & d \end{pmatrix}$, $a \equiv d \equiv 1(4)$, $b \equiv c \equiv 0(2)$, $ad - bc = 1$, as a prod-

uct of $\begin{pmatrix} 1 & 2 \\ 0 & 1 \end{pmatrix}$ and $\begin{pmatrix} 1 & 0 \\ 2 & 1 \end{pmatrix}$. Prove that his method must always

work by showing that these last two matrices generate $\overline{\Gamma}(2)$. It is enough

in doing this to show that exactly one product $\begin{pmatrix} 1 \pm 2 \\ 0 & 1 \end{pmatrix}\begin{pmatrix} a & b \\ c & d \end{pmatrix}$

and $\begin{pmatrix} 1 & 0 \\ \pm 2 & 1 \end{pmatrix}\begin{pmatrix} a & b \\ c & d \end{pmatrix}$ lowers $|a|$ if $|a| > |c|$, or $|c|$ if $|a| < |c|$

(why?). In fact, this observation shows that $\overline{\Gamma}(2)$ is the free product

of $\begin{pmatrix} 1 & 2 \\ 0 & 1 \end{pmatrix}$ and $\begin{pmatrix} 1 & 0 \\ 2 & 1 \end{pmatrix}$. Confirm your earlier calculations by

showing that

$$\begin{pmatrix} 9 & 22 \\ 2 & 5 \end{pmatrix} = \begin{pmatrix} 1 & 4 \\ 0 & 1 \end{pmatrix}\begin{pmatrix} 1 & 0 \\ 2 & 1 \end{pmatrix}\begin{pmatrix} 1 & 2 \\ 0 & 1 \end{pmatrix},$$

$$\begin{pmatrix} 5 & 6 \\ 4 & 5 \end{pmatrix} = \begin{pmatrix} 1 & 2 \\ 0 & 1 \end{pmatrix}\left[\begin{pmatrix} 1 & 0 \\ -2 & 1 \end{pmatrix}\begin{pmatrix} 1 & 2 \\ 0 & 1 \end{pmatrix}\right]^2,$$

$$\begin{pmatrix} 9 & 2 \\ 22 & 5 \end{pmatrix} = \begin{pmatrix} 1 & 0 \\ 2 & 1 \end{pmatrix}\begin{pmatrix} 1 & 2 \\ 0 & 1 \end{pmatrix}\begin{pmatrix} 1 & 0 \\ 4 & 1 \end{pmatrix},$$

$$\begin{pmatrix} 9 & 10 \\ 8 & 9 \end{pmatrix} = \left[\begin{pmatrix} 1 & 2 \\ 0 & 1 \end{pmatrix}\begin{pmatrix} 1 & 0 \\ -2 & 1 \end{pmatrix}\right]^4\begin{pmatrix} 1 & 2 \\ 0 & 1 \end{pmatrix}.$$

9. Every quadric Q in $\mathbf{C}P^3$ can be diagonalized so that it is described by an equation of the form $y_0 y_3 = y_1 y_2$. It follows that Q is the image of $\mathbf{C}P^1 \times \mathbf{C}P^1 \to Q \subset \mathbf{C}P^3$ by $[x_0, x_1] \times [x_2, x_3] \to [x_0 x_2, x_0 x_3, x_1 x_2, x_1 x_3]$. For a fixed $[x_2, x_3]$ the image is a projective line in Q, called an *A*-line; for a

fixed $[x_0, x_1]$ the image is a projective line in Q called a B-line. Show that no A-line meets any other A-line, nor any B-line another B-line, but that each A-line meets each B-line at a unique point.

Chapter V

Some Algebraic Curves

This chapter discusses a topic which was studied from various points of view throughout the nineteenth century and which presented itself in such different guises as: the 28 bitangents to a quartic curve, the study of a Riemann surface of genus 3 and its group of automorphisms, and the reduction of the modular equation of degree 8. These studies, which began separately, were drawn together by Klein in 1878 and proved crucial to his discovery of automorphic functions.

It is only possible to sketch the early developments of each part of this topic in the space available. I have written further on these topics in Gray [1989]. This treatment is divided schematically into two parts: the first on algebraic curves, particularly quartics, and Riemann surfaces; the second on the modular equation.

5.1 Algebraic curves, particularly quartics

An algebraic plane curve is, by definition, the locus in the plane[1] corresponding to a polynomial equation of some degree, n: $f(x, y) = 0$. For example $f(x, y) :=$ $x^3 y + y^3 + x = 0$ represents a quartic. The equation may be written in homogeneous coordinates $(x; y; z)$ by defining

$$F(x; y; z) = z^n f\left(\frac{x}{z}, \frac{y}{z}\right),$$

when the curve is considered to lie in the projective plane. The example above becomes $F(x; y; z) = x^3 y + y^3 z + z^3 x = 0$ in homogeneous form. In nineteenth century usage an algebraic curve of degree n was often called a C_n.

The study of higher plane curves, as C_ns were called when $n > 2$, goes back at least as far as Newton, who in 1667–68 made a thorough study of cubics (Newton [M. P. II, 10–89]), but for present purposes a start can be made with Plücker, who, in his *System der analytischen Geometrie* [1834, 264], showed that every C_n in the projective plane has in general $3n(n-2)$ inflection points. By "in general" he meant that the C_n has no multiple points or cusps. His argument was that at the inflection points of a curve $f(x, y) = 0$, a line $x = \kappa y + \gamma$ has a three-fold intersection with the curve. So

$$\frac{d^2}{dy^2} f(\kappa y + \gamma, y) = 0,$$

i.e.,

$$\frac{\partial^2 f}{\partial y^2}\kappa^2 + 2\kappa \frac{\partial^2 f}{\partial x \partial y} + \frac{\partial^2 f}{\partial x^2} = 0,$$

where κ is homogeneous of order $n-1$, since it satisfies

$$\frac{\partial f}{\partial y}\kappa + \frac{\partial f}{\partial x} = 0$$

(any three-fold intersection is automatically two-fold). Accordingly, the true inflection points lie at the intersection of $f(x, y) = 0$, of degree n, and

$$\frac{\partial^2 f}{\partial y^2} = 0$$

of degree $3n-4$. Of these he showed, however, that $2n$ are only points at infinity, so there are $3n(n-2)$ inflection points altogether. He also showed (p. 283) that the 9 inflection points of a cubic curve lie on four systems of three lines, each line containing three points, and pointed out that, as a result, only three of the inflection points can be real.[2]

Some of these theorems were subsequently proved again in Hesse [1844] in a way which enables the geometric significance of the Hessian to be explained; it was in this context indeed that Hesse introduced it.[3] Adjacent normals to a curve will meet at the appropriate centre of curvature, and Hesse argued, as had Plücker, that at a point of inflection the adjacent normals will be parallel, and the corresponding radius of curvature infinite. Its reciprocal, the mean curvature, is therefore zero, but this just is

$$\frac{\partial^2 f}{\partial x^2}\frac{\partial^2 f}{\partial y^2} - \left(\frac{\partial^2 f}{\partial x \partial y}\right)^2.$$

Hesse introduced homogeneous coordinates x_1, x_2, x_3 to simplify the treatment of the points "at infinity" (p. 131), and so found that the Hessian

$$\left| \frac{\partial^2 F}{\partial x_i \partial x_j} \right|$$

was of degree $3(n-2)$, if f is of degree n, and the Hessian of f meets f in its $3n(n-2)$ inflection points.

Poncelet had suggested in his [1832] that a C_n could have only finitely many bitangents (lines which touch the curve in two distinct places). It seems that this theorem was first proved by Jacobi [1850] and the number found to be $\frac{1}{2}n(n-2)(n^2-9)$ in general. This calculation solved an intriguing problem raised by Plücker in his *Theorie der algebraischen Curven* [1839, Ch. 4], as Jacobi explained. If $F(x;y;z) = 0$ is the equation of a curve, C_n, of degree n, and $(x_0;y_0;z_0) = p$ is a point on the curve, then the tangent to the curve at that point has the equation

$$x\frac{\partial F}{\partial x}(p) + y\frac{\partial F}{\partial y}(p) + z\frac{\partial F}{\partial z}(p) = 0.$$

Conversely, a tangent through the point $(x_1;y_1;z_1)$ touches the curve at a point p which satisfies the same equation,

$$x_1\frac{\partial F}{\partial x}(p) + y_1\frac{\partial F}{\partial y}(p) + z_1\frac{\partial F}{\partial z}(p) = 0.$$

Given F and (x_1, y_1, z_1), the locus of points p satisfying this equation is called the first polar of F with respect to the given point; it is a curve of degree $n - 1$. As such, it meets C_n in $n(n-1)$ points, so one immediately obtains the result that through any given point there are in general $n(n-1)$ tangents to a given curve. Following Möbius and Poncelet, nineteenth century geometers invoked a principle of duality, so that the tangents were regarded as points in a dual projective space. Algebraically this can be done by regarding projective space as made up of lines and interpreting each equation as determining an envelope. Geometrically this can be done by picking a circle in the plane and replacing each line by its polar with respect to the circle. Either way, the $n(n-1)$ tangents become $n(n-1)$ points on the dual curve, to be defined, and because the tangents all passed through the point $(x_1;y_1;z_1)$, the $n(n-1)$ points lie on a line (the dual of $(x_1;y_1;z_1)$). So the dual curve will have degree $n(n-1)$. To obtain it one makes the original point $(x_1;y_1;z_1)$ run along the curve, and considers the tangents at each point; their polars define the dual curve. Plücker's paradox is this: plainly the dual curve of the dual curve is the original curve. Yet the degree of the dual to a C_n is $n(n-1)$, so the degree of the dual of the dual is $n(n-1)[(n(n-1))-1]$, which is not n unless $n = 2$.

Plücker's solution rested on two observations. First, a tangent to a curve is simply a line meeting it at two coincident points, so any line through a double point is a tangent, and the first polar therefore passes through the double point.

Figure 5.1

Consequently, the number of lines through a given point which are truly tangent to a curve is diminished by 2 for each double point, for if the curve is regarded as two curves there are two spurious tangents:

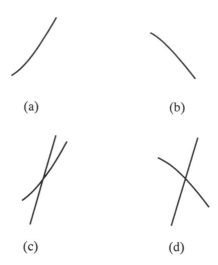

Figure 5.2

Second, if the original curve has a cusp, then the first polar not only passes through the cusp but it is tangent there. So each inflection point reduces the degree of dual by 3. Plücker argued accordingly that the dual curve should have α double points and β cusps, where $2\alpha + 3\beta = n(n-1)[n(n-1)-1] - n = n^3(n-2)$. Since the dual of a double point is a bitangent and of a cusp is an inflection point, Plücker could also speak of the α bitangent points and β inflection points of the original curve. He then stated that

$$\alpha = \frac{1}{2}n(n-2)(n^2-9),$$
$$\beta = 3n(n-2), \qquad\qquad\qquad [1839, \S330]$$

arguing that β was already known to be $3n(n-2)$. These formulae connecting the degree of the dual curve with the degree of the old curve and the number of its double points and cusps are nowadays known as Plücker's formulae.

Jacobi regarded the formula for α as more of a conjecture in need of a proof, and proved it directly using the condition that the equation which a line must satisfy in order to be a tangent (to the given C_n) has a repeated root when the line is a bitangent. This condition yields an equation of degree $(n-2)(n^2-9)$, and hence a curve meeting the original C_n in $n(n-2)(n^2-9)$ points, the points of contact of $\frac{1}{2}n(n-2)(n^2-9)$ bitangents.[4] So, for example, there are 28 bitangents to a quartic curve, a result obtained earlier by Hesse [1848, §3].

Plücker had made a study of the bitangents to a quartic in his [1839, Chapter V]. If two of the bitangents are chosen as axes, the equation of the curve may be written in the form $pqrs - \mu\Omega_2^2 = 0$, where $p, q, r,$ and s are linear terms, Ω_2 is a quadratic term, and μ is a constant. Plücker claimed that this reduction may be performed in

$$\frac{28.27.26}{2.3.4} = 819$$

ways, but went on to deduce incorrectly the number of conics meeting the quartic at the eight bitangent points associated to p, q, r and s. He correctly established that all 28 bitangents may be real [1839, §115] by considering deformations of the curve

$$\Omega_4 = (y^2 - x^2)(x - 1)\left(x - \frac{3}{2}\right) - 2(y^2 + x(x - 2))^2 = 0$$

to $\Omega_4 \pm \kappa = 0$.

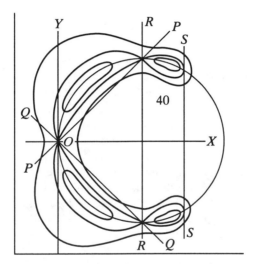

Figure 5.3 [Plücker, *Theorie*, 247, Fig 40]

The curve made up of the 4 meniscas has 28 real bitangents. He also showed that quartics may be found having 16 or 8 real bitangents, but no other values (greater than 8) (§122).

Jacob Steiner took up the study of quartics in 1848. He was an enthusiast for synthetic methods; Kline [1972, 836] records that he threatened to stop submitting articles to Crelle's *Journal* if his friend Crelle continued to publish Plücker's analytic papers. He also preferred to withhold the proofs of his theorems, leaving them as challenges for his colleagues. On this account, Cremona, who responded particularly to Steiner's work on cubic surfaces, called him "this Celebrated Sphinx".

In his papers [1848] and [1852] Steiner considered the interrelationships of the 28 bitangents, and stated several conclusions. These include the following two:[5]

The $\frac{1}{2}(28.27) = 378$ pairs of bitangents may be grouped into 63 groupings (*Gruppen*) of 6 distinct pairs so that the 6 intersection points of each pair lie on a conic; and the eight contact points of any two pairs of bitangents in the same grouping lie on a conic, so there are $\frac{1}{2}.6.5 = 15$ conics associated to each grouping. There are $\frac{1}{3}.63.15 = 315$ such conics altogether, since each is counted 3 times (4 bitangents yield 3 sets of two pairs).

The study of the 28 bitangents brought to light a great deal of rich mathematics which even the complexity of this survey can scarcely suggest. Their intimate connection with the 27 lines on a cubic surface was first made plain by Geiser [1869], who argued that given a cubic surface S and a point P not on it, the tangents through P to S form a cone of degree 6 (any plane through P cuts S in a cubic curve, to which there are $n(n-1) = 6$ tangents from P). If P is on S, the cone becomes the tangent plane, E, at P to S and a quartic surface, Q. Any plane through Q cuts it in a quartic curve — take the plane E and consider the curve $C = E \cap Q$. If g is one of the 27 lines on S, then any plane e through g meets S again in a conic which meets g at two points, P_1 and P_2. The lines PP_1 and PP_2 lie in Q, and define a plane which, furthermore, cuts E in a line that is a bitangent to C. In this way the 27 lines on S give rise to 27 bitangents on C, the 28th is obtained from the principal tangents to S at P. A more complicated argument enabled Geiser to show that the 27 lines can be obtained from the 28 bitangents.[6]

The group-theoretic aspects of this connection were explored by Jordan in his *Traité* [1870, 329–333]. These configuration of special points on a plane quartic were of much interest in the 1850s and 1860s, and it was possible to look forward to a rewarding study of higher plane curves if only suitably powerful enough techniques could be developed. The next crucial development was to come from Riemannian function theory, with the introduction of the generalization of elliptic functions (which are appropriate to cubic equations) to Abelian functions. An indication of how this development was connected by Jordan to the work of Hesse is given in Exercises 15 and 16 at the end of the chapter.

5.2   Function-theoretic geometry

The studies of Hesse and Steiner showed how intricate the geometry of plane curves could be. At about the same time, Riemann was developing the function theory of such curves along the lines of his Inaugural dissertation. The basic problem in the subject was to understand the integral of a rational function $R(x, y)$,

$$\int_0^z R(x, y)dx,$$

on an algebraic curve $F(x, y) = 0$. Jacobi had shown [1834] that if F is a non-singular curve of degree greater than 3, then there will be more than 2 periods to such an integral, and so it does not define a tractable function of its upper end point. In the simplest case, where $F(x, y) = y^2 - f(x)$, and f is of degree 5 or 6, he had suggested taking pairs of integrals and pairs of end points together. This approach was successfully carried out by Göpel [1847] and Rosenhain [1851] independently, and was then generalised magnificently by Weierstrass [1853, 1856] to the case

$F(x, y) = y^2 - R(x) = 0$, $R(x) = (a_0 - x) \ldots (a_{2n} - x)$ being of degree $2n + 1$, and the a_is being constant. By analogy with the elliptic integrals

$$\int \frac{dx}{\sqrt{\text{cubic}}};$$

this case was called hyperelliptic; Weierstrass' analysis of it did much to secure his invitation to Berlin. His method was quite novel: he used Abel's theorem and a system of simultaneous differential equations to study n functions each of n variables obtained by inverting sums of n integrals, i.e.,

$$U_i = \int_{a_1}^{x_1} \phi_i(x)dx + \int_{a_3}^{x_2} \phi_i(x)dx + \cdots + \int_{a_{2n-1}}^{x_n} \phi_i(x)dx,$$

where

$$\phi_i(x) = \frac{(x - a_1)(x - a_3) \ldots (x - a_{2n-1})}{2(x - a_{2i+1})\sqrt{R(x)}}, \quad 1 \le i \le n.$$

The integrands only vary from row to row. As Weierstrass showed, the special case of elliptic functions looks like this in his approach. The integral

$$u = \int_0^x \frac{dx}{[(1 - x^2)(1 - k^2 x^2)]^{1/2}}$$

defines the function $x = sn(u)$, which satisfies the differential equation

$$\frac{d^2 \log x}{du^2} = k^2 x^2 - \frac{1}{x^2}.$$

Conversely, starting from the differential equation, the substitution $x = P_1/P$ results in the equation

$$\frac{d^2 \log P_1}{du^2} - \frac{d^2 \log P}{du^2} = k^2 \frac{P_1^2}{P^2} - \frac{P^2}{P_1^2},$$

which can be factorized to give the two equivalent equations

$$\frac{d^2 \log}{du^2} P_1 = \frac{-P^2}{P_1^2}, \quad \frac{d^2 \log}{du^2} P = -\frac{k^2 P_1^2}{P^2}.$$

These equations, Weierstrass showed, can be solved to yield uniformly convergent power series expansions of P and P_1 on some suitable domain, and if one fixes the initial conditions at $u = 0$:

$$P = 1, \quad \frac{dP}{du} = 0, \quad P_1 = 0, \quad \frac{dP_1}{du} = 1,$$

then indeed

$$snu = \frac{P_1}{P}.$$

[Weierstrass 1856 = *Werke*, I, 297].

When R is of degree $2n + 1$, the differential equations gave Weierstrass uniformly convergent power series in n variables for n functions which he called $Al_i(x_1, \ldots, x_n)$, $1 \le i \le n$, in honour of Abel. The mathematical world was greatly excited by Weierstrass' work, which went a long way to solve a problem that had been outstanding for a generation. In only the next year, Riemann was able to solve the problem completely, but his methods were so novel that they did not command general assent. Until the work of Hilbert and Weyl, other methods were preferred to those of Riemann for dealing with the new functions. Riemann, his students, and Clebsch, also sought to use the new functions to understand the geometry of algebraic curves. Weierstrass himself was so struck by Riemann's paper that he withdrew a paper of his own, which he had nearly finished, on Abelian functions, and did not feel able to lecture on it satisfactorily until 1869.[7]

Riemann divided his paper [1857c] into two halves. In the first part, as he said, he discussed algebraic functions and their integrals without using theta-series. In §§1–5 he used his version of Dirichlet's principle to define algebraic functions in terms of their branching behaviour and their poles; in §§6–10 he expressed algebraic functions and their integrals as rational functions on a (Riemann) surface; in §§11–13 he looked at equivalent ways of expressing a given function (introducing the ideas of birational equivalence and moduli) and at the simplest expressions for a given function. In §14–16 he discussed Abel's addition theorem as a solution to a system of differential equations. In the second half of the paper he used theta functions to solve the general Jacobi inversion problem, thus generalizing Weierstrass' treatment of the hyperelliptic case.

To understand the geometrical implications of this work, it is enough to know (1) that Riemann succeeded in defining the genus, p, of a surface given by an algebraic equation in purely topological terms,[8] and in showing that such a surface has a linear space of holomorphic differentials of complex dimension p (this is the paradigm case of an index theorem.) Furthermore (2) he solved the problem of Jacobi inversion, which is:

> given the values $e_1, \ldots, e_p$ of p sums of p integrals of p linearly independent holomorphic differentials ω_i, $1 \le i \le p$,

$$\left(\sum_i \int_{a_i}^{x_i} \omega_1, \ldots, \sum_i \int_{a_i}^{x_i} \omega_p \right) \equiv (e_1, \ldots, e_p)$$

which are defined modulo periods, to find the corresponding values of $x_1, \ldots, x_p$. This inverts the integrals $u_j = \int \omega_j$ and makes $x_1, \ldots, x_p$ functions of $u_1, \ldots, u_p$. His solution introduced the infinitely many-valued function $\theta(u_1 - e_1, \ldots, u_p - e_p)$ of p variables on the (Riemann) surface, which was quasi-periodic and had a well-defined set of p zeros on the surface. These zeros, which came to be known classically as the theta-null values, were, he showed, the sought-after values of $x_1, \ldots, x_p$.[9]

This formulation, while not providing an explicit expression for $(x_1, \ldots, x_p)$ as a function of $(e_1, \ldots, e_p)$, at least connects Jacobi inversion for arbitrary curves with Jacobi inversion for elliptic curves, and reduces the general problem to the vanishing of a scalar function on the Riemann surface. It thus gives a hold on the pre-image of a specified point in $\mathbf{C}^p$ (mod periods) which Riemann was able to exploit. It is comparable to the use of the symmetric functions in solving the problem of finding three numbers x, y and z, when $x + y + z$, $xy + yz + zx$, and xyz are given, and one can write down a cubic equation.

The responses to Riemann's paper were various, but most mathematicians agreed that it was extremely important and difficult. The high degree of generality and the use of a transcendental method for constructing functions militated against its immediate application, and for a while it was only Clebsch and Riemann's own students who advanced his ideas. Foremost among them was Gustav Roch, who died in 1866 of tuberculosis, at the age of 29.

The geometric study of curves based on the theory of algebraic functions formed part of Riemann's lectures during February 1862, and some of this material was published in the *Werke* (second edition, no. XXXI) based on notes taken by Roch. More was published in the *Nachträge* [*Werke*, 1–66]. Riemann considered functions that have $p - 1$ second order infinities and $p - 1$ double zeros on a curve $F(s, z) = 0$, which he obtained as follows. A basis for the holomorphic integrands on the curve is

$$\left\{ \frac{\phi_i(s, z)}{\dfrac{\partial F}{\partial s}} \right\}_{1 \leq i \leq p},$$

and he had shown in [1857c, §10)] that each $\phi_i(s, z)/\phi_j(s, z)$ is a function with $2p - 2$ simple infinities and $2p - 2$ simple zeros.

Now he showed that it was possible to find a finite number of ϕ_i such that their quotients ϕ_i/ϕ_j have $p - 1$ double zeros and $p - 1$ double poles, as required. His construction involved associating characteristics to the theta-function, that is, expressions of the form

$$\begin{pmatrix} \varepsilon_1, \dots, \varepsilon_p \\ \varepsilon_1', \dots, \varepsilon_p' \end{pmatrix}.$$

The εs are integers, and could be taken to be either 0 or 1 without loss of generality. Riemann said the characteristic

$$\begin{pmatrix} \varepsilon_1, \dots, \varepsilon_p \\ \varepsilon_1', \dots, \varepsilon_p' \end{pmatrix}$$

was even if $\varepsilon_1 \varepsilon_1' + \cdots + \varepsilon_p \varepsilon_p' \equiv 0 \pmod 2$, and otherwise odd. The zeros of the theta-function are a p-tuple of values of integrals of the holomorphic integrands, whose upper end points are the zeros of the functions ϕ_i. A complicated argument enabled Riemann to show that the relation between the two sets of zeros was such that the odd characteristics corresponded to the ϕ_i with repeated zeros. There are $2^{p-1}.(2^p - 1)$ odd characteristics and $2^{p-1}.(2^p - 1)$ such functions accordingly, so when $p = 3$, there are 28 interesting odd characteristics (see Table 5.1).

On the Bitangents of a Quartic, by Professor Cayley.

The equations of the 28 bitangents of a quartic curve were obtained in a very elegant form by Riemann in the paper "Zur Theorie der Abelschen Functionen für den Fall $p = 3$," Werke, Leipzig, 1876, pp. 456—472; and see also Weber's "Theorie der Abelschen Functionen vom Geschlecht 3," Berlin, 1876. Riemann connects the several bitangents with the characteristics of the 28 odd functions, thus obtaining for them an algorithm which it is worth while to explain, but they will be given also with the algorithm employed p. 231 et seq. of the present work, which is in fact the more simple one. The characteristic of a triple θ-function is a symbol of the form

$$\alpha\ \beta\ \gamma,$$
$$\alpha'\ \beta'\ \gamma',$$

where each of the letters is $= 0$ or 1; there are thus in all 64 such symbols, but they are considered as odd or even according as the sum $\alpha\alpha' + \beta\beta' + \gamma\gamma'$ is odd or even; and the numbers of the odd and even characteristics are 28 and 36 respectively; and, as already mentioned, the 28 odd characteristics correspond to the 28 bitangents respectively.

We have x, y, z trilinear coordinates, α, β, γ, α', β', γ' constants chosen at pleasure, and then α', β', γ' determinate constants, such that the equations

$$x + y + z + \xi + v + \zeta = 0,$$
$$\alpha x + \beta y + \gamma z + \frac{\xi}{\alpha} + \frac{v}{\beta} + \frac{\zeta}{\gamma} = 0,$$
$$\alpha' x + \beta' y + \gamma' z + \frac{\xi}{\alpha'} + \frac{v}{\beta'} + \frac{\zeta}{\gamma'} = 0,$$
$$\alpha'' x + \beta'' y + \gamma'' z + \frac{\xi}{\alpha''} + \frac{v}{\beta''} + \frac{\zeta}{\gamma''} = 0,$$

are equivalent to three independent equations; this being so, they determine ξ, v, ζ each of them as a linear function of (x, y, z); and the equations of the bitangents of the curve $\sqrt{(x\xi)} + \sqrt{(yv)} + \sqrt{(z\zeta)} = 0$ (see Weber, p. 100) are

16	111 111	$z = 0,$
38	001 011	$y = 0,$
33	011 001	$z = 0,$
32	010 010	$\xi = 0,$
13	100 110	$v = 0,$
12	110 100	$\zeta = 0,$
48	101 100	$x + y + z = 0,$
14	010 011	$\xi + v + z = 0,$
65	100 101	$\alpha x + \beta y + \gamma z = 0,$
15	011 010	$\frac{\xi}{\alpha} + \frac{v}{\beta} + \frac{\zeta}{\gamma} = 0,$
66	110 010	$\alpha' x + \beta' y + \gamma' z = 0,$

16	001 101	$\frac{\xi}{\alpha'} + \beta' y + \gamma' z = 0,$
78	010 110	$\alpha'' x + \beta' y + \gamma' z = 0,$
17	101 001	$\frac{\xi}{\alpha''} + \beta' y + \gamma' z = 0,$
24	100 111	$x + v + z = 0,$
34	110 101	$x + y + \zeta = 0,$
25	101 110	$\alpha x + \frac{v}{\beta} + \gamma z = 0,$
35	111 100	$\alpha x + \beta y + \frac{\zeta}{\gamma} = 0,$
26	111 001	$\alpha' x + \frac{v}{\beta'} + \gamma' z = 0,$
36	101 011	$\alpha' x + \beta' y + \frac{\zeta}{\gamma'} = 0,$
27	011 101	$\alpha'' x + \frac{v}{\beta''} + \gamma' z = 0,$
37	001 111	$\alpha'' x + \beta' y + \frac{\zeta}{\gamma''} = 0,$
67	100 000	$\frac{x}{1-\beta\gamma} + \frac{y}{1-\gamma\alpha} + \frac{z}{1-\alpha\beta} = 0,$
57	110 011	$\frac{x}{1-\beta'\gamma'} + \frac{y}{1-\gamma'\alpha'} + \frac{z}{1-\alpha'\beta'} = 0,$
56	010 111	$\frac{x}{1-\beta''\gamma''} + \frac{y}{1-\gamma''\alpha''} + \frac{z}{1-\alpha''\beta''} = 0,$
45	001 001	$\frac{\xi}{\alpha(1-\beta\gamma)} + \frac{v}{\beta(1-\gamma\alpha)} + \frac{\zeta}{\gamma(1-\alpha\beta)} = 0,$
46	011 110	$\frac{\xi}{\alpha'(1-\beta'\gamma')} + \frac{v}{\beta'(1-\gamma'\alpha')} + \frac{\zeta}{\gamma'(1-\alpha'\beta')} = 0,$
47	111 010	$\alpha''(1-\beta''\gamma'') x + \beta''(1-\gamma''\alpha'') y + \gamma''(1-\alpha''\beta'') z = 0.$

The whole number of ways in which the equation of the curve can be expressed in a form such as $\sqrt{(x\xi)} + \sqrt{(yv)} + \sqrt{(z\zeta)} = 0$ is 1260; viz. the three pairs of bitangents entering into the equation of the curve are of one of the types

12.34,	13.24,	14.23	▨ No. is	70
12.34,	13.24,	56.78	□ ∥ "	630
12.35,	14.34,	16.35	◈ "	560
				1260

and it may be remarked that selecting at pleasure any two pairs out of a system of three pairs the type is always □ or ∥, viz. (see p. 231) the four bitangents are such that their points of contact are situate on a conic.

From Salmon's Higher Plane Curves
387-389.

Table 5.1

In his study of curves $F(s, z) = 0$ of genus $p = 3$, Riemann took new coordinates $\xi := \phi_1/\phi_3$ and $\eta := \phi_2/\phi_3$. Since ξ and η each take every value $2p - 2 = 4$ times, the curve is now expressed as a quartic $F(\xi, \eta) = 0$. The integrands

$$\phi \Big/ \frac{\partial F}{\partial \xi}$$

are everywhere finite, so ϕ is of degrees $2p - 5 = 1$ in ξ and η, and is therefore a linear function containing 3 constants, i.e., the dimension of the space of holomorphic integrands is 3. But then F is non-singular for, if it had r singular points, then a line $\phi = 0$ through a singular point would have a double zero there, and so the space of lines having double zeros on F would have dimensions $p + r$. But then $p + r = 3$ and $p = 3$, so $r = 0$. By construction each ϕ vanishes to the second order on F, so $\phi = 0$ represents a bitangent to the curve. Also, the restriction of a rational function σ in (s, z)-space to a curve $F(s, z) = 0$ gives a function on a curve, and the zero locus of σ is itself a curve which meets F where $\sigma(s, z) = 0$. Repeated zeros on F of the function correspond to points where σ touches F.

Riemann then introduced homogeneous coordinates x, y, z ($\xi = x/z, \eta = y/z$), so each ϕ was of the form $cx + c'y + c''z$. A further linear change of variable allowed him to consider x, y, and z as bitangents, in which case he showed (see the Exercises) that $F(x, y, z) = f^2(x, y, z) - xyzt$, where t is linear in x, y and z.

Riemann then argued that the equation can be reduced to the form $f^2 - xyzt$ in 6 different ways once x and y are given. For $F = f^2 - xyzt = \psi^2 - xypq$ implies $(f + \psi)(f - \psi) = xy(zt - pq)$, so xy divides, say, $f - \psi$, i.e., $\psi - f = \alpha xy, \alpha$ constant. Then $\alpha(f + \psi) = pq - zt$, or, eliminating ψ, $a\alpha f + \alpha^2 xy + zt = pq$. The left hand side is reducible, so its discriminant must vanish, and its discriminant is of degree three in its coefficients, and thus of degree 6 in α.

Accordingly, there are 6 decompositions of F in the form $f^2 - xyzt$; $\alpha = 0$, giving precisely the decomposition in which $pq = zt$, and $\alpha = \infty$ the decomposition $pq = xy$.

Salmon [1879, 227] gave a pleasing geometric interpretation of this. When the equation of the curve is written in the form $xyU = (z^2 + ayz + by^2 + cxz + dx^2)^2$, as it may always be by the above argument, it may thus be written in the form

$$xyU = V^2,$$

or

$$xy(\lambda^2 U + 2\lambda V + xy) = (xy + \lambda V)^2$$

where $U = 0$ and $V = 0$ represent conics. Six conics in the family $\lambda^2 U + 2\lambda V + xy = 0$ are reducible, i.e., are line pairs. One, $\lambda = 0$, gives the line pair $xy = 0$; the other 5 other line pairs, $zt = 0$ such that the equation of the curve becomes, on setting $xy + \lambda V = f$, $xyzt = f^2$.

Thus, "through the four points of contact of any two bitangents, we can describe five conics, each of which passes through the four points of contact of two other bitangents", and, since there are $\frac{1}{2}28.27 = 378$ pairs of bitangents, but each conic is counted six times, "there are in all $\frac{5}{6}(378)$ or 315 conics, each passing through the points of contact of four bitangents of a quartic". This result was given by Hesse, Steiner and, incorrectly, by Plücker.

Riemann now confronted two tasks: first, given a curve to find the equation of its 28 bitangents; second, to express the mutual relationships of the bitangents. In reverse order, the second task was accomplished by means of the notation for the characteristics, and then the symmetries between the bitangents enabled Riemann to find their equations. This is an attractive piece of work, and it was taken up by Clebsch in an important paper of 1864. Other papers in this spirit were written by Roch and Weber. It is interesting to note, however, that Klein did not work in this tradition when he turned to the geometry of quartic curves and sought to connect that subject with his earlier work on modular transformations. It is very likely that he was unaware of Riemann's lectures, which were not published until 1892, and he may not have understood the papers by Clebsch and Roch, which inevitably appealed to Riemann's obscure theory of Jacobi inversion. Even the self-appointed heir to Riemann was to admit he found the master very difficult.

5.3   Klein

In 1878 Klein began to publish a series of papers generalizing his earlier work to equations and transformations of higher degree than five. These works mark a considerable development in the theory of Riemann surfaces, and are the start of the systematic study of modular functions, Klein's greatest contribution to mathematics. They also form, as has been suggested, an important stage in the development of Galois Theory, being the origin of that part of the subject which concerns fields of rational functions on an algebraic curve. Klein distinguished in his papers between the function-theoretic and the purely algebraic directions in which his research was proceeding, concentrating chiefly on the former, which will be dwelt on accordingly.[10]

The most important paper on the function-theoretic side is Klein's "Über die Transformation siebenter Ordnung der elliptischen Functionen" [1878/79]. Klein described the path he took in this paper as going from a thorough description of a group of 168 linear substitutions

$$\omega' = \frac{\alpha\omega + \beta}{\gamma\omega + \delta}$$

which permute the roots of the modular equation, to a study of a certain function η as a branched function of J, where it turns out that η forms a surface of genus 3. This leads to the study of a curve in the projective plane having 168 symmetries which is, indeed, a quartic. Klein devoted quite some time to describing the curve as vividly as possible.

The group of the modular equation of order 7 is, as Galois had known, made up of the 168 linear substitutions

$$\omega' = \frac{\alpha\omega + \beta}{\gamma\omega + \delta}$$

with coefficients in the integers reduced mod 7, and determinant $+ 1$. [For, the column

$$\begin{pmatrix} \alpha \\ \gamma \end{pmatrix}$$

in one of the two matrices representing the transformation

$$\omega' = \frac{\alpha\omega + \beta}{\gamma\omega + \delta}$$

can have any of 8.6 different entries, the number of vectors through the origin in the field $\mathbf{Z}/7\mathbf{Z}$, and

$$\begin{pmatrix} \beta \\ \delta \end{pmatrix}$$

can lie in one of 7 other directions. So there are $8.6.7 = 336$ matrices, and 168 linear transformations.] The group was usually referred to the group of order 168, and later denoted G_{168}, as it will be here. It is the only simple group of that order,[11] and is isomorphic to $PSL\,(2;\mathbf{Z}/7\mathbf{Z})$. Klein described its elements according to their conjugacy class, which essentially determines their geometric character, referring to conjugacy by the word "*gleichberechtigt*": S_1 and S_2 are conjugate if there is an element S of G_{168} such that $S_1 = S^{-1}S_2S$. He found there were:

(a) 21 self-conjugate substitutions of period two, such as

$$\frac{-1}{\omega},$$

   characterised by the fact that $\alpha + \delta = 0$ [i.e., they have zero trace];

(b) 28 conjugate pairs of substitutions of period 3, such as

$$\frac{-2\omega}{3} \quad \text{and} \quad \frac{-3\omega}{2},$$

   characterised by $\alpha + \delta = \pm 1$;

(c) 48 substitutions of period 7 coming in 8 conjugate sets of 6, such as $\omega + 1, \omega + 2, \ldots, \omega + 6$ characterised by $\alpha + \delta = \pm 2$ and excluding the identity $\omega' = \omega$; and

(d) 21 conjugate pairs of substitutions of period 4 corresponding to each substitution of period 2, such as

$$\frac{2\omega + 2}{-2\omega + 2}, \frac{2\omega - 2}{2\omega + 2}$$

corresponding to

$$\frac{-1}{\omega},$$

characterised by $\alpha + \delta = \pm 3$.

There are accordingly the following subgroups of G_{168}:

(1) The identity;

(2) 21 G_2s of order 2

(3) 28 G_3s of order 3;

(4) 21 G_4s of order 4;

(5) 8 G_7s of order 7;

(6) 14 G'_4s of order 4, each containing two substitutions of order two which commute with (*sind vertauschbar ... mit*) a given substitution of order 2, for example

$$\omega, \quad \frac{-1}{\omega}, \quad \frac{2\omega + 3}{3\omega - 2}, \quad \frac{3\omega - 2}{-2\omega - 3}$$

in two families of 7 conjugates;

(7) 28 conjugate G'_6s of order 6, containing the three substitutions of order 2 commuting with a given G_3;

(8) 21 conjugate G'_8s of order 8, the centralizers of the G_4s;

(9) 8 conjugate G'_{21}s of order 21, the centralizers of the G_7s;

(10) 14 G''_{24}s of order 24 in two families of 7 conjugates, arising from the families of G'_4s, which are octahedral groups (this paper, §14).

Klein claimed that this list was complete, but in footnote 6 in the *Werke*, he mentions that each G''_{24} contains a normal subgroup G''_{12} of order 12, isomorphic to the even permutations on 4 things. The G'_4s introduced casually here have come to be known as examples of Klein's four-group.

Klein next considered the subgroup of $PSL(2; \mathbf{Z})$ consisting of those maps

$$\omega' = \frac{\alpha\omega + \beta}{\gamma\omega + \delta}$$

which are conjugate to the identity mod 7 [in his later work he called this group the principal congruence subgroup of level 7; it is usually denoted $\overline{\Gamma}(7)$] and introduced a single-valued function η on the upper half plane with the property that

$$\eta(\omega) = \eta\left(\frac{\alpha\omega + \beta}{\gamma\omega + \delta}\right) \quad \text{if and only if} \quad \omega' = \frac{\alpha\omega + \beta}{\gamma\omega + \delta} \quad \text{is in} \quad \overline{\Gamma}(7).$$

The fundamental region for η is 168 copies of the fundamental region for J, since $\overline{\Gamma}(7)$ has index 168 in $PSL(2; \mathbf{Z})$, and $PSL(2; \mathbf{Z})/\overline{\Gamma}(7) \simeq G_{168}$. So for each value of $J(\omega)$, there are 168 different values of $\eta(\omega)$. As for the branching of η over J, it followed from an earlier paper of Klein's that only $J = 0$, 1, and ∞ can be branch points. It follows from the classification of the substitutions and their fixed points that, at $J = 0$, the branching is of order 3 and the leaves hang in 56 cycles; at $J = 1$ the branching is of order 2 in 84 cycles; and at $J = \infty$, the order is 7 (typically $\omega' = \omega + 1$). The genus of the surface is thus

$$p = \frac{1}{2}(2 - 2.168 + 2.56 + 1.84 + 6.24) = 3,$$

and η forms a Riemann surface of genus 3 admitting 168 conformal self-transformations. [$\overline{\Gamma}(7)$ moves the fundamental region around en bloc, and so provides the identifications of the sides to yield a closed surface; G_{168} provides the self-transformations of the surface.]

To understand this surface better, Klein introduced certain special points corresponding to the branching. The 24 points a correspond to $J = \infty$; the 56 points b to $J = 0$; the 84 points c to $J = 1$. Since each a is fixed by a G_7, of which there are 8, each G_7 fixes 3 as. Likewise, each G_3 fixes 2 bs, and each G_2 four cs. No G_4 fixes any point.

To obtain an equation for the curve, Klein (quoting Clebsch–Gordan [1866] p. 65 and Clebsch–Lindemann [1876] pp. 687, 712, but only in the *Werke*) argued that the equation would be either a plane quintic with triple points if the curve was hyperelliptic, or a non-singular quartic. The curve in question could not be hyperelliptic [because G_{168} is simple], so it can be represented by a non-singular quartic, C_4, and G_{168} becomes a group of plane collineations. Klein remarked here that G_{168} had seemingly been omitted by Jordan in his list of finite subgroups of $PGL(3, \mathbf{C})$ [1877/78], and only included in the Correction [1878]. But now the as, bs, and cs can be immediately connected with distinguished point sets on a quartic that correspond to certain projective properties. The as are the 24 inflection points; the 28 pairs of bs the points of contact of the 28 bitangents; the 21 quadruples of cs the sextactic points, at which a cubic has threefold contact with the quartic (a sextactic point is one where a conic has 6-fold contact with a curve, the term is due to Cayley [1859b]).

Furthermore, the triples of as (inflection points) preserved by a G_3 can be regarded as follows. Each inflection tangent to a C_4 meets it again at 1 point, so there are 24 thus distinguished points, which must be the inflection points themselves (since they form the only distinguished set of 24 points). No inflection tangent is a bitangent, so each inflection tangent meets the curve at a different inflection point. A collineation fixing a given a must also fix its inflection tangent, and so the new inflection point. By iterating this argument, one sees that the triple as preserved by the G_3 are the vertices of a triangle whose sides are inflection tangents; , and there are 8 such inflection triangles.

The 21 quadruples of cs are invariant under collineations of order 2, i.e., perspectivities, which thus occur having 21 centres and 21 axes, each quadruple lying on an axis. There are 4 axes through each centre, 4 centres on each axis (since each transformation of order two fixes 4 cs and 4 other points). Each bitangent carries 3 centres, through each of which pass 4 bitangents. Each G''_{24} permutes a distinguished set of 4 bitangents.

If an inflection triangle is taken as a triangle of reference, so the sides are $\lambda = 0$, $\mu = 0$, $\nu = 0$, it is soon clear (§4) that the equation of the curve becomes $f(\lambda, \mu, \nu) = \lambda^3\mu + \mu^3\nu + \nu^3\lambda = 0$. Klein gave a series of explicit calculations for the equations of bitangents with respect to this triangle of reference and others[12] adapted to display the symmetry and so the invariance with respect to particular subgroups of G_{168}. He also gave a representation of the group G_{168} as projective collineations of the following forms:

$$\lambda' = A\lambda + B\mu + C\nu$$

(1) $\mu' = B\lambda + C\mu + A\nu$

$$\nu' = C\lambda + A\mu + B\nu$$

where

$$A = \frac{\gamma^5 - \gamma^2}{\sqrt{-7}}, \quad B = \frac{\gamma^3 - \gamma^4}{\sqrt{-7}}, \quad C = \frac{\gamma^6 - \gamma}{\sqrt{-7}}$$

and $\gamma = e^{2\pi i/7}$ so $\gamma + \gamma^4 + \gamma^2 - \gamma^6 - \gamma^3 - \gamma^5 = \sqrt{-7}$, or

(2) $\lambda' = \mu, \mu' = \nu, \nu' = \lambda$, or

(3) $\lambda' = \gamma\lambda, \mu' = \gamma^4\mu, \nu' = \gamma^2\nu$, and compounds of (1), (2), and (3).

The associated invariants and covariants of the curve are fairly easy to study geometrically. Its Hessian $\nabla = 5\lambda^2\mu^2\nu^2 - (\lambda^5\nu + \nu^5\mu + \mu^5\lambda)$ of course meets it at the 24 inflection points, and since this is the smallest set of distinguished points, there is no invariant polynomial of degree less than 6, and, up to constant multiples, only the Hessian of degree 6. Similarly, the next invariant polynomial will be of degree 14 and its intersection with the curve will be the bitangent points. There are many such polynomials, but all are sums of $f^2\nabla$ and any one which is not a constant multiple of $f^2\nabla$. Klein chose

$$C = \frac{1}{9} \begin{vmatrix} \dfrac{\partial^2 f}{\partial\lambda^2} & \dfrac{\partial^2 f}{\partial\lambda\partial\mu} & \dfrac{\partial^2 f}{\partial\lambda\partial\nu} & \dfrac{\partial\nabla}{\partial\lambda} \\[2mm] \dfrac{\partial^2 f}{\partial\mu\partial\lambda} & \dfrac{\partial^2 f}{\partial\mu^2} & \dfrac{\partial^2 f}{\partial\mu\partial\nu} & \dfrac{\partial\nabla}{\partial\mu} \\[2mm] \dfrac{\partial^2 f}{\partial\nu\partial\lambda} & \dfrac{\partial^2 f}{\partial\nu\partial\mu} & \dfrac{\partial^2 f}{\partial\nu^2} & \dfrac{\partial\nabla}{\partial\nu} \\[2mm] \dfrac{\partial\nabla}{\partial\lambda} & \dfrac{\partial\nabla}{\partial\mu} & \dfrac{\partial\nabla}{\partial\nu} & 0 \end{vmatrix}.$$

For an invariant of degree 21, he chose the functional determinant K, of f, ∇, and C:

$$K = \frac{1}{14} \begin{vmatrix} \dfrac{\partial f}{\partial \lambda} & \dfrac{\partial \nabla}{\partial \lambda} & \dfrac{\partial C}{\partial \lambda} \\[2mm] \dfrac{\partial f}{\partial \mu} & \dfrac{\partial \nabla}{\partial \mu} & \dfrac{\partial C}{\partial \mu} \\[2mm] \dfrac{\partial f}{\partial \nu} & \dfrac{\partial \nabla}{\partial \nu} & \dfrac{\partial C}{\partial \nu} \end{vmatrix}.$$

Since K is the only invariant polynomial of degree 21, it must represent the 21 axes of the perspectivities. Finally, to represent the 168 corresponding points on $f = 0$, he considered the family of curves $\nabla^7 = kC^3$, where k is a constant. There is a linear relation between ∇^7, C^3, and K^2 precisely as in the icosahedral case. In this case a comparison of coefficients in the explicit representations of ∇, C, and K in terms of λ, μ, ν shows that

$$(-\nabla)^7 = \left(\frac{C}{12}\right)^3 - 27\left(\frac{K}{216}\right)^2,$$

i.e.,

$$J : J - 1 : 1 = \left(\frac{C}{12}\right)^3 : 27\left(\frac{K}{216}\right)^2 : -\nabla^7,$$

or, in terms of the invariants of elliptic integrals

$$g_2 = \frac{C}{12}; \quad g_3 = \frac{K}{216}; \quad \sqrt[7]{\Delta} = -\nabla,$$

and f, ∇, C, K represent a complete system of covariants[13] of f.

The equation of degree 168 is the resolvent of the modular equation. Since G_{168} has subgroups of indices 7 and 8, the equation has resolvents of degrees 7 and 8, which Klein also found. He connected the resolvent of degree 8 to the 36 systems of contact cubics of even characteristic via the 8 inflection triangles. This led him to an explicit solution to the equation of degree 168 in terms of elliptic functions. He remarked in footnote 21 "[The equation] must also be solvable by means of a linear differential equation of the third order; how can one construct it?" This problem was solved by Hurwitz and Halphen.[14]

Klein then turned, in the concluding sections of the paper, to the task of describing these resolvents "as graphically as possible with the aid of Analysis Situs". The function J maps one half of its fundamental region onto the upper half-plane, the other onto the lower half-plane. Let the triangle mapping onto the upper half-plane be shaded. Then the fundamental region for μ contains 168 shaded and 168 unshaded triangles. The vertices of these triangles, in accordance with the branching of μ, are the a, b, and c points; at a point a, 14 triangles meet, at a b, 6 triangles, and at a c, 4 triangles. As originally presented, the points a are at infinity, and the vertical angle, i.e., 0 whereas the vertical angles at the bs are $\pi/3$, and at the cs are $\pi/2$. Klein found it more attractive to work with triangle having angles $\pi/7$, $\pi/2$, $\pi/3$.

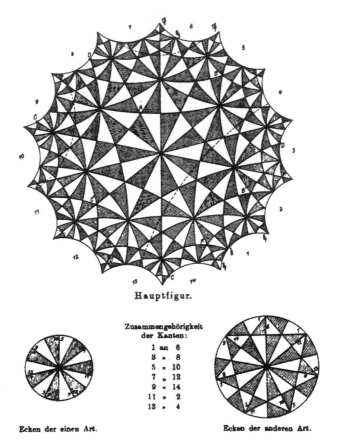

Hauptfigur.

Zusammengehörigkeit
der Kanten:

1	an	6
3	„	8
5	„	10
7	„	12
9	„	14
11	„	2
13	„	4

Ecken der einen Art.   Ecken der anderen Art.

Figure 5.4

He did not explain how this transition could be made; his student M. W. Haskell wrote a thesis on this curve in 1890 and pointed out that the Schwarzian s-function

$$s\left(\frac{1}{3}, \frac{1}{2}, \frac{1}{7}, J\right)$$

can be used [Haskell, 1890, p. 9n]. An examination of elements in $PSL(2; \mathbf{Z})$, congruent to the identity mod 7, provides the identification of the edges on the boundary needed to make the curvilinear 14-gon into a closed surface: if the edges are numbered anti-clockwise, then 1 is identified with 6, 3 with 8, 5 with 10, and so on, until 13 is identified with 4.

The beautiful figure of 168 shaded and 168 unshaded triangles not only displays the inflection points, bitangent points, and sextactic points of the curve as the vertices of the triangles. It also contains several interesting lines, which run straight through the as, bs, and cs in this order:

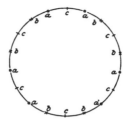

Figure 5.5

They are the curves on the surface on which J is real-valued. On the closed surfaces, the lines are closed curves, i.e., projective lines. There are 28 symmetry lines, and they can be traced on the figure. Some care is needed at the vertices of the 14-gon, which are of two kinds. For example, the vertical line $0_1 0_8$ continues as the edge 5 which is, of course, identified now with the edge 10. Klein remarked (§13), "These symmetry lines are for many purposes the simplest means of orientation on our surface; I will use them here to define the mutually corresponding points a, b, c on the surface. It is then easy to picture the corresponding self transformations of our surface". The seven symmetry lines emanating from an a meet again at the two corresponding as. The three symmetry lines through a point b meet at the corresponding b; and the two through a c meet again at a c. Furthermore, if the surface is cut open along the seven lines through a given a, then, because the genus is 3, it appears as a simply-connected region with boundary curve: the 14-gon is recovered. Two corresponding bs give rise to three symmetry lines which meet another symmetry line through two cs, so there is a one-to-one correspondence between the bs and the symmetry lines. Finally, the two lines through a pair of corresponding cs meet all the other symmetry lines but two, which themselves meet in two cs, giving rise to the quadruples of c-points.

Klein wanted to display the figure in as regular a way as possible, but he knew that there is no solid in three-dimensional space whose symmetry group is G_{168}. However, he pointed out, G_{168} contains several octahedral subgroups G''_{24}. Each permutes a set of 4 pairs of b-points. It is possible to surround a b-point with 3 14-gons (strictly 7-gons, because of the adjacent vertical angles of $\pi/2$) made up of the 14 triangles meeting at an a-point, thus dividing the figure so that it is seen as containing $4.2.(3.14) = 336$ triangles. The G''_{24} acts then on a true octahedron, with a b point at the mid point of each face, the a points as vertices, and some c-points as mid edge points, provided diametrically opposite points are identified.

The aspects of the figure which cannot be realized in three-space have this interpretation: one imagines the octahedron as composed of three hyperboloids of one sheet with axes crossing at right angles, and with opposite edges identified at infinity, thus representing a surface of genus 3. The axes of the hyperboloids may be said to pass through the vertices of the octahedron.

Klein concluded the paper with a discussion of the real part of the curve, in

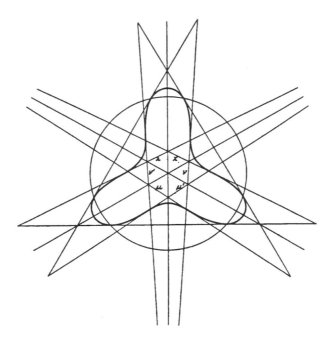

Figure 5.6

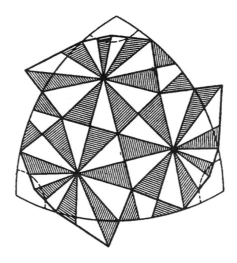

Figure 5.7

keeping with an earlier interest, and supplied the following attractive figure shown opposite.

This paper must have convinced Klein, if he needed convincing, of the power of his new methods. A year later he reported on this progress in a paper [1880/81] suggestively entitled, "Towards a theory of elliptic modular functions". In this work he set himself the task of showing how the different forms in which one encountered the modular equation were very special cases of a simple general principle. The theory should begin, he said, with an analysis of all the subgroups of $PSL(2; \mathbf{Z})$. He proposed in this paper only to deal with subgroups of finite index, which he admitted was a great restriction. He pointed out that this use of groups as a classifying principle went beyond Galois in that it invoked infinite as well as finite groups — to be precise, infinite normal (*ausgezeichneten*) subgroups of $PSL(2; \mathbf{Z})$. His second classifying principle, introduced on empirical grounds, was that the subgroups should be defined arithmetically by means of congruences, but, he pointed out sternly, "it must be clearly understood that not all subgroups are congruence groups". Thirdly, he argued that a fundamental polygon should be introduced for each group, and the function theory of the corresponding Riemann surface brought into consideration. In particular, the genus of the surface should be evaluated, above all to see whether or not it is zero.

For a given subgroup, he considered the functions invariant under the substitutions of the subgroup; such functions he called *Moduln*, here translated as modules (moduli would also be confusing). When $p = 0$, a module can be found taking each value precisely once on the fundamental polygon. For $p > 0$, two modules are needed to specify a point, and they will be related by an algebraic equation of degree p. The case $p = 0$ includes, he pointed out, the νth roots of Legendre's modules κ^2 and of $\kappa^2 \kappa'^2$ for integers ν. Up until, now only the following had been studied in the usual theory: $\kappa^2, \kappa, \sqrt{\kappa}, \sqrt[4]{\kappa}, \kappa^2 \kappa'^2, \kappa\kappa', \sqrt{\kappa\kappa'}, \sqrt[4]{\kappa\kappa'}, \sqrt[3]{\kappa\kappa'}, \sqrt[6]{\kappa\kappa'}, \sqrt[12]{\kappa\kappa'}$, and this was, he said, because only these are congruence modules.

For a modular transformation $w' = w/n$ or $w' = -n/w$, Klein proposed considering the equation between $J(w)$ and $J(w')$ as the prototype of all modular equations, and to investigate it by finding its degree, its Galois group, and subgroups thereof.

One notices very clearly here the role of invariant functions and the sense in which Galois groups appear as indices of the relation between two fields of functions. It is in this sense that Klein and also Gordan have been described in Chapter III as extending or going beyond the Galois theory of their day.[15] Also, Klein was clearly aware of how many groups are excluded from his approach by the arithmetic restriction. It was to be his chief concern for the next two years to find some way of eliminating it from the theory of Riemann surfaces and modular functions, as will be discussed in the final chapter.

Exercises

The first 13 exercises concern a fascinating paper by Hesse, which I briefly describe. The geometry of plane curves is very rich. They may be touched not only by lines but also by curves of various degrees. Hesse took up the crucial notion of systems of curves touching a given C_n in his paper [1855a]. He defined two C_{n-1}'s which touch a given C_n to be in the same system if their contact points lay on a common C_{n-1}. He claimed (§8) there were 36 systems of cubics touching a quartic in six different points, but deferred the proof until his [1855b]. For these cubics the six points of contact do not lie on a conic, but he also found 28 systems of cubics whose contact points do lie in 6's on conics (§9), and showed that any conic through two bitangent points meets the C_4 again at 6 points, which can be taken as the contact points of the C_4 and a C_3. He also found there are 63 systems of conics which touch a given C_4 (§10).

To discuss the bitangents themselves Hesse, in his [1855b], passed to the consideration of figures in space, observing that eight points in general position in space give rise to $\frac{1}{2}.8.7 = 28$ lines. He gave a thorough treatment of the geometry of a plane quartic Δ in terms of the geometry of a related sextic curve in space, κ; in particular he showed that a line joining any two of the points cuts κ at two points, which correspond to bitangent points on Δ. So the 28 bitangents correspond to 28 lines in space. These 28 lines through 8 points have various pleasing interrelationships, which formed the subject of Sections 9–12 and 15–16 of Hesse's paper, and which will be convenient to list.

Three bitangent pairs give rise to 6 lines in space. If these pairs of lines are the opposite edges of a tetrahedron, then the four contact points of one pair and the four intersection points of the remaining two pairs lie on a conic (§9).

Four bitangents give rise to 8 points on Δ, which lie on a conic if the four corresponding lines in space either form a quadrilateral (§9) or join the eight points $a_1, \ldots, a_8$ in pairs (§12).

If four points on Δ are the points at which a conic, C_2, touches Δ, then any other conic through those points meets Δ at four more points which may also be taken as the points in which Δ touches a conic, say C_2'. C_2 and C_2' belong to the same system of conics touching Δ (§10).

Three bitangents to Δ, whose corresponding lines in space form a triangle or meet at a common point, give rise to 6 bitangent points (§10, 12).

The 12 contact points of 6 bitangents lie on a cubic if the corresponding lines in space form either two disjoint triangles or two triangles with at most a point in common (§10).

In Sections 15 and 16 Hesse counted the configurations of each kind which can arise, and found:

(i) There are 2016 triples of bitangents whose 6 points do not lie on a conic, and 1260 triples whose 6 points do.

(ii) There are 315 quadruples of bitangents whose 8 points lie on a conic.

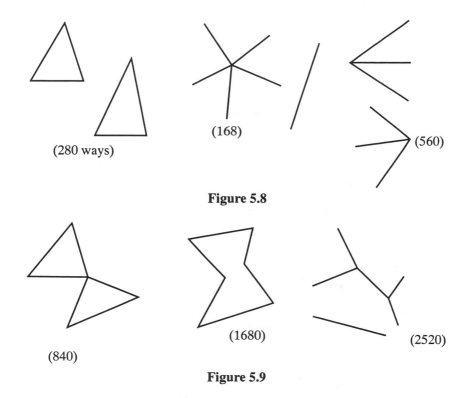

Figure 5.8

Figure 5.9

(iii) There are $1008 + 5040$ sextuples of bitangents whose 12 points lie on a cubic, of which the 1008 arrangements contain no tuples whose 6 points lie on a conic. These occur when the corresponding lines in space form one of the following configurations:

The $5040 (= 7!)$ arrangements each yield 12 points lying on 4 conics, each pair of conics cutting Δ in 2 more bitangent pairs, so producing 4 more bitangents and 8 points lying on a conic. The corresponding arrangements of lines in space are:

Hesse showed that there are 36 different ways (i.e., unrelated by linear transformations) of representing the same quartic D. These are obtained by dividing the eight points $a_1, \ldots, a_8$ up into two sets of 4, which can be done in $\frac{8.7.6.5}{4!2} = 35$ ways; each division will produce a new representation, so there are $35 + 1 = 36$ in all. He listed the corresponding arrangement of bitangents in a table at the end of §14. There are therefore 36 systems of cubics associated to a given quartic, whose contact points do not lie on a conic, as Hesse had claimed in the earlier paper.

1. Suppose $F(x, y, z) = ax^2 + by^2 + cz^2 + 2fyz + 2gzx + 2hxy = 0$ represents a conic in the projective plane. Let $p_1 = [x_1, y_1, z_1]$ and $P_2 = [x_2, y_2, z_2]$

be two points, so any point on the line $\gamma x + \mu y + \nu z = 0$ joining them has coordinates $[\ell_1 x_1 + \ell_2 x_2, \ell_1 y_1 + \ell_2 y_2, \ell_1 z_1 + \ell_2 z_2]$. Show that the line meets the conic where $\ell_1^2 F_1 + 2\ell_1 \ell_2 F_{12} + \ell_2^2 F_2 = 0$, where

$$F_1 = F(p_1), \ F_2 = F(p_2) \text{ and}$$

$$F_{12} = ax_1 x_2 + by_1 y_2 + cz_1 z_2 + f(y_1 z_2 + y_z z_1) +$$
$$g(z_1 x_2 + z_2 x_1) + h(x_1 y_2 + x_2 y_1).$$

2. Show that, if P_1 lies on the conic, then the line meets the conic where

$$F_2 x_1 - 2F_{12} x_2 = 0, \ F_2 y_1 - 2F_{12} y_2 = 0,$$

and

$$F_2 z_1 = 2F_{12} z_2 = 0.$$

Deduce that the equation of the tangent to the conic at z_1 is $F_{12} = 0$, interpreted as a locus in x_2, y_2, z_2.

3. Recall that if the two tangents from a point q to a conic meet it in points lying on a line ℓ, then ℓ is called the polar of q and q the pole of ℓ. Deduce that the polar of the point (x_2, y_2, z_2) is the line $f_{12} = 0$, interpreted as an equation in (x_1, y_1, z_1)

4. Show that the pole of $\lambda x + \mu y + \nu z = 0$ is the point (x_1, y_1, z_1) which satisfies $F_{x_1} = \lambda, \ F_{y_1} = \mu, \ F_{z_1} = \nu$, where $F_{x_1} = \dfrac{\partial F}{\partial x}_{(x_1, y_1, z_1)}$ and F_{y_1} and F_{z_1} are defined similarly.

5. Show that if $F = 0$ and $G = 0$ are the equations of two conics through four points, then any other conic through those points has an equation of the form $\alpha F + \beta G = 0$.

6. Show, given a line ℓ and the family on conics through four points form a quadrilateral in a plane, that the locus of the poles of ℓ with respect to the conics is itself a conic which passes through the diagonal points of the quadrilateral.

7. Generalise this to obtain Chasles' theorem: given a plane and the (one-parameter) family of quadrics through 8 points in general position, the locus of the pole of the plane is a space curve of the third order. This curve will be denoted σ in the sequel.

8. Obtain Hesse's theorem, that the locus of the pole of a plane with respect to the two parameter family of quadrics through 7 points is a cubic surface, $\sum$.

9. Consider the one-parameter family of cones through 7 points, and show that the locus of their vertices is a space sextic, K.

10. The surface $\sum$ depends on the choice of plane in (8), but K does not. Deduce that any two $\sum$ meet in K and a variable cubic curve in space.

11. Show that K corresponds to a plane quartic as follows. Let f_1, f_2, and f_3 be any three quadrics through 7 points. Then any other quadric through those points has equation $\lambda_1 f_1 + \lambda_2 f_2 + \lambda_3 f_3 = 0$. Let $2F = \sum u_{ij} x_i x_j = 0$ be a cone in this family, with vertex (x_1, x_2, x_3, x_4). The u's are linear in the λ's. Eliminate x_1, x_2, x_3, x_4 from $2F = 0$ and obtain $\Delta = \det(u_{ij}) = 0$, which is to be thought of as a quartic in $\lambda_1, \lambda_2, \lambda_3$.

12. Show that a line meets Δ at four points corresponding to four points on K, which are the vertices of 4 cones meeting in a common curve, and establish the converse.

13. Deduce from (12) that two lines meeting Δ yield 8 points on K, which can be taken as the 8 intersection points of 3 quadrics.

14. Verify Klein's calculation that the following octahedral G'_{24} is contained in $PSL(2; Z/7Z)$. Take a Klein four-group $G'_4 = \left\{ w, -\dfrac{1}{w}, \dfrac{2w+3}{3w-2}, \dfrac{3w-2}{-2w-3} \right\}$, find 6 elements of order 4 whose squares are in G'_4. Find 6 elements of order 2 which do not lie in G'_4 but commute with one of its elements, and lastly, find 4 pairs of commuting elements of order 3.

Klein gave this solution:

$$\frac{2w+2}{-2w+2} \text{ and } \frac{2w-2}{2w+2} \text{ whose squares are } -1/w,$$

$$\frac{w+1}{w+2} \text{ and } \frac{-2w+1}{w-1} \text{ whose squares are } \frac{2w+3}{3w-2}$$

$$\frac{3w-3}{-3w+1} \text{ and } \frac{w+3}{3w+3} \text{ whose squares are } \frac{3w-2}{-2w-3}$$

$$\frac{2w-3}{-3w-2} \text{ and } \frac{3w+2}{2w-3} \text{ which commute with } -1/w$$

$$\frac{-w+1}{-2w+1} \text{ and } \frac{w+2}{-w-1} \text{ which commute with } \frac{2w+3}{3w-3}$$

$$\frac{3w-1}{3w-3} \text{ and } \frac{-3w-3}{w+3} \text{ which commute with } \frac{3w-2}{-2w-3};$$

and the following commuting pairs of elements of order 3:

$$\frac{-3w-1}{2}, \frac{-2w-1}{3}; \frac{2w}{w-3}, \frac{3w}{w-2}; \frac{2}{3w+1}, \frac{-w+2}{3w}; \frac{-w+3}{2w}, \frac{-3}{2w+1}.$$

Show that the G''_{24} exhibited above permutes the following pairs projective points in the projective line over the field of 7 elements:

$$\{0, -2\}, \{1, 2\}, \{3, -1\}, \{\infty, -3\}.$$

Show that $\dfrac{2w - 3}{-3w - 2}$ switches $\{1, 2\}$ and $\{-1, 3\}$ but fixes $\{0, -2\}$, $\{\infty, -3\}$;

that $\dfrac{-w + 1}{-2w + 1}$ switches $\{1, 2\}$ and $\{0, -2\}$ but fixes $\{-1, 3\}$ and $\{\infty, -3\}$; and

that $\dfrac{3w - 1}{3w - 3}$ switches $\{1, 2\}$ and $\{\infty, -3\}$ but fixes $\{-1, 3\}$ and $\{0, -2\}$. This means that the group generated by those three elements certainly contains S_4. Show that they generate precisely S_4 by verifying that they satisfy the appropriate relations.

15. Interpret the 64 characteristics introduced by Riemann as vectors in the 6 dimensional vector space over the field of 2 elements, and show how the symplectic form corresponding to the 6×6 matrix $\begin{pmatrix} 0 & I \\ -I & 0 \end{pmatrix} \equiv \begin{pmatrix} 0 & I \\ I & 0 \end{pmatrix}$ mod 2 picks out the odd characteristics. Hence, relate the group of the bitangents to $Sp(6, Z/2Z)$. Hesse's analysis of the bitangents as rederived by Clebsch enabled Jordan in his *Traité* (§332) to show that the group of the bitangents is actually isomorphic $Sp(6; Z/2Z)$.

16. By considering generators and relations, show that G_{168} is also isomorphic to $SL(3, Z/2Z)$, a result Weber (*Lehrbuch der Algebra*, 1896, II p. 539) tells us Kronecker discovered while working on the modular equation at the prime 7. This isomorphism well exhibits the action of G_{168} on the bitangents.

17. By using techniques similar to those in Exercise 14, prove Hermite's theorem that $PSL(2; Z/5Z)$ is isomorphic to A_5, the alternating group on 5 symbols.

Chapter VI

Automorphic Functions

This final chapter is, naturally, concerned with the triumphant accomplishments of Poincaré: the creation of the theory of Fuchsian groups and automorphic functions. These developments brought together the theory of linear differential equations and the group-theoretic approach to the study of Riemann surfaces, so this account draws on all of the preceding material. It begins with a significant stage that is intermediate between the embryonic general theory and the developed Fuchsian theory: Lamé's equation.

6.1 Lamé's Equation

Lamé's equation[1] may be written as either

$$\frac{d^2y}{dx^2} + \frac{1}{2}\left[\frac{1}{x-e_1} + \frac{1}{x-e_2} + \frac{1}{x-e_3}\right]\frac{dy}{dx} - \frac{(n(n+1)x+B)y}{4(x-e_1)(x-e_2)(x-e_3)} = 0, \tag{53}$$

which exhibits it algebraically as an equation of the Fuchsian type,

$$\frac{d^2y}{du^2} - (n(n+1)\mathcal{P}(u) + B)y = 0 \tag{54}$$

which may be called the Weierstrassian form, or as

$$\frac{d^2y}{dv^2} - (n(n+1)k^2sn^2v + A)y = 0 \tag{55}$$

which may be called the Jacobian form. Form (1) is essentially how it was introduced by Lamé [1845] in connection with a study of triply orthogonal coordinates in space. The substitution $x = \mathcal{P}(u)$, where $\mathcal{P}$ is Weierstrass' elliptic function with half-periods e_1, e_2, and e_3 transforms (1) into (2); in this form the equation was

studied by Halphen [1884]. The substitution $r = u(e_1 - e_3)^{1/2}$ transforms (2) into
(3), where $A = \dfrac{B + e_3 n(n + 1)}{(e_1 - e_3)}$ and $k = \left(\dfrac{e_2 - e_3}{e_1 - e_3}\right)^{1/2}$; in this form it was studied
by Hermite [1877] and Fuchs [1878].

As an equation of the Fuchsian type it has four regular singular points: e_1, e_2, and e_3, at which the exponents are 0 and $\frac{1}{2}$ in each case, and infinity, at which the exponents are $-\frac{1}{2}n$, $\frac{1}{2}(n + 1)$. Its properties, however, come out more clearly when it is written in either of the elliptic forms. Hermite [1877 = *Oeuvres* III, 266] summarized the work of his predecessors as follows. Lamé had shown that for suitable values of the constant, A, in (55) one solution can be written as a polynomial of degree n in snx. Liouville [1845] and Heine [1845] independently had studied the second solution and Heine established a connection, via what he called higher order Lamé functions, with spherical harmonics. In a series of papers Hermite [1877–1882 = *Oeuvres* III, 266-418] considered the case when the constant A may be arbitrary, and showed that the solutions were always elliptic functions of the second kind. These are functions F such that

$$F(x + 2K) = \mu F(x)$$

$$F(x + 2iK') = \mu' F(x) \tag{56}$$

for some constants μ and μ'. Hermite showed that they may always be written in terms of Jacobi's functions Θ and H. The designation of the second kind was introduced by Picard [1879], it conforms with Legendre's classification of elliptic integrals. Hermite gave many examples of how elliptic functions, of both the old and the new kinds, could solve problems in applied mathematics, and initiated a considerable amount of research in that area.[2]

Fuchs, in his [1877c], observed that the explicit forms of the solution follow from the elementary fact that, if y_1 is one solution of an equation $\dfrac{d^2y}{dx^2} + p_1\dfrac{dy}{dx} + p_0 y = 0$, then there is another, linearly independent, solution of the form $y_2 = y_1 \displaystyle\int \dfrac{e^{-\int p_1\, dx}}{y_1^2}\, dx$. Hermite's work connected with cases studied by Fuchs [1876a, §13], when working on the all-solutions-algebraic problem, of equations integrable by elliptic or Abelian functions: the equation $\dfrac{d^2y}{dx^2} = Py$ has a solution of the form

$$y = \phi(z)^{1/2} e^{\left(\sqrt{-\frac{\lambda}{4}} \int \frac{dz}{\phi(z)}\right)},$$

where $\phi^2(z)$ is rational in z, λ is a constant, and $P = \frac{1}{4}\left(\dfrac{d}{dz}\log\phi\right) + \frac{1}{2}\dfrac{d^2}{dz^2}\log\phi - \dfrac{\lambda}{4\phi^2}$. Fuchs observed [1878a] that $\lambda = 0$ led to Heine's Lamé functions of higher order, and that integrals are elliptic if $\phi^2(z) = (1 - z^2)(1 - k^2 z^2)$, when the differential equation reduces to Lamé's.

In a third paper [1878c] Fuchs investigated what conditions must be imposed on the coefficient P so that $\dfrac{d^2y}{dx^2} = Py$ has a basis of solutions consisting of two elliptic

functions of the second kind whose poles are the poles of P. Since $P = \dfrac{1}{y_i}\dfrac{d^2 y_i}{dx^2}$ for
any solution y_i of the equation, P must be a single-valued doubly periodic function
of x, which, Fuchs went on to show, can be written explicitly as a sum of Jacobian
functions H.

The converse to this theorem was established by Picard, who showed [1879b,
1880a, b] that every differential equation $\dfrac{d^2 y}{dx^2} + p\dfrac{dy}{dx} + qy = 0$ of the Fuchsian
class with doubly periodic coefficients and single-valued solutions, has as its general solution a sum of two doubly periodic functions of the second kind, with an
analogous result for equations of higher order. He also showed how the solutions
can in general be written explicitly; the exceptional cases were treated by Mittag–
Leffler [1880]. Picard's [1881] presented his results to a German audience with
extensions to systems of first-order linear equations. The result about the basis of
solutions followed easily from Picard's observations that, if it is false, then there are
equations of the form

$$f(x + 4K) = A f(x) + B f(x + 2K)$$

$$f(x + 4iK') = A' f(x) + B' f(x + 2iK'),$$

where f is a solution of the differential equation and $2K$ and $2iK'$ are the periods
of p and q. But then a suitable linear combination of $f(x)$ and $f(x + 2K)$ can be
found, say $\alpha f(x) + \beta f(x + 2K)$, such that

$$\alpha f(x + 2K) + \beta f(x + 4K) = \lambda(\alpha f(x) + \beta f(x + 2K))$$

for some λ, and so $\alpha f(x) + \beta f(x + 2K)$ is the sought-after solution. Once it
has been found, and because $\dfrac{e^{-\int p\,dx}}{f^2(x)}$ is doubly-periodic since the general solution is single-valued, the function $\dfrac{e^{-\int p\,dx}}{f^2(x)}$ is also doubly-periodic of the second
kind. There is then an independent doubly periodic solution (of the second kind):
$f(x)\displaystyle\int \dfrac{e^{-\int p\,dx}}{f^2(x)}$. The quasi-periods of the solutions are the periods of the coefficients. It may, of course, happen that the solutions are themselves doubly-periodic
functions of the first kind.

A thorough study of equations with doubly-periodic coefficients was made
by Halphen in 1880 in his prize-winning essay (published as [1884] and again in
[1921]). He showed (p. 55) that any linear differential equation with single-valued[3]
doubly-periodic coefficients and a pair of independent solutions whose ratio is only
undefined at infinity can be transformed into one of the same form for which the solutions are single-valued. The periods of the coefficients will in general be changed
by this transformation. Halphen drew on the earlier results of Hermite and Picard,
and on Weierstrass' theory of elliptic functions, as presented in [Kiepert, 1874]
and [Mittag–Leffler, 1876].[4] Halphen also showed how to solve the hypergeometric

equation when its solutions are elliptic functions, and how to solve Lamé's equation:
e.g., (93, equation 44)

$$\frac{1}{y}\frac{d^2y}{du^2} = \frac{3}{4}\mathcal{P}(u)$$

has the solutions

$$y = \left(\mathcal{P}'\left(\frac{u}{2}\right)\right)^{-1/2} \text{ and } y = \left(\mathcal{P}'\left(\frac{u}{2}\right)\right)^{-1/2} \quad \mathcal{P}(u/2).$$

Halphen's presentation also showed how one might proceed from equations with
rational coefficients to those with elliptic coefficients and then to those with more
general algebraic coefficients via the theory of invariants (Chapter IV above). Nev-
ertheless, for the most part he confined himself to cases that are integrable by elliptic
functions. The successful treatment of the more general cases was inspired by the
work of Fuchs which will now be discussed, and presented to the Academy in an
essay which took second place to Halphen's.

In a series of paper [1880 *a, b, c, d, e*, 1881 *a, b, c*] (this summary follows
[1880 *a, b*]) Fuchs took the equation

$$\frac{d^2y}{dz^2} + P(z)\frac{dy}{dz} + Q(z)y = 0, \tag{57}$$

where P and Q are rational functions of z, took functions $f(z)$ and $\phi(z)$ as a basis
of solutions for it, and considered, by analogy with Jacobi inversion, the equations

$$\int_{\zeta_1}^{z_1} f(z)\,dz + \int_{\zeta_2}^{z_2} f(z)\,dz = u_1$$

$$\int_{\zeta_1}^{z_1} \phi(z)\,dz + \int_{\zeta_2}^{z_2} \phi(z)\,dz = u_2 \tag{58}$$

as defining functions of u_1 and u_2 : $z_1 = F_1(u_1, u_2)$, $z_2 = F_2(u_1, u_2)$. Plainly by
varying the paths of integration, one obtains

$$F_i(\alpha_{11}u_1 + \alpha_{12}u_2 + \alpha_1 c, \alpha_{21}u_1 + \alpha_{22}u_2 + \alpha_{22}c) = F_i(u_1, u_2), i = 1, 2,$$

for integers α_{ij} which describe the monodromy of u_1 and u_2 as the paths cross the
cuts joining the singularities of (57) to ∞ and α_1 and α_2 are analogous to the periods
of an elliptic integral. What follows is obscure in Fuchs.

He let $a_i, i = 1, \ldots, n$ and ∞ be the singular points of the differential equation,
and took the roots of the associated indicial equations to be $r_1^{(i)}, r_2^{(i)}, i = 1, \ldots, n$,
and s_1 and s_2. He let a and b be two distinct points, possibly infinite, and let u_1,
u_2 be such that $z_1(u_1, u_2) = a$ and $z_2(u_1, u_2) = b$ and let either a, b, or both
be singular points, but insisted that $f(z_2)\phi(z_1) - f(z_1)\phi(z_2) \neq 0$. (Call these
conditions A.) Then the four derivatives $\dfrac{\partial z_i}{\partial u_j}$ are holomorphic functions of z_1, z_2
near $z_1 = a, z_2 = b$. Furthermore, every value $(z_1, z_2) \in \mathbf{C}^2$ is to be attainable
with finite $(u_1, u_2) \in \mathbf{C}^2$.

For this it is necessary and sufficient that at each finite singular point a_i, $r_1^{(i)}$ and $r_2^{(i)}$ satisfy $r_1^{(i)} = -1 + \dfrac{1}{n_i}$, $r_2^{(i)} = -1 + \dfrac{k_i}{n_i}$, $1 < k_i < n_i$, n_i, k_i positive integers and at ∞, $s_1 = 1 + \frac{1}{n}$, $s_2 = 1 + \frac{k}{n}$, $1 < k$; n, k, positive integers.

Extra conditions on the roots of the indicial equation ensure that $\dfrac{f(z)}{\phi(z)} = \zeta$ defines z as a single-valued functions of ζ and that the equation $f(z_1)\phi(z_2) - f(z_2)\phi(z_1) = 0$ has only the trivial solution $z_1 = z_2$. They are that either $r_2^{(i)} - r_1^{(i)} = 1$ or $\dfrac{1}{n_i}$ and either $s_2 - s_2 = 1$ or $\frac{1}{n}$, together with the extra condition that the solutions to the differential equation do not involve logarithmic terms (conditions B). If furthermore $r_2^{(i)} - r_1^{(i)} = \frac{1}{n_i}$ or $r_2^{(i)} = -r_1^{(i)} = \frac{1}{2}$, and $s_2 - s_1 = \frac{1}{n}$ or $s_2 = \frac{5}{2}$, $s_1 = \frac{3}{2}$ (conditions C) and if ∞ is not a regular point, but Fuchs is obscure here, then z_1, z_2 are roots of a quadratic equation, whose coefficients are single-valued holomorphic functions of u_1, u_2. In this case the number of finite singular points cannot exceed, but can equal, 6. In the example where 6 finite singular points occur the functions $F_1(u_1, u_2)$, $F_2(u_1, u_2)$ are necessarily hyperelliptic, but generally they will not be even Abelian functions, since the differential equation will not be algebraically integrable.

Fuchs' proofs of these assertions proceeded by a case by case analysis of each kind of singularity that could occur in terms of the local power series expansions of the functions. Poincaré pointed out (see below) that the analysis rapidly becomes confusing and was, in any case, incomplete.

The condition that no logarithmic terms appear in the solutions to the differential equation even though $r_2 - r_1 = 1$, an integer, is a strong restriction on the kind of branching that can occur. Fuchs also seems to be assuming, or perhaps is only interested, in the case when ζ takes every value in C, not merely in some disc.

The curious condition that there are at most six finite singular points was obtained as follows (§7). From the general theory of differential equations of the Fuchsian type, if the number of finite singular points is A, then

$$\sum_i (r_1^{(i)} + r_2^{(i)}) + s_1 + s_2 = A - 1.$$

If there are A' where $r_1 = \dfrac{-1}{2}$ and $r_2 = \frac{1}{2}$ and A'' others, where necessarily $r_1^{(i)} + r_2^{(i)} = -2 + \dfrac{k_i}{n_i}$, $1 < k_i < n_i$, then $\displaystyle\sum_{i=1}^{A''} \dfrac{3}{n_i} + s_1 + s_2 = A' + 3A'' - 1$ (summing only over the second kind).

But $s_1 + s_2 \leq 5$, and $\dfrac{1}{n_i} \leq \dfrac{1}{2}$ implies

$$3 \sum_1^{A''} \frac{1}{n_i} \leq \frac{3A''}{2},$$

so $A' + 3A'' - 1 \leq \dfrac{3A''}{2} + 5$ or

$$A + \frac{1}{2}A'' \leq 6,$$

so $A \leq 6$, as was to be shown.

This case can arise when ∞ is an ordinary point, $s_1 = 2$ and $s_2 = 3$. These were the conclusions drawn from conditions (C).

As an example of the case when there are 6 singular points, Fuchs (§8) adduced the hyperelliptic integrals

$$y_1 = \int \frac{g(z)}{\sqrt{\phi(z)}}, \quad y_2 = \int \frac{h(z)}{\sqrt{\phi(z)}}, \quad \text{where } \phi(z) = (z - a_1) \cdots (z - a_6)$$

and ∞ is not a singular point, so $s_1 = 2$, $s_2 = 3$ and $g(z)$ and $h(z)$ are linearly independent polynomials of degree 0 or 1 (say $g(z) := 1$, $h(z) := z$). In this case $z_1 = F_1(u_1, u_2)$ and $z_2 = F_2(u_1, u_2)$ are hyperelliptic functions of the first kind, but (§9) non-Abelian functions may arise. Fuchs gave as an example an equation with finite singular points

$$a_1, \text{ at which } r_1^{(1)} = -\frac{2}{3}, \quad r_2^{(1)} = -\frac{1}{3}$$

$$a_2, \text{ at which } r_1^{(2)} = -\frac{5}{6}, \quad r_2^{(2)} = -\frac{4}{6}$$

and

$$\infty, \text{ at which } s_1 = \frac{3}{2}, s_2 = 2.$$

(These satisfy conditions (C) with $n_1 = 3, n_2 = 6, n = 2$.)

The original differential equation $\dfrac{d^2 y}{dz^2} + P(z) \dfrac{dy}{dz} + Q(z)y = 0$ is transformed by the substitution of $y = (z - a_1)^{-1}(z - a_2)^{5/4}\omega$ into one with exponents $\frac{1}{3}$, $\frac{2}{3}$ at a_1, $\frac{5}{12}$, $\frac{5}{12}$ at a_2, and $\frac{-3}{4}$, $\frac{-1}{4}$ at ∞, and so is of the form $\dfrac{d^2\omega}{dz^2}$. But, by Fuchs' test (Chapter III, p. 83) the denominators of the exponents at a_2 are $12 > 10$, so the equation is not algebraically integrable unless its solutions (or a second degree homogeneous polynomial in the solutions) is the root of a rational function. But this is not possible either, since the denominators at a_1 and a_2 are neither 1, 2, nor 4.

Fuchs was chiefly concerned with studying the inversion of equations (6) and only slightly interested in the function $\zeta = \dfrac{f(z)}{\phi(z)}$. His obscure papers rather confused the two problems but they were soon to be disentangled.

On the 29th May 1880, a young Frenchman at the University of Caen wrote a letter to Fuchs expressing interest in the subject of his paper in the *Journal für Mathematik*, but seeking clarification of some points. The young man was very interested in the global theory of differential equations, whether first-order real or linear and complex. Within two years his work was to transform both subjects completely, opening up whole new aspects of research in the one, and in the other leaving little, it has been said, for his successors to do. His name was Jules Henri Poincaré.

6.2 Poincaré

Poincaré was born in Nancy on 29 April, 1854. His father was professor of medicine at the university there, his mother, a very active and intelligent women, consistently encouraged him intellectually, and his childhood seems to have been very happy.[5] He did not at first show an exceptional aptitude for mathematics, but towards the end of his school career his brilliance became apparent, and he entered the École Polytechnique at the top of his class. Even then he displayed what were to be life-long characteristics: a capacity to immerse himself completely in abstract thought, seldom bothering to resort to pen and paper, a great clarity of ideas, a dislike for taking notes so that he gave the impression of taking ideas in directly, and a perfect memory for details of all kinds. When asked to solve a problem he could reply, it was said, with the swiftness of an arrow. He had a slight stoop, he could not draw at all, which was a problem more for his examiners than for him, and he was totally incompetent in physical exercises. He graduated only second from the École Polytechnique because of his inability to draw, and proceeded to the École des Mines in 1875. In 1878 he presented his doctoral thesis to the faculty of Paris on the subject of partial differential equations. Darboux said of it that it contained enough ideas for several good theses, although some points in it still needed to be corrected or made precise. This fecundity and inaccuracy is typical of Poincaré; ideas spilled forth so fast that, like Gauss, he seems not to have had the time to go back over his discoveries and polish them. On the first of December 1879 he was at the École des Mines in Caen and in charge of the analysis course at the Faculty of Sciences.

1880 was a busy time for him. On the 22nd of March, he deposited his essay "Mémoire sur les courbes définies par une équation différentielle" with the Académie des Sciences as his entry in their prize competition. That essay considered first-order non-linear differential equations $\frac{dx}{X} = \frac{dy}{Y}$, where X and Y are real polynomial functions of real variables x and y, and investigated the global properties of their solutions. He later withdrew the essay, on 14 June 1880, without the examiners reporting on it, perhaps wanting to concentrate on the theory of complex differential equations.

Poincaré's question to Fuchs concerned the nature of the inverse function $z = z(\zeta)$. Fuchs had claimed that z is a meromorphic function of $\zeta = \frac{f(z)}{\phi(z)}$, whether z is an ordinary or a singular point of the differential equation. Indeed, z is finite at ordinary points and infinite at singular points. Poincaré observed[6] that z is meromorphic at $\zeta = \infty$, so $z = z(\zeta)$ seems to be meromorphic on the whole

ζ-sphere, and so is a rational function of ζ. But then the original differential equation must have all its solutions algebraic, which Fuchs had expressly denied. Poincaré suggested that there were three kinds of ζ value: those reached by $\dfrac{f(z)}{\phi(z)}$ as z traced out a finite contour on the z-sphere; those reached on an infinite contour, and those which are not attained at all. *A priori*, he said, all three situations could occur, and indeed the last two would if the differential equation did not have all its solution algebraic. Fuchs' proof would only work for ζ-values of the first kind; however, Poincaré went on, he could show that $z(\zeta)$ was meromorphic even if the other kinds occurred, and he was led to hypothesize that (1) if indeed all ζ-values were of the first kind, then z would be a rational function; (2) if there are values of only the first and second kinds but z is monodromic at the values of the second kind, then Fuchs' theorem is still true; (3) if z is not monodromic or (4) if the values of the third kind occur and so the domain of z is only a domain D on the ζ-sphere, then z is single-valued on D. In this case the ζ-values of

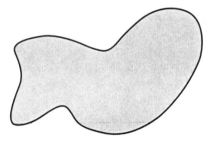

Figure 6.1

the first kind occur inside D, as shown. Those of the second kind lie on the boundary of D, and the unattainable values lie outside D. There is, finally, a fifth case, when all three kinds of ζ occur, but D has this form,

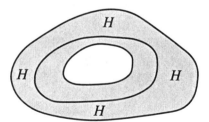

Figure 6.2

values of the first kind filling out the annulus. Now, said Poincaré, z will not return to its original value on tracing out a closed curve HHHH in D.

Fuchs replied on the 5th June: "Your letter shows you read German with deep understanding, so I shall reply in it". He admitted that his Theorem I was imprecisely worded, and suggested that the hypothesis of his earlier *Göttingen Nachrichten* articles were to be preferred, namely, that the exponents at the ith singular point satisfy either $r_2^{(i)} = r_1^{(i)} + 1$ or $r_1 - 1 + \dfrac{1}{n_i}$, $r_2 = -1 + \dfrac{2}{n_i}$, and the exponents at infinity satisfy $\rho_2 = \rho_1 + 1$ or $\rho_1 = 1 + \frac{1}{\nu}$, $\rho_2 = 1 + \frac{2}{\nu}$ for integers n_i, ν. He added a few words on the meaning: he excluded paths in which $f(z)$ and $\phi(z)$ both become infinite, and then the remaining ζ-values filled out a simply connected region of the ζ-plane with the excluded values on the boundary.

Poincaré replied on the 12th, apologising for the delay in doing so, but he had been away. He found some points were still obscure, and suggested the following argument. Suppose the singular points of the differential equation are joined to ∞ by cuts, and z moves without crossing the cuts. Then ζ traces out a connected region F_0. Let z cross the cuts, but no more than m times. Then the values of ζ fill out a connected region F_m. As m tends to infinity, F_m tends to the region Fuchs called F, and F will be simply connected if F_m is simply connected for all m. Now, asked Poincaré, "is that a consequence of your proof? One needs to add some explanation". He agreed that F_m could not cover itself as it grew in this fashion:

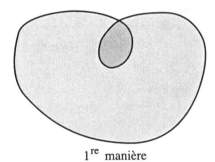

1re manière

Figure 6.3

but the proof left open this possibility:

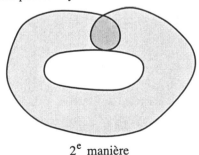

2e manière

Figure 6.4

He said that when there were only two finite singular points it was true that z was a single-valued function, "that I can prove differently", and he went on, "but it is not obvious in general. In the case that there are only two finite singular points I have found some remarkable properties of the functions you define, and which I intend to publish. I ask your permission to give them the name of Fuchsian functions". In conclusion he asked if he might show Fuchs' letter to Hermite.

Fuchs replied on the 16th, promising to send him an extract of his forthcoming complete list of the second order differential equations of the kind he was considering. This work, he said, makes any further discussion superfluous. He was very interested in the letters, and very pleased about the name. Of course his replies could be shown to Hermite.

The reply points to an interesting difference of emphasis between the two men. For Fuchs, the main problem was to study functions obtained by inverting the integrals of solutions to a differential equation, thus generalizing the Jacobi inversion. As a special case one might also ask that the inverse of the quotient of the solutions is single-valued, which imposed extra conditions. For Poincaré, interested in the global nature of the solutions to differential equations, it was only the special case which was of interest, and he gradually sought to emancipate it from its Jacobian origins. There is also some humour in the situation of the young man gently explaining about analytic continuation and the difference between single-valued and unbranched functions, to one who had consistently studied and applied the technique for fifteen years.

Poincaré's reply of the 19th June (one is struck by the efficiency of the postal service almost as much as by the rapidity of his thought) pointed up this difference of emphasis. Taking the condition on the exponents to be

$$r_1^{(i)} = -1 + \frac{1}{n_i}, r_2^{(i)} = -1 + \frac{2}{n_i} \text{ or } r_1^{(i)} = -\frac{1}{2} = -r_2^{(i)}, \text{ and}$$

$$s_1 = 1 + \frac{1}{n}, s_2 = 1 + \frac{2}{n} \text{ or } s_1 = \frac{3}{2}, s_2 = \frac{5}{2} \text{ (at infinity)},$$

he wrote that he had found that when the differential equation was put in the form $y'' + qy = 0$, the finite singular points with exponent difference 1 vanished. Thus he found that, at all the finite singular points, the exponent difference was an aliquot part of 1 and not equal to 1, and that there were no more than 3 singular points. If there was only one, then z was a rational function of ζ. If there were two, and the exponent differences were ρ_1, and ρ_2, and ρ_3 at infinity, then either $\rho_1 + \rho_2 + \rho_3 > 1$, in which case z is rational in ζ, or $\rho_1 + \rho_2 + \rho_3 = 1$, in which case z was doubly periodic. Even in this case there were difficulties, as he showed with this example.

Let $z = \wedge(u)$ be a doubly periodic function, and set $\eta = \left(\dfrac{dz}{du}\right)^{1/2} e^{\alpha u}$. Then η and $\left(\dfrac{dz}{du}\right)^{1/2} e^{-\alpha u}$ satisfy a second-order linear differential equation and $\zeta = e^{2\alpha u}$. So for z to be single-valued in ζ, it must have period $\dfrac{i\pi}{\alpha}$ as a function of u, which, he pointed out, was not the case in general. Finally, if there were three finite singular points, then the exponents would have to be $-\frac{1}{2}$ and 0, and at infinity they would be $\frac{3}{2}$, 2. But although these satisfied Fuchs' criteria, z was not a single-valued function of ζ, so the theorem is wrong. Poincaré proposed to drop the requirement that Fuchs' functions $z_1 + z_2$, $z_1 \cdot z_2$ be single-valued in u_1 and u_2. He went on to say that this gives a "...much greater class of equations than you have studied, but to which your conclusions apply. Unhappily my objection requires a more profound study, in that I can only treat two singular points". Dropping the conditions on $z_1 + z_2$, $z_1 \cdot z_2$ admits the possibility that the exponent differences ρ_1, ρ_2, and ρ_3 satisfy $\rho_1 + \rho_2 + \rho_3 < 1$. Now z is neither rational nor doubly periodic, but is still single-valued. "These functions I call Fuchsian, they solve differential equations with two singular points whenever ρ_1, ρ_2, and ρ_3 are commensurable with each other. Fuchsian functions are very like elliptic functions, they are defined in a certain circle and are meromorphic inside it". On the other hand, he concluded, he knew nothing about what happened when there were more than two singular points.

We do not have Fuchs' reply, but Poincaré wrote to him again on the 30th of July to thank him for the table of solutions "which lifts my doubts completely", although he went on to point out a condition on some of the coefficients of the differential equations which Fuchs had not stated explicitly in the formulation of his theorems. As to his own research on the new functions, he remarked that they

> ...present the greatest analogy with elliptic functions, and can be represented as the quotient of two infinite series in infinitely many ways. Amongst those series are those which are entire series playing the role of Theta functions. These converge in a certain circle and do not exist outside it, as thus does the Fuchsian function itself. Besides these functions there are others which play the same role as the zeta functions in the theory of elliptic functions, and by means of which I solve linear differential equations of arbitrary orders with rational coefficients whenever there are only two finite singular points and the roots of the three determinantal equations are commensurable. I have also thought of functions which are to Fuchsian functions as abelian functions are to elliptic functions and by means of which I hope to solve all linear equations when the roots of the determinantal equations are commensurable. Finally functions precisely analogous to Fuchsian functions will give me, I think, the solutions to a great number of differential equations with irrational coefficients.

Poincaré's last letter (20th March, 1881) merely announces that he will soon publish his research on the Fuchsian functions, which partly resemble elliptic functions and partly modular functions, and on the use of zeta Fuchsian functions to solve differential equations with algebraic coefficients. In fact, his first two articles on these matters already appeared in the *Comptes Rendus*, and these will be discussed below.

Poincaré had not had long to study Fuchs' work by the time he wrote him his first letter on 29th May 1880, for he had only received the journal at the start of the month. Yet, incredibly, he had already written up and presented his findings in the form of an essay for the prize of the Paris Académie des Sciences. Indeed, he had submitted this entry on the 28th May 1880, the day before he wrote to Fuchs. His essay was awarded second prize, behind Halphen's, but it was not published until Nörlund edited it for *Acta Mathematica* (vol 39, 1923, 58-93, in *Oeuvres*, I, 578–613). In awarding it second prize, Hermite said of the essay, which also discussed irregular solutions, that:

> " ...the author successively treated two entirely different questions, of which he made a profound study with a talent by which the commission was greatly struck. The second ... concerns the beautiful and important researches of M. Fuchs, ... The results ... presented some lacunas in certain cases that the author has recognized and drawn attention to in thus completing an extremely interesting analytic theory. This theory has suggested to him the origin of transcendents, including in particular elliptic functions, and which has permitted him to obtain the solutions to linear equations of the second order in some very general cases. A fertile path is there that the author has not entirely gone down, but which manifests an inventive and profound spirit. The commission can only urge him to follow up his researches in drawing to the attention of the Academy the excellent talent of which they give proof. (Quoted in Poincaré, *Oeuvres*, II, 73.)

In the essay Poincaré considered when the quotient $z = \dfrac{f(x)}{\phi(x)}$ of two independent solutions of the differential equation $\dfrac{d^2y}{dx^2} = Qy$ defines, by inversion, a meromorphic function x of z. He found Fuchs' conditions were not necessary and sufficient. It was necessary and sufficient for x to be meromorphic on some domain that the roots of the indicial equation at each singular point, including infinity, differ by an aliquot part of unity (i.e., $\rho_1 - \rho_2 = 1/n$, for some positive integer n). If the domain is to be the whole complex sphere then this condition is still necessary, but it is not longer sufficient. He found that Fuchs' methods did not enable him to analyse the question very well, as special cases began to proliferate, and sought to give it a more profound study, beginning with Fuchs' example ([Fuchs 1880b, 168 = 1906, p. 210]) of a differential equation in which there are two finite singular points a_1 and a_2, where the exponent differences are $\frac{1}{3}$ and $\frac{1}{6}$, and the exponent difference at ∞ is $\frac{1}{2}$. In this case he found the change in z was of the form

$$z \rightarrow z'', \quad \frac{z'' - \alpha}{z'' - \beta} = e^{2\pi i/3}\left(\frac{z - \alpha}{z - \beta}\right) \quad \text{upon analytic continuation around } a_1,$$

$$z \rightarrow z'', \quad \frac{z'' - \gamma}{z'' - \delta} = e^{\pi i/3}\left(\frac{z - \gamma}{z - \delta}\right), \quad \text{upon analytic continuation around } a_2,$$

and $\frac{1}{z} \rightarrow -\frac{1}{z}$ (around ∞).

Accordingly x is a meromorphic single-valued function of z mapping a parallelogram composed of eight equilateral triangles onto the complex sphere, and $z = \infty$ is its only singular point, so z is an elliptic function. The differential equation, Poincaré showed, has in fact an algebraic solution $y_1 = (x - a_1)^{1/3}(x - a_2)^{5/2}$ and a non-algebraic solution y_2 such that

$$\frac{y_2}{y_1} = \int (x - a_1)^{-2/3}(x - a_2)^{-5/6}\,dx.$$

This result agrees with Fuchs' theory.

Poincaré next investigated when a doubly-periodic function can give rise to a second order linear differential equation, and found that one could always exhibit such an equation having rational coefficients for which the solution was a doubly periodic function having 2 poles. If furthermore the periods, h and K, were such that

$$2i\pi = 0 \ (\mathrm{mod}\ h, K)$$

then x would be a monodromic function of z with period $2i\pi$.

His reasoning is too lengthy to reproduce, but the condition on h and K derives essentially from the behaviour of z under analytic continuation: if z is monodromic, it must reproduce as $z \to z' = \dfrac{az + b}{a'z + b'}$ (for some a, b, a', b'), and thus can be written as

$$\frac{z' - \alpha}{z' - \beta} = \lambda \frac{z - \alpha}{z - \beta},$$

and $\lambda^n = 1$. After a further argument Poincaré concluded (p. 79) that (i) there were cases when one solution of the original differential equation was algebraic, and then Fuchs' theory was correct, but (ii) there were cases when the differential equation had four singular points, elliptic functions were involved, and then extra conditions were needed.

However, it might be that the domain of x could not be the whole z-sphere. Poincaré showed that this could happen even when the differential equation had only two finite singular points. For example, if the exponent differences were $\frac{1}{4}$, $\frac{1}{2}$ and $\frac{1}{6}$ at ∞, then as long as x crosses no cuts, z stays within a quadrilateral (Figure 2, p. 86) $\alpha 0 \alpha' \gamma$.

Furthermore, however x is conducted about in its plane, z cannot escape the circle HH'. Poincaré described the quadrilateral as 'mixtiligne', the circular-arc sides meet the circle HH at right angles. This geometric picture is quite general; curvilinear polygons are obtained with non-re-entrant angles and circular arc sides orthogonal to the boundary circle. Thus the domain of x is $|z| < OH$, and Poincaré then investigated whether x is meromorphic. This reduces to showing that, as x is continued analytically, the polygons do not overlap. This does not occur if the angles satisfy conditions derived from Fuchs' theory, unless the overlap is in the form of an annular region (see Figure 6.6):[7]

However, if the angles are not re-entrant, this cannot happen, and so x is meromorphic. Poincaré's proof of this is of incidental interest. He projected the circle

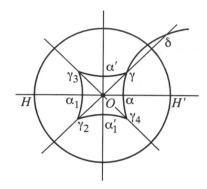

Figure 6.5

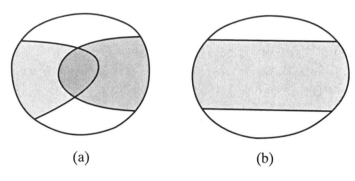

(a)   (b)

Figure 6.6

HH' stereographically onto the southern hemisphere of a sphere, and then projected the image orthogonally back onto its original plane. The circular arcs orthogonal to HH' become straight lines, which renders the theorem trivial.

This result virtually concluded Poincaré's essay. As he said in his letter to Fuchs, his understanding was limited essentially to the case of two finite singular points. But one notices in this last argument that the final image of the disc is the Beltrami picture of non-Euclidean geometry. Poincaré did not recognize it at this stage, but he soon did.

Poincaré himself has left us one of the most justly celebrated accounts of the process of mathematical discovery, which concerns exactly his route to the theory of Fuchsian functions. Although it is very well known, it is not apparent from the usual sources (Darboux [Poincaré, *Oeuvres* II, 1vii, 1viii], Hadamard [1954, 12–15]) precisely how and when it connects with the correspondence with Fuchs; even the year, 1880, is left unstated. Poincaré gave this account in a lecture he gave to the Société de Psychologie in Paris 1908, and it was later published as the third essay in his volume *Science et Méthode* [1909]. In 1980, I discovered that there is still more evidence about Poincaré's work in 1880, which does not appear to have been considered before. The *Comptes Rendus* record three anonymous supplements to the essay bearing the motto 'Non inultus premor', received by the Académie on 28th June, 6th September, and 20th December. Prize essays were submitted anonymously and only identified by a motto. This motto identifies the supplements as Poincaré's (it is in fact the motto of his home town of Nancy). The supplements are to be found in the Poincaré dossier in the Académie des Sciences, but for some reason Nörlund did not publish them when he published the essay in *Acta Mathematica*, nor have they been included in Poincaré's *Oeuvres*. The supplements confirm and greatly amplify what Poincaré said in the lecture 28 years later, and have recently been published (Poincaré [1997].)[8]

Poincaré began by doubting that Fuchsian functions could exist, but shortly came to the opposite view. He tells us in the lecture that:

"For two weeks I tried to prove that no function could exist analogous to those I have since called the Fuchsian functions: I was then totally ignorant. Every day I sat down at my desk and spent an hour or two there: I tried a great number of combinations and never arrived at any result. One evening I took a cup of coffee, contrary to my habit; I could not get to sleep, the ideas surged up in a crowd, I felt them bump against one another, until two of them hooked onto one another, as one might say, to form a stable combination. In the morning I had established the existence of a class of Fuchsian functions, those which are derived from the hypergeometric series. I had only to write up the results, which just took me a few hours." ([1909, p. 50])

This account is consistent with his knowledge of Fuchs' problem as presented in the prize essay, and plainly marks his realization of the fundamental invariance properties of the new functions. It is most likely therefore that it refers to the period between 29 May and 12 June 1880 (the date of the second letter). He may also have accomplished the second stage of his discoveries by that time:

"I then wanted to represent the functions as a quotient of two series; this idea was perfectly conscious and deliberate; the analogy with elliptic functions guided me. I asked myself what must be the property of these series, if they exist, and came without difficulty to construct the series that I called theta-Fuchsian". ([1909], p. 51.)

His probable ignorance of Riemann's theory of theta functions may well have been a blessing to him here, for the analogy is with the Jacobian theory of the theta-function of a single variable.

Next comes his marvellous, almost Proustian, donné:

"At that moment I left Caen where I then lived, to take part in a geological expedition organized by the École des Mines. The circumstances of the journey made me forget my mathematical work; arrived at Coutances we boarded an omnibus for I don't know what journey. At the moment when I put my foot on the step the idea came to me, without anything in my previous thoughts having prepared me for it; that the transformations I had made use of to define the Fuchsian functions were identical with those of non-Euclidian geometry. I did not verify this, I did not have the time for it, since scarcely had I sat down in the bus than I resumed the conversation already begun, but I was entirely certain at once. On returning to Caen I verified the result at leisure to salve my conscience." ([1909], 51, 52.)

The letter of 12th June records that he had been away from Caen, and speaks of the remarkable properties of these new functions. One may surely conclude that this alludes to the connection with non-Euclidean geometry. It would be possible that the reference to being away is misleading, and that the crucial bus journey took place between the 12th and the 19th. This would explain the more independent tone of the third letter, and its explicit reference to the domain of the functions being a circle. Nonetheless it is the second letter which speaks of the dramatic nature of the discoveries, whereas the third is more of a methodical reworking of the material guided by the new insights. It is hard to account for the excitement of the second letter if the progress is only arriving 'without difficulty' at the theta-Fuchsian series. That noone had had such an idea before does not seem to have been a particular source of joy to Poincaré, who happily lacked the personally competitive streak of some mathematicians.

The first supplement sheds considerable light on Poincaré's grasp of Fuchsian functions and non–Euclidean geometry at this time. It is 80 pages long, and was received by the Académie on 28th June 1880. Poincaré began by reviewing the tessellation of the disc by 'mixtiligne' quadrilaterals obtained by successively operating on one, which he called Q, by transformations M and N. He observed (p. 9) that these transformations form a group, and remarked:

> "There are close connections with the above considerations and the non-Euclidean geometry of Lobachevskii. In fact, what is a geometry? It is the study of a *group of operations* formed by the displacements one can apply to a figure without deforming it. In Euclidean geometry the group reduces to *rotations* and *translations*. In the pseudogeometry of Lobachevskii it is more complicated."

Indeed, the group of operations formed by means of M and N is *isomorphic* ('*isomorphe*') to a group contained in the pseudogeometric group. To study the group formed by means of M and N is therefore *to do the geometry of Lobachevskii*. Pseudogeometry will consequently provide us with a convenient language for expressing what we will have to say about this group. (Poincaré's emphasis.)

Poincaré's realization on boarding the bus at Coutances can be described very simply. He realized that the straighted version of the 'mixtiligne' figures described at the end of his Prize essay were identical with the figures in Beltrami's description of non-Euclidean geometry; that therefore the original figures were conformally accurate representations of non-Euclidean figures; and finally that this meant the transformations formed from M and N were non-Euclidean isometries. Beltrami's detailed discussion of the non-Euclidean differential geometry of the disc (described in Appendix 4) enabled Poincaré to give a new meaning to his previously analytical transformations. Consequently, on p. 20, he remarked that:

"The Fuchsian functions are to the geometry of Lobachevskii what the doubly periodic functions are to that of Euclid."

Fuchsian functions only illuminate the study of a differential equation if they can be defined independently of the equation. This Poincaré did by introducing the Fuchsian and theta-Fuchsian series. He attempted to prove their convergence by an ingenious argument which again appealed to non-Euclidean geometry. The theta–Fuchsians are perhaps easier to understand. Poincaré took an arbitrary rational function, H, and considered $\sum_K H(zK)\dfrac{d}{dz}(zK)^m$, where the summation is taken over all elements K in the group under consideration. The convergence argument involved considering the partial sum over n elements $K_1, \ldots, K_n$, and estimating the non-Euclidean area and Euclidean perimeter of the region $QK_1 \cup \cdots \cup QK_n$. Poincaré showed that the sum converged if $m > 1$, but was unable to show that his series defining the Fuchsian functions converged. However, he remarked (p. 66):

"The quotient of two theta-Fuchsian series (corresponding to the same value of m) is a rational function of a Fuchsian function."

He also introduced zeta-Fuchsian functions (p. 49) to solve differential equations where the exponents are arbitrary rationals, and not simply reciprocals of integers. All this will be discussed in more detail below.

Poincaré remained stuck on the case of the hypergeometric equation at least until his fourth letter, 30th July. Liberation came from an unexpected source, arithmetic, just as arithmetical considerations had earlier enriched Klein's work. But here the response was to be entirely different:

"I then undertook to study some arithmetical questions without any great result appearing and without expecting that this could have the least connection with my previous researches. Disgusted with my lack of success, I went to spend some days at the seaside and thought of quite different things. One day, walking along the cliff, the idea came to me, always with the same characteristics of brevity, suddenness, and immediate certainly, that the arithmetical transformations of ternary indefinite quadratic forms were identical with those of non-Euclidean geometry."

"Once back at Caen I reflected on this result and drew consequences from it; the example of quadratic forms showed me that there were Fuchsian groups other than those which correspond to the hypergeometric series; I saw that I could apply them to the theory of theta-Fuchsian series, and that, as a consequence, there were Fuchsian functions other than those, which derived from the hypergeometric series, the only ones I knew at that time. I naturally proposed to construct all these functions; I laid siege systematically and carried off one after another all the works begun; there was one however, which still held out and as the chase became involved it took pride of place. But all my efforts only served to make me know the difficulty better, which was already something. All this work was quite conscious." ([1909], 52, 53.)

The second supplement, of 23 pages, was received by the Académie on 6th September 1880. Most of it is given over to a more rigorous description of non-Euclidean geometry, and to tessellations of the disc by polygons with angles π/m for integers m. When the polygon is a triangle, he also discussed more carefully the ways of constructing Fuchsian functions in this case and was led to conjecture a result which he said he was not yet in any state to prove — the Riemann mapping theorem! Then, on p. 17, he abruptly stated the connection with the theory of quadratic forms, a subject he had studied under Hermite.

He let T be a matrix ('substitution') with integer coefficients which preserved an indefinite ternary quadratic form Φ and S be a substitution sending $\zeta^2 + \eta^2 - \zeta^2$ to Φ. Then STS^{-1} preserves $\xi^2 + \eta^2 - \zeta^2$ and sends (ξ, η, ζ) to (ξ', η', ζ'), say. The quantities

$$z = \frac{\xi}{\zeta} + \sqrt{-1}\frac{\eta}{\zeta}, \quad z' = \frac{\xi'}{\zeta'} + \sqrt{-1}\frac{\eta'}{\zeta'}$$

are related by transformation $z' = zK$ of the non-Euclidean plane provided $\xi^2 + \eta^2 - \zeta^2 < 0$. He did not prove that a sheet of the hyperboloid of two sheets provides a model of non-Euclidean geometry[9] — which is easy enough to prove (see p. 300) — and remarked only that (p. 19):

"All the points zK are the vertices of a polygonal net obtained by decomposing the pseudogeometric plane into polygons pseudogeometrically equal to each other".

Next he observed that the corresponding Fuchsian functions (obtained as quotients of theta-Fuchsian series as before) can be used to solve differential equations with algebraic coefficients (see below). It becomes clear that Poincaré has not heard of the Schwarzian derivative.

The third supplement, of only 12 pages, was received on 20 December. Its main result is the extension of the method of polygonal decomposition to include cases where the angles are zero, and the roots of the indicial equation differ by integers. The notable example is Legendre's equation. Poincaré's method is to push the polygons outwards until one or more vertices are 'at infinity', i.e., are on the boundary of the disc, and the corresponding angles vanish. Since the polygons hitherto studied had angles which were only rational multiples of π, Poincaré's argument relies heavily on its geometrical plausibility.

The unpublished work makes abundantly evident the astounding clarity of Poincaré's mind, coupled to an almost equally dramatic ignorance of contemporary mathematics. There is no mention of the work of Schwarz on the hypergeometric equation, so naturally and even pictorially connected with the crucial case $\rho_1 + \rho_2 + \rho_3 < 1$. Nor is there any mention of the work of Dedekind or Klein, and even Hermite's work on modular functions, which he must have known, seems to have been forgotten. We shall see that these omissions are not mere oversights; Poincaré genuinely did not then know the German work. The contrast with the deliberately well-read Felix Klein could not be more marked.

Klein

Although Klein was far from idle in 1880, he did not make the dramatic progress that Poincaré did. He reworked his 1879 paper on eleventh order transformations of elliptic functions, connecting it, and his work on the transformations of orders 5 and 7, with division values of theta functions, in a paper he finished on January 3, 1881, [1881]. He reported to the *Akademie der Wissenschaften* in Munich on the significance of his work for the study of the normal forms of elliptic integrals of the first kind, and concerned himself with the work of his students Gierster, Dyck, and Hurwitz. Bianchi stayed in Munich during the year autumn 1879 to autumn

1880, and Klein [1923, p. 6] described the work he did with him as "more transient ..., but as very essential for me", for it showed him how to connect his work with the developments of the Weierstrassian school, including that of his friend Kiepert. In the autumn of 1880 he moved to become professor at the University of Leipzig, and gave his inaugural address on the 25th of October on the relationship of the new mathematics to applications, a theme dear to his heart. Carl Neumann stimulated his interest in applied mathematics, and he wrote a short paper in mid-January 1881 on the Lamé functions. Even so he was, he said later, so deeply immersed in geometric function theory that he began to lecture on it as soon as he arrived in Leipzig. He gave two series of lectures; the second has become very well-known in the slightly reworked form of his book *Über Riemanns Theorie der algebraischen Functionen und ihre Integrale*".

Klein wanted to conceive of an algebraic curve $F(s, z)$ not as spread out over the z-plane, but as a closed surface in its own right, and a complex function as a pair of flows with singularities on the surface. In so doing, he argued that he was only following Riemann's own approach, and he sought out Riemann's students and colleagues to ask them how true it was that Riemann had thought this way. At first Prym agreed with Klein, but later correspondence between Prym and Betti makes it clear that Riemann had not,[10] so it seems that Klein's description must be regarded not as historically accurate, but as an inspired response to reading Riemann. Although his presentation of Riemann's topological ideas is highly attractive, it can perhaps be doubted if it helped Klein with his *"old problem ... : to construct all discontinuous groups in one variable, esp. the corresponding automorphic functions"* ([1923a], p. 581, italics Klein's) for it avoids the problem of how the corresponding Riemann surfaces are best constructed. One sees here very clearly how Klein wanted to advance a particular view of mathematics, on the one hand visual and intuitive, on the other naturally algebraic, group-theoretic or invariant-theoretic, and linked to projective geometry. The unity of these domains and the connection with number theory delighted him, and also I shall suggest, bound him too closely to one view of the problem.

Klein had been working on Riemann's ideas since 1874, when he began a series of four papers which showed how to display Riemann surfaces in a more intuitively intelligible way. He only began to study Riemann's work more intensively in his later years in Munich. This brought him up against a style of mathematics that was new to him, for Riemann's approach to geometry was much more metrical, even topological, and markedly less algebraic than Klein's. Klein later wrote [1923a, p. 477] that although Clebsch had introduced him to Riemann's results: "the geometrical-physical considerations by which Riemann had come to his conclusions were properly not seen to be adequate or extendable. In this respect the development which led me to Riemann was in contradiction with the tradition in which I was brought up". Klein took the chief importance of Riemann's ideas to be their implications for complex function theory, and although Klein had been called to Leipzig to lecture specifically on geometry, he took the opportunity to deepen his understanding of function theory. He wrote in his autobiography [1923b, 20f] (for an English translation see Rowe and Gray (forthcoming)):

> "But I did not conceive of the word geometry one-sidedly as the subject
> of objects in space, but rather as a way of thinking that can be applied
> with profit in all domains of mathematics. Correspondingly, despite
> many contradictions, I began my Leipzig professorship with a lecture
> on geometric function theory in which I went further with the ideas that
> had stirred me in Munich .... At the time I began a cycle of lectures on
> geometry that comprised analytic geometry, projective geometry, and
> differential geometry."

It is only on arrival at Leipzig that Klein began to don the Riemannian mantle. At that time, as he was willing to admit, he found understanding Riemann's work extremely difficult, and we should be careful not to see him in any simple way as Riemann's true successor.

1881

Klein's attention was first drawn to Poincaré when he read Poincaré's three notes "Sur les fonctions Fuchsiennes" in the *Comptes Rendus* which will now be described. The papers, and the ones which followed them, are a jumble of promises, allusions, examples, and sometimes, mistakes, within which various themes can be detected. The main object of study is what Poincaré called Fuchsian functions, functions defined on a disc or half plane and invariant under certain subgroups of $SL(2, \mathbf{R})$. The main method of study is the use of fundamental regions or polygons, in the sense of the non-Euclidean geometry intrinsic to the disc. The main results concern the relation between Fuchsian functions associated to a given group, and the use of these related functions to solve linear differential equations with rational or algebraic coefficients.

In the first paper (14 February 1881) Poincaré announced the discovery of a large class of functions generalizing elliptic function and permitting the solution of a differential equations with algebraic coefficients, which he proposed to call Fuchsian "in honour of M. Fuchs whose works have been very useful to me in these researches". These functions were invariant under discontinuous subgroups of the group $z \to \dfrac{az+b}{cz+d}$ which leaves a "fundamental" circle fixed; discontinuous meaning no z is infinitesimally near any transform of itself. As a result, the group, called by Poincaré a Fuchsian group, can be studied by looking at how it transforms a region R bounded by arcs of circles perpendicular to the fundamental circle. He said non-Euclidean geometry would be helpful here, but he did not explain how. Instead he gave an example of a triangular polygon $R_0 = ABCD$ vertical angles $B\hat{A}C = B\hat{D}C = \frac{\pi}{\alpha}, C\hat{B}A = C\hat{B}D = \frac{\pi}{\beta}, B\hat{C}A = B\hat{C}D = \frac{\pi}{\gamma}$ where α, β, γ are positive integers or ∞ and $\frac{1}{\alpha} + \frac{1}{\beta} + \frac{1}{\gamma} < 1$. The transforms of such a region would provide an example of a Fuchsian group, and Poincaré asked if any polygon would do. As other examples he gave the special case $a = 2, b = 3, \gamma = \infty$, which gives $SL(2; Z)$, and any linear substitution group preserving an indefinite ternary form $(px^2 - qy^2 - rz^2)$. Finally he introduced functions he called theta-Fuchsians, by analogy with the theta functions occurring in the theory of elliptic functions. Suppose z is a point inside the fundamental circle, K_i an element of the Fuchsian group and zK_i the transform of z by K_i. The single-valued function θ is a theta function if, for some integer m, it satisfies $\theta(zK_i) = \theta(z) \left(\dfrac{dzK_i}{dz} \right)^{-m}$, or equivalently if $\theta \left(\dfrac{az + b}{cz + d} \right) = \theta(z)(cz + d)^{2m}$, assuming $ad - bc = 1$.

The existence of theta-Fuchsian functions follows, he said, from the convergence of the series

$$\sum_{i=1}^{\infty} H(zK_i) \left(\frac{dzK_i}{dz} \right)^m, \; m \text{ an integer greater than 1,}$$

where H is a rational function of z. Two cases arise, one in which every point of the fundamental circle is an essential singular point of the theta-Fuchsian, the other in which the essential singular points, although infinite in number, are isolated. Only in this case can the function be extended to the whole plane.

Next week's installment, so to speak, [1881b], began by pointing out that the quotient of two theta-Fuchsian functions corresponding to the same Fuchsian group was a Fuchsian function. Poincaré then observed that between any two Fuchsian functions corresponding to the same group there existed an algebraic relation, but he gave no proof, and that every Fuchsian function F permitted one to solve a linear differential equation with algebraic coefficients.[11] For, if $x = F(z)$ is a Fuchsian function, then $y_1 = \left(\frac{dF}{dz} \right)^{\frac{1}{2}}$ and $y_2 = z \left(\frac{dF}{dz} \right)^{\frac{1}{2}}$ are the solutions of the equation $\dfrac{d^2 y}{dx^2} = y\phi(x)$, where ϕ is algebraic in y. He gave the hypergeometric equation as an example, and pointed out that one variant of that,

$$\frac{d^2 y}{dx^2} = y \left[\frac{x(x - 1) - 1}{4x^2(x^2 - 1)} \right],$$

gives the periods of $\sin am$ as functions of the square of the modulus. Zeta-functions (to be defined below) were also introduced, and Poincaré claimed that such functions provided a basis of solutions to any linear differential equation having rational coefficients and two finite singular points.

The third paper [1881c] suggested a connection between Fuchsian functions $x(z)$ and $y(z)$ corresponding to the same group and abelian integrals $u(x, y)$. Regarding u as a function of z, and operating on z by transformations in the given group which send (x, y) round a cycle, Poincaré was led to relate the number of generators, $2p + 2$, of the group to the periods of the abelian integral, and consequently obtained an upper bound, p, for the genus of the algebraic equation connecting x and y. Only a truly great mathematician can achieve this degree of visionary imprecision.

Klein read these notes on 11th June 1881 and wrote to Poincaré the next day. He had, he said,[12] considered these topics deeply in recent years, and had written about elliptic modular functions in several articles, which "naturally are only a special case of the relations of dependence considered by you, but a closer comparison would show you that I had a very general point of view". After listing these works, and referring to related work of Halphen and Schwarz, Klein went on to say that the task of modern analysis was to find all functions invariant under linear transformations, of which those invariant under finite groups and certain infinite groups, as the elliptic modular functions, were examples. He had talked to other mathematicians about these questions, without coming to any definite result, and now he wondered if he should not have been in touch earlier with Poincaré or Picard; he hoped his letter would start a correspondence. He admitted that at the moment other duties kept him from working on the problem, but he would return to it soon, for he was to lecture on differential equations in the winter.

Klein was rather exaggerating his achievements, and seems a little concerned to impress Poincaré with what he had already done, perhaps even to over-awe him. We have seen that he was, in fact, at an impasse with his research in this direction, and had moved off onto other topics. Now that a rival had entered the field, he would return to it by the winter.

Poincaré's reply (15 June) was characteristically more modest, and he was more willing to admit to ignorance, even quite astounding ignorance. He immediately conceded priority over certain results to Klein, but said he was not at all surprised "... for I know how well you are versed in the study of non-Euclidean geometry which is the veritable key to the problem which occupies us". He would do justice in that matter when he next published his results, meanwhile he would try to find the relevant *Mathematische Annalen*, which were not in the library at Caen. But, since that would take time, could Klein please explain some things straight away. Why speak of modular functions in the plural, when there was only one, the square of the modulus as a function of the periods? And had Klein found all the circular-arc polygons which give rise to discontinuous groups and found the corresponding functions?

Klein got the letter on the 18th and replied the next day, enclosing reprints of his own articles and promising to send those of his students Dyck, Gierster, and Hurwitz. Then he warmed to the theme of using Fuchs' name. All such research, he said, was based on Riemann. His own was closely related to that of Schwarz, which he urged Poincaré to read "if you do not already know it". The work of Dedekind had shown how modular functions could be represented geometrically, which had already become clear to him (Klein), whereas Fuchs' work was ungeometric. He did not criticize the rest of Fuchs' work on differential equations, but here he had made a fundamental mistake which Dedekind had had to correct. On the subject of polygons Klein pointed out that any polygon was equivalent to a half plane, so repeated reflection or inversion in the sides generated a group and invariant functions in the manner of Schwarz and Weierstrass, without the need to return to general Riemannian principles. But some polygons gave rise to discontinuous groups which did not preserve a fixed circle, so "the analogy with non-Euclidean geometry (which is in fact very familiar to me) does not always hold". He gave this polygon as an example:

Figure 6.7

Poincaré replied on the 22nd, before the reprints had arrived, seeking permission to quote the passage about the group in his next publication and defending the name 'Fuchsian function' on the grounds that, even if "... the viewpoint of the geometric savant of Heidelberg is completely different from yours and mine, it is also certain that his work served as a point of departure ..." and so it was only just that his name should stay attached to those functions.

Klein's brief reply (25 June) demurred, directing Poincaré back to Fuchs' original publication in the *Journal für Mathematik* for the purposes of comparison.

Poincaré wrote on the 27th to say that finally the reprints had arrived, having been sent via the Sorbonne and the Collège de France even though they were correctly addressed (so the postal service was not always so efficient). He now admitted that "I would have chosen a different name [for the functions] had I known of Schwarz's work, but I only knew of it from your letter after the publication of my results, ..." and he could not change the name now without insulting Fuchs. As to the mathematics, had Klein determined the fundamental polygons for all the principal congruence subgroups? And what, in that connection, was the *Geschlecht* in the sense of Analysis situs? Was it the same as the *genre* that he, Poincaré, had defined, for he only knew that they both vanished simultaneously. Poincaré asked for a definition of the topological genus, or, if it was too long to give in a letter, a reference to where it could be read. As for the polygon, presumably its sides should not meet when extended, and finally, he asked what Klein understood by general Riemannian principles?

On July 2nd Klein answered these questions as well as he could. Dyck had established the polygons for congruence subgroups for prime n; composite n had not been considered. Genus in the sense of the analysis situs was the maximal number of closed curves that can be drawn without disconnecting the surface, and was *materially the same number* (Klein's emphasis) as the genus of the algebraic equation representing the surface. He went on, "I have only conjecturally a freer representation of a Riemann surface and the definition of p based on it".

The furthest he had gone on the question of connection of polygons and curves was this. Map a half-plane onto a polygon such as this one,

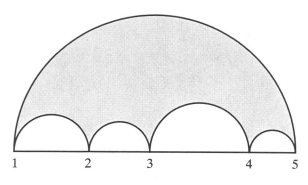

Figure 6.8

and let the points corresponding to 1, 2, 3, 4, 5 be I, II, III, IV, V respectively, which can be arbitrary. Let I, II etc. be the branch points of an algebraic function, $w(z)$, which can have no other branch points. So w and z are single-valued functions of the right kind, and if all the branch points lie on a circle in the z-plane, there is nothing more to be said. But in the other cases, and here the unsymmetric polygon of the last letter is relevant, reflections generated a fundamental space. But did they together cover only one part of the plane? "I find myself already brought to a halt on this difficulty for a long time". In this conclusion Schottky's study of inversion in families of non-intersection circles was interesting.

Riemann's principles he went on, do not tell you how to construct a function, so for that reason these consequences are somewhat uncertain, and Weierstrass and Schwarz have done a lot of work on circular arc polygons. Riemann's principles map a many-leaved surface to a given polygon and enable one to prove the existence of a function having prescribed infinities and real parts to their periods. "This theorem, which by the way I have only completely understood recently, includes, so far as I can see, all the existence proofs of which you speak in your notes as special cases or easy consequences".

Finally, referring to another of Poincaré's papers ("Sur les fonctions abeliennes", Comptes rendus 92, 18 April 1881) Klein asked why the moduli were presumed to be $4p + 2$ in number when they are really only $3p - 3$. "Haven't you read the relevant passage of Riemann? And is the entire discussion of Brill and Noether...unknown to you?"

Poincaré's reply of 5 July apologized profusely for asking about the topological genus when it was defined "on the next page of your memoir". Not having had a refusal of his request to quote Klein, he had done so, taking silence for consent. As for the branch points of algebraic functions, that had been proved by him too, he said, and published on May 23rd, but where was it in the work of his predecessors? Finally, the number $4p + 2$ had only been needed as an upper bound and was easy to obtain. He made no mention of his deficient reading.

This part of the correspondence closed with Klein's letter of 9 July, in which he pointed out that Riemann's use of Dirichlet's principle was not conclusive, but that one could find stronger proofs in, e.g., Schwarz's work. By now Poincaré had published several more short papers, and these should be described.

The notes of 18 April, 23 and 30 May [1881e,g,h], published before the correspondence with Klein had begun, dealt in more detail with the role of circular-arc polygons in generating groups. Poincaré supposed the fundamental circle to be bisected by the real axis, O_x, and considered the polygon with sides $C_1, \ldots, C_n$ as follows:

(i) All the C_i meet the fundamental circle at right angles;

(ii) C_1 meets 0_x at α_1 and β_1, making an angle of $\dfrac{\lambda_1}{2}$;

(iii) C_i meets C_{i-1} at α_i and β_i, making an angle of λ_i:

(iv) C_n meets O_x at α_{n+1} and β_{n+1}, making an angle of $\dfrac{\lambda_{n+1}}{2}$;

where he supposed each λ_i is an aliquot part of 2π and $\lambda_1 + 2\lambda_2 + \cdots + 2\lambda_n + \lambda_{n+1} < 2\pi(n-1)$ (so that the construction is possible).

The transformation sending z to z_j by

$$\frac{z_j - \alpha_j}{z_j - \beta_j} = e^{i\lambda_j} \left(\frac{z - \alpha_j}{z - \beta_j} \right) \quad j = 1, \ldots, n+1$$

leaves α_j and β_j fixed, and rotates the family of coaxal circles surrounding α_j and β_j by λ_j. Since one may suppose α_j is inside the fundamental circle and β_j outside, this has the effect of rotating the polygon about the vertex α_j so that its new position lies alongside its old one. Even so it is not true, as Poincaré claimed, that one obtains a discontinuous group in this way unless the sides of the polygons have the same length (i.e., vertex α_{j-1} is rotated onto α_{j+1}). The idea is clear nonetheless; successive transformations move the polygon around crabwise and a function defined on the original domain extends to (meromorphic) Fuchsian function accordingly, defined on the interior of the unit disc, and satisfying $F(z) = F(z_1) = \cdots = F(z_{n+1})$, so F is invariant under all composites of the above transformations.

Since two such functions x and y will be related algebraically, one can consider the genus of the equation connecting them, $f(x, y) = 0$. Should it be zero, so f is a rational function, then taking $x = F(z)$, $y = \left(\frac{dF}{dz} \right)^{\frac{1}{2}}$. One obtains

$$\frac{d^2 y}{dx^2} = y\phi(x),$$

in which ϕ is also a rational function, and the singular points of the equation, being the infinities of ϕ, are at $F(\alpha_1), \ldots, F(\alpha_n)$. Suppose these to be all arbitrary but real, and $F(z)$ to be real along the sides $C_1, \ldots, C_n$ of the polygon. Then the case considered is that of an algebraic function with arbitrary real branch points, as discussed in the correspondence. Suppose furthermore that $\lambda_1 = \ldots \lambda_{n+1} = 0$, so $\alpha_1, \ldots \alpha_{n+1}$ lie on the fundamental circle itself. Then $F(z)$ fails to take the values $F(\alpha_1), \ldots, F(\alpha_{n+1})$. So if one has a linear differential equation with coefficients rational in x and real singular points at $x = F(\alpha_1), \ldots, F(\alpha_{n+1})$, then one sets $x = F(z)$ and the solutions of the equation are zeta fuchsian functions of z. In this way a large class of differential equations are solved.

On 30th May Poincaré clarified this a little: when $\lambda_{n+1} = 0$, the appropriate $F(z)$ is the limit of the original $F(z)$ as the λ's tend to 0, and it is only in this case that the values of $F(\alpha_1), \ldots, F(\alpha_{n+1})$ can be arbitrary. In this paper he broached a continuity argument to show that if a given differential equation has $2n$ singular points, these can perhaps be allowed to tend to $2n$ real points and the solutions to the real case allowed to deform into those in the general case:

"If I succeed in showing that these equations always have a real solution, I will have shown that all linear equations with algebraic coefficients can be solved by Fuchsian and zeta-fuchsian transcendents."

On the 27th June, after he had heard from Klein, Poincaré [1881i] returned to the description of non-Euclidean polygons, this time defined in the upper half plane (which can be obtained from the disc by an inversion). He took a and b in the upper half plane, with conjugates $\bar{a}$ and $\bar{b}$, and defined

$$(a, b) := \frac{(a - \bar{a})(b - \bar{b})}{(a - \bar{b})(b - \bar{a})}.$$

Since this is invariant under the projective group $PSL(2 : \mathbf{R})$ which preserves the upper half plane, it can play the role of a metric invariant in the sense of non-Euclidean geometry. In particular there will be a transformation sending a to c and b to d if and only if $(a, b) = (c, d)$. If this condition is stipulated for each of n pairs of sides of a $2n - gon$ whose angles are, as before, aliquot parts of 2π, and whose vertices are made to correspond only if they are either both above the real axis or both on it — or both segments of it, Poincaré admitted the infinite case,

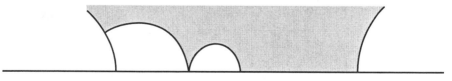

Figure 6.9

then indeed a discontinuous group is obtained. Poincaré added, without a hint of a proof, that every Fuchsian group can be obtained in this way. Turning to Klein's example of an unsymmetric polygon, which does not preserve a fundamental circle, and to his generalization of that to a region bounded by $2n$ circles exterior to one another and possibly touching externally, he said that successive reflections again generated a discontinuous group and so also invariant functions. Making amends for

his earlier choice of names, and surely with a twinkle in his eye, Poincaré added that
"... I propose to call [these functions] Kleinian functions, because it is to M. Klein
that one owes the discovery. There will also be theta-Kleinian and zeta-Kleinian
functions, analogous to the theta-Fuchsian functions."

In the paper of 11th July entitled, "Sur les groupes Kleinéens" [1881j] Poincaré
delightfully restored the analogy between discontinuous groups and non-Euclidean
polygons which Klein had said his example of unsymmetric polygons broke. His
insight was to take the (x, y)-plane as the boundary of the space (x, y, z), $z \geq$
0, and to regard the polygon as bounded by arcs which were the intersections of
hemispheres centre $(x, y, 0)$ with the (x, y)-plane. Thus the groups generated by
reflections on the sides is regarded as acting on 3 dimensional non-Euclidean space,
realized as the space above the (x, y)-plane in this way:

A pseudogeometric plane (Poincaré referred to the pseudogeometry of
Lobachevskii) is a hemisphere; a pseudogeometric line is the intersection of two
such planes; the distance along a line from p to s is half the logarithm of the cross-
ratio $pscd$, where c and d are the points where the line ps meets the xy plane; and
the angle between two intersecting lines is their usual geometric angle.

This is the first appearance in print of Poincaré's conformal model of non-
Euclidean geometry. Oddly enough, it is three-dimensional, just as the original
versions of Lobachevskii and Bolyai were.

Papers continued to stream out of Caen. In another (8 August = [1881k]) he
claimed that he had been able to show by a simple polynomial change of variable
that his earlier work (30 May) on differential equations led to the conclusions that

(i) Every linear differential equation with algebraic coefficients can be solved by
zeta-Fuchsians;

(ii) The coordinates of points on any algebraic curve can be expressed as Fuchsian
functions of an auxiliary variable.

The second point is the uniformization theorem for algebraic curves: if $f(u, v) = 0$
is the equation of an algebraic curve, then it is claimed that there are Fuchsian
functions $\phi(x)$ and $\psi(x)$ such that $u = \phi(x)$, $v = \psi(x)$, $f(\phi, x), \psi(x)) = 0$. (For
a modern account of this important theorem, see Abikoff [1981]; for a history, see
Gray [1994].)

Klein meanwhile had concluded his summer lectures on geometric function
theory, and on 7th October he sent the manuscript of *Über Riemanns Theorie der
algebraischen Funktionen und ihrer Integrale* to Teubner in Leipzig. Now he could
return to his old problem. He wrote to Poincaré on the 4th of December, congrat-
ulating him on solving the general differential equation with algebraic coefficients
and the uniformization problem, and asking for an article on this work for *Mathematische Annalen*, of which he was the editor. He, Klein, would add a letter to it,
connecting it with his work on modular functions. Publication would serve two purposes: acquainting the readership of the *Annalen* with his work, and explaining its
relationship to Klein's (one wonders who would gain most by this). Poincaré, who
had moved to Paris by now agreed, and sent a manuscript to Klein on December

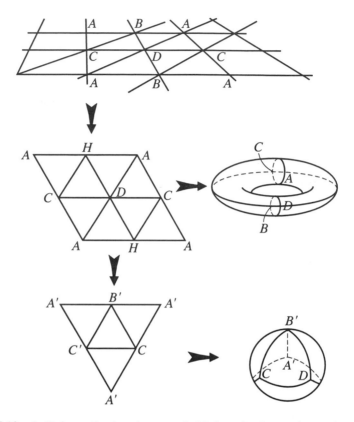

Figure 6.10   In Poincaré's view the upper half plane (top), acted upon by a discrete subgroup (a lattice), has a quotient space which is a torus. In the view of Riemann and Klein the algebraic curve $y^2 = $ quartic in x is a double cover of the x-sphere branched over four points.

17th, which made the deadline for the last part of volume 19. In content it much resembles Poincaré's report presented to the *Mémoirs de l'Académie Nationale des Sciences, Arts et Belles-Lettres de Caen*, written and published shortly before, but it goes into more details on the topics Klein had requested. Before discussing these two memoirs, which represent Poincaré's first attempts to survey what he had done in that memorable year, one must turn to the sourer discussion of priorities which Klein's letters provoked.

Klein again objected to the name Fuchsian functions, on the grounds that Fuchs had published nothing on the domain of Fuchsian functions, whereas Schwarz had. Nor had he, Klein, done anything on Kleinian functions except to bring one special case to Poincaré's attention; in this case Schottky deserved more of a mention, as did Dyck for the group-theoretic emphasis in his work. Klein had warned Poincaré by letter (13 January 1882) that he would protest against the names as much as was in him, and Poincaré indicated that he would like space in the *Annalen* to defend his choice. So, in the next issue he pointed out that it was not ignorance of Schwarz that made him select Fuchs' name, but the impossibility of forgetting the remarkable discoveries of Fuchs, based on the theory of linear differential equations. Furthermore, Schwarz was only a little interested in differential equations, and Klein indeed dwelled more on elliptic functions, whereas Fuchs had brought forward a new point of view on differential equations "which had become the point of departure for my researches". As for Kleinian functions, the name belonged rightly to the man who had "stressed their principal importance", said Poincaré, turning Klein's observation on Schottky in his own work [1882a] against him.

Now that Klein had criticized Fuchs publicly, Fuchs came to hear of the dispute and could reply himself, which he did in the *Journal für Mathematik* [1882] and also privately in a letter to Poincaré.[13] Stung by the allegation that he had published nothing on invariant functions, he cited not only those works of his which had inspired Poincaré, but also his earlier study [1875] of differential equations with algebraic solutions. He observed that Klein himself, in the *Annalen*, vol II, had said that these works had led to his own developments. Fuchs was not quite correct. Klein had said rather that he should, at the conclusion of a work devoted to differential equation having algebraic solutions, refer to Fuchs' [1875], "which likewise concerns the subject matter here presented, the study of which I was recently induced to do, even to derive the simple method I substitute here" [Klein, 1876a, = 1922, 305]. In that paper he invoked Schwarz's work at the start, and his own work on binary forms. It is entirely plausible that this was his route to the problem, and that his acquaintance with Fuchs' studies was slight and scarcely influential.

It is a story he was to repeat in his book *Vorlesungen über das Ikosaeder* [1884], although towards the end of his life he conceded that Fuchs had been a particular impulse [1922, 257]. But he was eager to see that Poincaré harboured no doubts on the matter. His next letter to Paris (3 April 1882) contains a lengthy restatement of the historical priorities as viewed from Düsseldorf. In Klein's view his study of linear transformations of a variable went back to 1874. When in 1876 he solved the problem thrown up by Fuchs, the facts were contrary to what Fuchs had said: "I did not take the ideas from his work, rather I showed that his theme must be handled with *my* ideas" (Klein's emphasis). He would welcome the new class of functions of several variables being called Fuchsian, but by the way, were they single-valued or only unbranched? Then Klein returned to mathematical comments. This remark is the most interesting one: uniformization permits one to show, what is only claimed as probable in the lectures on Riemann's Theory of Algebraic Functions and their Integrals: that a Riemann surface of genus $p > 0$ cannot have infinitely many discrete self transformations. [Klein meant $p > 1$; see his section XIX].

Klein hoped his letter would close the debate, and Poincaré at once agreed (4 April). He wrote that he had not started the debate and he would not prolong it. As for the question of "Kleinian" he had known of no right to the property before Klein's letter. As for Fuchs' functions of two variables, these divided into three classes, two effectively uniform and one only unbranched, and he directed Klein's attention to the smaller works of Fuchs which had followed the main paper. He concluded by wishing the debate about names to be over: *"Name is Schall and Rauch"* he wrote in German ("Name is sound and smoke").[14] Happily, no more was said, at least on this occasion, but Fuchs remained acutely sensitive to plagiarism, once accusing Hurwitz of it, in 1887, once P. Vernier, and once, incorrectly, Loewy (in 1896). In private he had indeed already registered a similar wound in 1881. He wrote (30 March 1881) to Casorati complaining that Brioschi had referred only to Klein and not to him on the algebraic solutions question, although his work preceded Klein's [1876a] (letter quoted by E. Neuenschwander [1978]).

Poincaré's two memoirs on Fuchsian functions have been mentioned already, and shall now be described, dwelling on the *Annalen* paper, which subsumes the Caen one. Poincaré hardly ever supplied proofs in these articles; statements were given bald, supported only by reference to long calculations or even obscurely qualified by vague remarks about simple cases. Some of these evasions are unimportant, others are more interesting.

A Fuchsian group, he said, is a group of substitutions of the form $S_i = \left(\zeta, \dfrac{\alpha_i \zeta + \beta_i}{\gamma_i \zeta + \delta_i} \right)$, where α_i, β_i, γ_i, δ_i are all real and $\alpha_i \delta_i - \beta_i \gamma_i = 1$, which move a region R_0 of the upper half plane around so that the various $S_i R_0$ form a sort of web. The $S_i R_0$ are to be non-Euclidean polygons, not necessarily finite in extent, fitting together along their common edges, and whose angles are aliquot multiplies of 2π.

The theta-Fuchsian function $\theta(\zeta)$ is defined as follows: let $H(\zeta)$ be any rational function. Then the series

$$\theta(\zeta) = \sum_{i=1}^{\infty} H\left(\frac{\alpha_i \zeta + \beta_i}{\gamma_i \zeta + \delta_i} \right) \frac{1}{(\gamma_i \zeta + \delta_i)} 2m$$

converges for each integer m greater than 1 and defines a theta-Fuchsian function. Consideration of $\int \dfrac{\theta'(\zeta)d\zeta}{\theta(\zeta)}$ over the boundary of R_0 shows that the number of its zeros and its infinities in R_0 is finite. A quotient of two theta-Fuchsians is a Fuchsian function $F(\zeta)$; it satisfies

$$F\left(\frac{\alpha_i \zeta + \beta_i}{\gamma_i \zeta + \delta_i} \right) = F(\zeta).$$

If R_0 has a segment of its boundary in common with the real axis, F can be extended analytically into the lower half plane, but otherwise, he remarked, it is only defined on the upper half plane.

If the substitutions of the group have all these properties except that of preserving a half-plane or circle, then Poincaré called the group Kleinian, and the functions theta-Kleinian and Kleinian respectively, remarking that their study required processes derived from three dimensional non-Euclidean geometry. He gave no reason for the name in the *Annalen*; in the Caen essay he said that Klein had been the first to give an example of such groups.

Poincaré stated four main results for the theory of Fuchsian functions and groups. The first concerned the genus of the algebraic equation connecting two Fuchsian functions belonging to the same groups. He called two edges of R_0 conjugate if they were identified by some S_i, and said that vertices which were identified belonged to the same cycle. In the case of a finite polygon, if there were $2n$ sides to R_0 and p cycles, then genus of the equation connecting any two functions was $\dfrac{n+1-p}{2}$. For a polygon with "sides" along the real axis, the formula is $n - p$.

The second theorem concerned the connection between Fuchsian functions and second order differential equations. If x and y are two Fuchsian or Kleinian functions corresponding to the same group, then

$$v_1 = \sqrt{\frac{dx}{d\zeta}} \text{ and } v_2 = \zeta \sqrt{\frac{dx}{d\zeta}}$$

are certainly solutions of

$$\frac{d^2 v}{dx^2} = v\phi(x, y),$$

where ϕ is an algebraic function. Poincaré claimed that he could state very complicated conditions on the coefficients of ϕ under which the converse was true, and the solutions of $\dfrac{d^2 v}{dx^2} = v\phi(x, y)$ were Fuchsian functions, but he did not do so. He gave instead one example of an equation not of the Fuchsian type, in which if the coefficients satisfied certain (unspecified) inequalities the solutions were Kleinian functions, and if they satisfied certain (unspecified) equalities the solutions were Fuchsian. This brings in the "notorious" accessory parameters.

He had claimed in [1881m] that an equation with specified singularities is Fuchsian for exactly one choice of parameters, a claim he proved in his [1884a]. See below, p. 218.

The third theorem was uniformization, derived from the existence of a Fuchsian function $F(\zeta)$ omitting $n + 1$ values $a_1, a_2, \ldots, a_n, \infty$ from its range, which rested in turn on a long numerical calculation, which was not given. If $f(x, y) = 0$ is an algebraic equation with singularities at $a_1, a_2, \ldots, a_n$, and $x = F(\zeta)$, then y is also a Fuchsian function and the curve $f(x, y) = 0$ has been uniformized.

The fourth theorem concerned zeta-fuchsian functions, i.e., functions

$$\phi_1(\zeta), \phi_2(\zeta), \ldots, \phi_n(\zeta)$$

which satisfy

$$\phi_\lambda \left(\frac{\alpha_i \zeta + \beta_i}{\gamma_i \zeta + \gamma_i} \right) = A^i_{1\lambda} \phi_1(\zeta) + A^i_{2,\lambda} \phi_2(\zeta) + \cdots + A^i_{n,\lambda} \phi_n(\zeta)$$

for constants A such that $\det (A^i_{j,\lambda}) = 1$ for all i. Poincaré claimed that his earlier results could be extended to show that every nth order linear differential equation with algebraic coefficients can be solved by Fuchsian and zeta-Fuchsian functions, or by Kleinian and zeta-Kleinian ones.

6.5 Klein's response

The same volume of the *Annalen* also contained Klein's first response to Poincaré's ideas. Excited, but also challenged by the sudden emergence of the younger man, he felt driven to use methods which even he regarded as "to some extent irregular" and to publish his results before he had managed to put them in order. This was a departure from his usual practice, which might indeed gloss over special cases and prefer geometric reasoning to the arithmetizing of, say, Weierstrass, but is nonetheless quite precise. The contrast between Klein and the Berlin school shows itself in Klein's preference for generic situations to exceptional cases; however, the generic situation is always treated carefully. The contrast between Klein and Poincaré is between the clear-sighted and the visionary. Poincaré's genius was paradoxically helped by his strange lack of mathematical education; he was making it up as he went along. Klein was reformulating his considerable store of existing knowledge, bringing to the new problems not only techniques he knew had worked in the past but a unifying view of mathematics to which, he felt, the new subject must conform. The result of each comparison is oddly similar. The thoroughness of the Berlin school revealed hidden riches in known fields, whereas Klein's study of modular functions was almost a new field of study; Klein's thoroughness in turn lacks the vivid recreation of the theory of Riemann surfaces and differential equations one finds in Poincaré.

In his first brief paper [1882b] Klein asserted that for each Riemann surface of genus greater than 1, there is always a unique function h which is single-valued on the simply-connected version of the surface, on crossing a cut changes to a function of the form $\dfrac{\alpha\eta + \beta}{\gamma\eta + \delta}$, and so by analytic continuation, maps the cut surface without overlaps onto a $2p$-connected region of the η-sphere. The (suppressed) argument to the existence was a continuity one; the uniqueness derives from the manner of the analytic continuation. It has the uniformization of the Riemann surface as a corollary, and Klein gives a more general description of the generation of the discontinuous group in this context. Poincaré always at this time used infinite polygons in his discussion of uniformization, as Freudenthal [1954] pointed out. Klein regarded the analytic continuation of η as $\dfrac{\alpha\eta + \beta}{\gamma\eta + \delta}$ as moving the image of the cut surface around on the η-sphere. Furthermore, these generating substitutions, although they must satisfy certain inequalities, determined by the shape of the fundamental region, and contain $3p - 3$ complex variable parameters. He explained this count as follows. The image of the cut surface may be taken to be a polygon whose $2p$ sides, the images of the cut, are identified in pairs. Each identification, being a fractional linear transformation, contains three parameters, so $3p$ parameters specify the group, but then 3 may be chosen arbitrarily since the group does not depend on the initial position of the polygon. Klein's description of the decomposition of a Riemann surface by means of cuts is confused: $2p$ cuts are needed to render a surface of genus p simply connected, when it becomes a polygon of $4p$ sides. In the "Neue Beiträge" he counted more carefully: there are $2p$ lengths and $4p$ angles in such a polygon, so $6p - 3$ independent real coordinates, and then, using the upper half plane model, the coefficients of η are real and η may be replaced by $\dfrac{\alpha\eta + \beta}{\gamma\eta + \delta}$, so three parameters are inessential, and the function η depends on $6p - 6$ real or $3p - 3$ complex parameters. Now, the space of moduli for Riemann surfaces has complex dimension $3p - 3$ and Klein made the audacious claim that every Riemann surface corresponds according to a unique group. As Freudenthal [1954] remarked, the direct route to uniformization is the only point at which Klein surpassed Poincaré. One might even make this point a little more strongly: it is only Klein who invoked the birational classification of surfaces and so made it all precise how the correspondence between discontinuous groups and curves might be expected to work.

Klein went on to describe symmetric surfaces and surfaces whose equations have real coefficients. In this way he encountered as a special case the figure bounded by $p + 1$ non-intersecting circles, which, he said, Schottky had already described "without emphasizing its principal importance".

The next developments were vividly described by Klein himself in a lecture of 1916, reprinted in his *Werke* (III, 584) and his *Entwicklung* (p. 379). He wrote: "In Easter 1882 I went to recover my health to the North Sea, and indeed to Nordeney. I wanted to write a second part of my notes on Riemann in peace, in fact to work out the existence proof for algebraic functions on a given Riemann surface in a new form. But I only stayed there for eight days because the life was too miserable, since violent storms made any excursions impossible and I had severe asthma. I decided to go back as soon as possible to my home in Düsseldorf. On the last night, 22nd to 23rd March, when I needed to sit on the sofa because of my asthma, suddenly the 'Grenzkreis theorem' appeared before me at about two thirty as it was already quite properly prefigured in the picture of the 14-gon in volume 14 of the *Mathematische Annalen*. On the following morning, in the post wagon that used to go from Norden to Emden in those days, I carefully thought through what I had found once more in every detail. I knew now that I had a great theorem. Arrived in Düsseldorf I wrote all of it at once, dated it the 27th of March, sent it to Teubner and allowing for corrections to Poincaré and Schwarz, and for example to Hurwitz." So Klein too had a sleepless night of inspiration.

Hurwitz immediately recognized the importance of the theorem, but when Poincaré replied on 4th April he made no mention of it, as we have seen. In his next note for the *Comptes Rendus*, however, Poincaré modestly remarked that the results "have been obtained by Klein on other grounds". Schwarz's response was more complicated. At first he denied the validity of the *Grenzkreis* theorem. Then he accepted it, and proposed that it be seen as a conformal representation of an infinite sheeted covering of the Riemann surface on a disc. Later still he derived it directly by showing how the concept of distance can be defined on the surface corresponding to the non-Euclidean metric on the disc (see Klein [1923a, 584]).

The Grenzkreis theorem (or boundary circle theorem) is the assertion that every Riemann surface of genus greater than one can be represented in an essentially unique way by an invariant function without branch points, defined on a disc-like region (Klein called it a spherical skull-cap bounded by a plane). What is novel in Klein's theorem is the idea that the function η in the *Ruckkehrschnitt* theorem can be so chosen that the corresponding image of the Riemann surface on the sphere is a circular arc polygon of $4p$ sides all perpendicular to a common circle. When this is done, the images of the surface under analytic continuation of η fill out the region bounded by the circle, which is a limit that they approach indefinitely without overlapping. This gives a magnificent correspondence: the space of all groups and the space of all curves correspond one-to-one via the function η. What Klein saw in the figure of the 14-gon was a way of depicting any Riemann surface within a disc-like region upon which a group acts discontinuously. This dispenses entirely with Riemann's approach, which started from the equation, considered to define a surface spread out over the complex sphere. This surface was rendered simply connected by means of cuts, and then functions on the surface were studied in terms of their behaviour at singularities and on crossing the cuts; in short, the surface is given intact and sits above the z-sphere. In the new view it is given opened out and sits underneath the η-domain. Analytic continuation of functions is given by the group action simultaneously with the identification of the sides — in modern jargon the surface is in Riemann's view a total space and, in the new view, a quotient or base space.[15]

Seen in this light one may say that this was Poincaré's attitude all along. Totally ignorant of Riemann's work (he did not know even of Dirichlet's principle, as his question (27 June 1881) about "general Riemannian principles" makes clear) he had come to construct Riemann surfaces naturally from discontinuous groups. In his mind this connected with linear differential equations, a topic much less interesting to Klein. This insight of Poincaré, so painfully gained by Klein, testifies to the strong hold the idea of mathematical unity had upon Klein. The paradox is that Klein, who had done so much to further non-Euclidean geometry in the 1870s, did not appreciate it here. This derives from his view of geometry as essentially projective. He had been able in 1871 to ground non-Euclidean geometry in projective geometry; the Klein model is a projective model. Thereafter his attitude to geometry had been connected with invariant theory or with line geometry, again a projective idea, or else with topological or visual properties of figures, questions of reality, etc. He does try to discuss intrinsic metrical ideas on a Riemann surface in his lectures on Riemann's algebraic functions, but the discussion is clumsy. Differential geometry was not his forte, and did not fit centrally into his view of mathematics. On the other hand, Poincaré went directly to a conformal model of non-Euclidean geometry, and viewed the transformations as isometries; Klein missed the metrical force of the idea of reflection and kept it confined to a complex analytic circle of ideas. This is one way in which Poincaré's ideas are more flexible and better adapted to the problem at hand. Klein for once was too committed to his picture of the overall nature of mathematics. It is a cruel irony to penetrate further into Riemann's way of doing things than anyone else had done at just the moment when epoch-making progress is being made on Riemannian questions in a non-Riemannian way. One can only admire the heroic and largely successful effort Klein made to grasp the new ideas against the current of his recent work.

The fruits of his labour were published in his long paper "Neue Beiträge zur Riemannschen Functionentheorie", finished in Leipzig 2 October 1882 and published in vol. 21 of the *Mathematische Annalen*, 1882/83. It will shortly be compared with Poincaré's two long papers in volume I of Mittag-Leffler's *Acta Mathematica*, also published in 1882; but first some more of Poincaré's short notes and the correspondence of the two men must be pursued.

Poincaré had published two notes on discontinuous groups and Fuchsian functions, while Klein wrestled with the Grenzkreis Theorem. The one on discontinuous groups (27 March 1882) made explicit the connection he had been so pleased to discover between number theory and non-Euclidean geometry. In it he pointed out that substitutions $(x, y, z; ax + by + cz, a'x + b'y + c'z, a''x + b''y + c''z)$ with real coefficients, preserving x^2-xy, correspond to substitutions

$$\left(x, y, \frac{ax + by + c}{a''x + b''y + c''}, \frac{a'x + b'y + b'}{a''x + b''y + c''} \right)$$

and then to substitutions

$$\left(t, \frac{\alpha t + \beta}{\gamma t + \delta} \right),$$

where $\alpha, \beta, \gamma, \delta$ are real and t is a complex quantity such that $xt^2 + 2zt + y = 0$. To make t lie in a half-plane he assumed $z^2 - xy$ was negative. Accordingly, discontinuous subgroups G of the initial group correspond to discontinuous subgroups G' of the final group, and the same is true for other forms in x, y, z. In particular one might start by assuming $a, b, \ldots, c''$ were integers. Furthermore, the polygonal region R' appropriate to G' (when G' is thought of as transforming the interior of a conic, say a circle) corresponds very simply to the polygon R moved around by G. Place a sphere so that stereographic projection from the plane to the sphere lifts the circular region on which G' acts up to a hemisphere. Now project the image vertically down onto the plane. The image of R' so obtained is R, and is rectilinear. This is the now standard process (discussed above) for converting the conformal description of non-Euclidean geometry of the projective one.

In the paper on Fuchsian groups (10 April) he looked again at nth order linear differential equations whose coefficients $p_i(x, y)$ are rational in x and y, and may become infinite at some points, where x and y satisfy an algebraic equation $f(x, y) = 0$ of genus p. He claimed that if there exists two Fuchsian functions $F(z)$ and $F_1(z)$ defined inside some fundamental circle, such that $x = F(z)$ and $y = F_1(z)$ satisfy $f(x, y) = 0$, and the only singular points of the equation are such that the roots of the determinant equation are multiples of $\frac{1}{n}$, and F, F_1 and their first $n - 1$ derivatives vanish at the singular points, then the equation is solvable by zeta-Fuchsian functions of z. Furthermore the fundamental region has $4p$ sides and either opposite sides are identified, or the sides are identified $4q + 1$ with $4q + 3$ and $4q + 2$ with $4q + 4$.

Klein was struck by the close resemblance of this paper of Poincaré's to his own, and wrote to him on the 7th of May to say he found the methods interchangeable. He went on: "I prove my theorems by continuity in that I assume the two lemmata: 1. that to any "groupe discontinu" there corresponds a Riemann surface, and 2. that to a single necessarily dissected Riemann surface (under the restrictions of the present theorem) there can correspond only *one* such group (in so far as a group does generally correspond to it". (Klein's emphasis.) He offered Poincaré a new version of his general theorem to accompany the two notes in the *Annalen*, for which he said he had very little time. Let $p = \mu_1 + \mu_2 + \cdots + \mu_m$, where the μ's are integers greater than 1. Take m points on the Riemann surface, $0_1, \ldots, 0_m$, and draw $2\mu_i$ boundary cuts from 0_i. On the η-sphere draw m circles lying outside one another and draw a circular-arc polygon in the regions they bound having $4\mu_1$ sides each perpendicular to the first circle, then another of $4\mu_2$ sides standing on the second circle, and so on until an m-fold connected polygon is obtained. Order the sides of the boundary A_1, B_1, A_1^{-1}, B_1^{-1}, A_2, B_2, ... Then the corresponding linear substitutions satisfy $A_1 B_1 A_1^{-1} B_1^{-1} \ldots A_{\mu_i}^{-1} B_{\mu_i}^{-1} = 1$. Then, he said "there is always one and only one analytic function which represents the dissected Riemann surface on the circular-arc polygon described in this way".

In his paper [1884a, 331–2] Poincaré gave this criticism of Klein's argument. Suppose S and S' are two manifolds of the same dimension, that to each s in S there corresponds a unique s' in S', and that to each s' in S' there corresponds at most one point in S. If this correspondence is analytic, then the condition that S is closed (i.e., without boundary) forces the map from S to S' to be an onto map. That is, to each s' in S' there always corresponds a unique s in S. But if S has a boundary, the conclusion no longer follows. For Klein's argument to be valid, it is necessary to show that the space of groups has no boundary, which said Poincaré, is "not at all evident *a priori*". The question is made more difficult by the fact that the same group may have many very different fundamental polygons, and the onto nature of the map must be established by a special discussion which Klein had failed to provide. "This is a difficulty one cannot overcome in a few lines" [1884a, 332]. But Poincaré only replied at the time (12 May) that their methods differed less in the general principle than in the details. As for the lemmas, he had established the first by considerations of power series, and the second "presents no difficulty and it is probable we will establish it in the same way". Once this was done, "and it is in effect there that I begin, like you I employ continuity, ...".

Klein replied by return of post (14 May). To prove the second lemma he considered the sets of all groups and of all Riemann surfaces as forming analytic manifolds with analytic boundaries in Weierstrass' sense of the term. The correspondence between these manifolds is analytic and $1 - x$ in different places, where x is 0 or 1, and the lemma follows from the result that the mapping from groups to surfaces has a nowhere vanishing functional determinant. But he admitted this argument was only advanced in principle, he expected that to carry it out would be a lot of trouble and it might have to be modified. He had also been to see Schwarz in Göttingen, and without Schwarz's permission to do so he offered Poincaré Schwarz's ideas on the problem. Schwarz imagined the dissected Riemann surface infinitely covered by leaves joined along the boundary cuts, so as to form a surface. This surface, at least for η-functions of the 'right' kind, would be simply-connected and simply bounded (in as much as infinite surfaces can be described in such ways at all) and so can be mapped onto the interior of a circle. As he said, "this Schwarzian line of thought is in any case very beautiful".

Poincaré agreed in that estimation (18 May) but admitted that he had tried and failed to extend Schwarz's ideas to more complicated problems, and hoped that Schwarz would have better luck.

As spring turned into summer both men were free to write up their research in extended form. Poincaré had been commissioned by Mittag-Leffler to publish his work in the first issue of the new *Acta Mathematica*,[16] and Klein had his own *Mathematische Annalen*. It is clear from their correspondence that they had each arrived at a convenient summit, offering a view over a considerable amount of new mathematical terrain, and with a hint of new peaks in the distance. In the event both men chose to survey their discoveries, rather than to continue their assault, and indeed the uniformization theorem and Kleinian groups were to prove markedly less tractable problems. As a result, it is possible to describe their long papers quite briefly.

Klein began his "Neue Beiträge zur Riemannschen Funktionentheorie" with a discussion of a Riemann surface as a closed surface in space, carrying a metric

$$ds^2 = E\,dp^2 + 2E\,dp\,dq + G\,dq^2$$

for which

$$\frac{\partial}{\partial p}\left(\frac{F\frac{\partial u}{\partial q} - G\frac{\partial u}{\partial p}}{\sqrt{EG - F^2}}\right) + \frac{\partial}{\partial q}\left(\frac{G\frac{\partial u}{\partial p} - E\frac{\partial u}{\partial q}}{\sqrt{EG - F^2}}\right) = 0\,.$$

This is the approach taken in the earlier "Riemann's Theorie der algebraischen Funktionen und ihrer Integrale". It leads naturally to a local theory of Riemann surfaces and then to the global study of the patches of surfaces obtained from the above equation. Dirichlet's principle as proved by Schwarz is the crucial tool for representing surfaces. Klein quoted an extensive passage from a letter Schwarz wrote to him on 1st February 1882 in which the non-simply connected case is discussed. Various patches are described including circular arc-triangles and polygons, and the way in which they might fit together to form surfaces is discussed, following the lines of the their correspondence. This line of argument is extended in the second section of the paper to a general discussion of the analytic continuation of a function by means of tilings or polygonal nets (*Bildungsnetz*) which leads to the central idea of the paper: single-valued functions with linear transformations to themselves.

These functions and their groups are discussed carefully in section III. Klein pointed out in §5 that even among discontinuous groups there are those which have no fundamental region, the fundamental points are everywhere dense in the complex plane. Then there are groups which arise when the complex plane is divided up by several natural boundaries, and those which have no connected fundamental domain. So, said Klein, the idea of a discontinuous group must be supplemented by that of the fundamental domain before it becomes workable. He then devoted several paragraphs to examples where the fundamental domains are well understood: the regular solids, the parallelogram lattice for elliptic functions, and the circular arc triangles of Schwarz. In the last case he introduced considerations of non-Euclidean geometry as Poincaré had done. He went on to look at more general examples where the group preserves a fundamental circle, and showed how Riemann surfaces may be produced by drawing a certain fundamental polygon and identifying the sides. In §14 he showed that a bounded Riemann surface of genus p with n branch points may be produced by identifying the sides of a polygon of $4p + 2n$ sides. To each boundary cut correspond 2 sides A_i and A_1^{-1}, B_i and B_i^{-1}. There are 8 real choices to be made to specify A_1 and A_1^{-1}, B_1 and B_1^{-1}; 6 for each subsequent quadruple. The branch points each require two sides (which are identified by the group, the curve so obtained joins each branch point to the initial point 0 of A_1) so there are $2n$ choices to be made; thus there are seemingly $6p + 2n + 2$ choices in all. But the polygon must be closed, so three choices are lost, the position of 0 is arbitrary, losing two more, and η may be replaced by $\dfrac{\alpha\eta + \beta}{\gamma\eta + \delta}$ with real α, β, γ, δ, so the decomposition depends on $6p + 2n - 6$ real parameters.

Fairly general remarks followed, regarding the situation when there is no fundamental circle, and the process of continuously varying the parameters. This gave way to the fourth and penultimate section of the paper, on the so-called fundamental theorem. This theorem asserted that to every Riemann surface of genus p, with n branch points joined by lines ℓ_k to a common point, there corresponds one and only one normal function η which maps the polygon onto the dissected surface. The theorem, said Klein, had the *Grenzkreis* theorem and the *Ruckkehrschnitt* theorem (to give them their later names) as special cases. Its proof rested on a 1–1 correspondence between two manifolds, M_1 consisting of all Riemann surfaces of type (p, n, ℓ_k) and M_2 of all the normalized functions η. M_1 and M_2 have the same dimension, $6p + 2n - 6$. Klein argued that each element of M_2 can be associated to exactly one element of M_1. For, given η, one obtained a Riemann surface of a certain type, but any two which could be obtained could be deformed continuously into one another since the dissected surfaces can be mapped onto discs, and their images in the discs deformed continuously into one another. This was, he said, intuitively evident. Conversely, to each element of M_1 there corresponds a unique element of M_2, i.e., given a Riemann surface one can find a unique function η. For, Klein argued, suppose these were two: η and η^1. Then analytic continuation of the same fundamental region would yield two equal regions, which, together with any isolated singularities and their boundary points, may be mapped onto one another by a harmonic function; this in turn establishes a linear relationship between η and η^1. Finally Klein invoked a continuity argument to establish that the correspondence between M_1 and M_2 was analytic. He admitted that the arguments in this section constituted only grounds for a proof and not a rigorous proof, which is indeed a fair understatement of what still had to be done.

Klein concluded his paper by remarking on how the theory of elliptic functions fitted into this framework. The group in this case is commutative, and the transformation theory concerns the passage from η-functions for one modulus to η-functions invariant under a subgroup.

As Klein himself said, the price he had to pay for the work was extra-ordinarily high. In autumn 1882, while working on his "Neue Beiträge" his health broke down completely. A sad passage in his *Werke* [II, 258] records that he had to rest for a long time, and was never able to work again at the same high level; elsewhere he spoke of the centre of his productive thought being destroyed. The competitive element in his personality might be supposed to have driven him too hard, and much has been made of the 'race' between the two men, but the true story is clearly more complicated.

It was not simply, or even chiefly, a race. The generosity of the two men towards each other made it more of a cooperative effort. As has been made clear, doubts and insights are shared in the letters, even to the point of divulging private discussions with other mathematicians. Klein never withheld anything, and cannot be accused of delay in reporting his ideas. But, although Klein had a profound grasp of mathematics, he was not the innovator that Poincaré was, and did not bring to the quest for automorphic functions the same depth of vision that Poincaré did. As a contest it was unequal, but Klein was too fiercely ambitious not to feel it as a challenge.

In 1880 Klein was the dominant figure of his generation. Only Schwarz could rival him in Germany, there was no one to compare with him in France or England, and there can be no doubts about his ambition.

The magisterial tone of at least his early letters to Poincaré, the ceaseless carping about Fuchs, the displays of erudition betray a man eager to be seen as the leader of his profession. The older generation in Berlin were out of reach, but their likely successors (Fuchs, Schwarz, Frobenius) he saw as rivals as much as colleagues. The familiar picture of Klein in his later years as an autocrat who could not be contradicted merely confirms the portrait of the young man who wanted to know all mathematics, and who felt he must shout at the Berlin seminars to put across his point of view. And yet it is a fine achievement to accompany Poincaré for a year as he invented the theory of automorphic functions, so often reformulating old ideas in ways that ran counter to the tradition Klein had just struggled for some years to master. Klein felt called upon to try his utmost to match Poincaré's achievements, and failed cruelly in the attempt. The mathematical community today shows fewer signs than ever of resisting comparing men by comparing their work. Klein's ambition was sustained by a hierarchical mathematical community which a hundred years later generates similar pressures; his grievous misfortune was not his fault.

Poincaré's two papers straddled Klein's. The first, on Fuchsian groups, was finished in July and printed in September, the second, on Fuchsian functions, was finished in late October and published at the end of November 1882. The papers gave the first detailed account of the theory since the resume published in *Annalen*, but, despite their greater length (62 and 102 pages), they are more reticent.

Poincaré's papers of 1882

The first paper describes how a discontinuous group of transformations of the upper half-plane H, $z' = \dfrac{az+b}{cz+d}$ $(ad - bc = 1, a, b, c, d \in \mathbf{R})$ may be considered geometrically. If $\alpha, \beta \in H$, and the circle through them perpendicular to the real axis meets the real axis, $\mathbf{R}$, at h, k, then the cross-ratio $\dfrac{(\alpha - h)\,(\beta - k)}{(\alpha - k)\,(\beta - k)} = [\alpha, \beta]$ is preserved by all elements of the group and is determined by a and b. It is multiplicative in that if g lies on the same circle, then $[\alpha, \beta][\beta, \gamma] = [\alpha, \gamma]$, and, for infinitesimal points z and $z + dz$ in H, Poincaré showed that

$$[z, z + dz] = 1 + \frac{|dz|}{y},$$

neglecting higher terms. So $\log([\alpha, \beta])$ may be taken as defining the distance between α and β, whence the element of arc length is $ds = \dfrac{|dz|}{y}$ and the element of area is $dS = \dfrac{dx\,dy}{y^2}$, setting $z = x + iy$, $dz = dx + idy$. Poincaré observed that these definitions implied that the geometry thus introduced into the upper half plane was non-Euclidean, but decided not to employ that terminology in order to avoid any confusion. He called the group 'discontinuous' if no substitution in it could be found, say f_i, for which $\log[z, f_i(z)]$ was infinitesimally small, and he called a discontinuous group of real substitutions a Fuchsian group (§3).

To each Fuchsian group he assumed he could associate a region R_0 of H such that the transforms of R_0 by the elements of the group covered it exactly once with overlaps only on the boundaries of regions. Two regions with a piece of boundary in common he called limitrophic (*limitrophes*). An edge of a region was a piece of boundary in the form of an arc of a circle perpendicular to the real axis, i.e., a non-Euclidean line segment; two edges of one region were said to be conjugate if one edge could be mapped onto the other by an element of the group. The interior of R_0 was mapped by f_i, an element of the group, onto the interior of $f_i(R_0)$. R_0 was, by definition, a region containing only one point in the set $\{\dfrac{a_i z + b_i}{c_i z + d_i} = f_i(z)\}$ as f_i ran through the group, for each point $z \in H$. Poincaré noted (§4) that this constraint fell far short of defining R_0 uniquely, and argued that one could always find an R_0 which was in one piece and without a hole. For, if R_0 has a hole, it also has a second piece exterior to it, S_0, and a transformation, f_i, exists which maps S_0 into the hole, so R_0 can be replaced by $R_0 + f_i(S_0) - S_0$. Furthermore, the region R_0 may be taken to be bounded by edges and to form a convex region, upon suitably adding and subtracting pieces to R_0 along conjugate sides. Poincaré admitted polygonal regions for which segments of the real axis formed part of the boundary (he called such segments vertices). Poincaré called a convex region R_0 bounded by edges a normal polygon, and he claimed that given such a polygon and the pairing of conjugate sides the group was determined completely. As Nörlund noted [1916, 126], this is true unless an edge AB has both vertices on the real axis, when the polygon determines an infinity of groups.

Poincaré was thus led to 7 families of groups, depending on the nature of the vertices of R_0. A vertex was of the first kind if it lay strictly in H, of the second kind if it was a point of **R**, of the third kind if it was a segment of **R**. Accordingly, the seven kinds of region were obtained by insisting that either (1) all vertices were of the first kind, or (2) all were of the second kind, or (3) all of the third kind, or (4) all were of the second and third kinds, or (5) all of the first and third kinds, or (6) all of the first and second kinds, or (7) were of all kinds.

He said vertices z and z' corresponded if an element of the group formed a cycle. The cycle was of the first category if it was closed under the process of leaving a vertex, tracing an edge and then its conjugate edge until a new vertex is reached (the edges being directed in a standard way in advance), and so on, and all the vertices were of the first kind. Cycles of the second category are closed and only contain vertices of the second kind. Cycles of the third category contain vertices of the second or third kind, and are open. For R_0 to generate a Fuchsian group, he observed that the angle at any vertex of the first kind must be an aliquot part of 2π, and that corresponding sides must have the same length (as they evidently will). Conversely, he showed (§6) that these conditions are also sufficient for R_0 to generate a Fuchsian group. At this stage in his argument Poincaré assumed that certain regions R_0 could be found for which the set of all $f_i(R_0)$ did not cover all of H but was a proper subset, which is false.

He then gave several examples before turning to the computation of the genus of the surface obtained by identifying corresponding sides of R_0. He found (by Euler's formula) that if R_0 had $2n$ edges all of the first kind and q closed cycles of vertices, then the genus, p, was given by

$$p = \frac{n+1-q}{2}.$$

Similarly, if R_0 has n edges of the second kind, then Euler's formula implied that the genus was

$$p = n - q.$$

Finally, if R_0 was of the third kind, the genus was necessarily 2.

Poincaré discussed how the generating polygon might be simplified, gave more examples and observed that a Fuchsian group has as many fundamental relations between its fundamental substitutions (those which transform R_0 to a limitrophic region) as it has cycles of the first kind. He observed that a Fuchsian group is obtained by letting a discontinuous group of complex substitutions preserve a fixed circle, but if a discontinuous group has no fixed circle, it is not Fuchsian but Kleinian. Finally he gave some brief historical notes which indicate how much he had learned from Klein while still paying generous tribute to Fuchs.

In his second paper, on Fuchsian functions, Poincaré began by showing that,

if $f_i(z) = \dfrac{\alpha_i z + \beta_i}{\gamma_i z + \delta_i}$, then $\dfrac{df_i(z)}{dz} = \dfrac{1}{(\gamma_i z + \delta_i)^2}$, and $\displaystyle\sum_{i=0}^{\infty} \left| \dfrac{df_i}{dz} \right|^m$ converges for any integer m greater than 1. His z-domain is now the unit circle, and he showed that the series converges inside and outside the circle, and at those points which are not limit points of a sequence $\left(-\dfrac{\delta_i}{\gamma_i}\right)$, i.e., z belongs to an edge of the second kind (is interior to a segment of the boundary of the unit circle which is part of the boundary of R_0). He gave two proofs of this result; the second rests essentially on the observation that there are not many points $f_1(z)$ within any circle centre $z = 0$ and radius $R < 1$. Furthermore, he said, if the parameters determining R_0 are varied without the angle sum changing, then R_0 in it changed form generate isomorphic groups and the series $\sum_i \left(\dfrac{df_i}{dz}\right)^m$ depends continuously on these parameters.

He defined a theta-Fuchsian function by the formula

$$\theta(z) = \sum H\left(\frac{\alpha_i z + \beta_i}{\gamma_i z + \delta_i}\right)(\gamma_i z + \delta_i)^{-2m}$$

where $H(z)$ is an arbitrary rational function of z having no pole on the fundamental circle, and m is an integer greater than 1. If H has poles at $a_1, \ldots, a_p$ inside the unit circle, then $\theta(z)$ is infinite at all the points $\dfrac{\alpha_i a_k + \beta_i}{\gamma_i a_k + \delta_i}$, but the series converges everywhere else. In Section V of the paper Poincaré investigated whether H could be chosen such that $\theta(z)$ vanished identically, and gave examples to show that this could happen. E.g., (no. 4) if α_r is a vertex of the polygon R_0 and α'_r its image in the fundamental circle, then the Fuchsian group contains the substitution $\left(\left(\dfrac{z - \alpha_r}{z - \alpha'_r}\right)\right.$, $e^{2\pi i/\beta_r}\left(\dfrac{z - \alpha_r}{z - \alpha'_r}\right)\right)$ for some integer β_r. If $H(z) := \left(\dfrac{z - \alpha_r}{z - \alpha'_r}\right)^p \dfrac{1}{(z - \alpha'_r)^{2m}}$ and $p + m \equiv 0 \bmod \beta_r$ then $\theta(z) = 0$ for all z. For the function θ may be written as $\theta(z) = \sum_i \dfrac{(f_i - \alpha_i)p}{(f_1 - \alpha'_r)^{p+2m}}\left(\dfrac{df_i}{dz}\right)^m$ and if the summation is taken first over the rotations around α_r and then over each coset the factor $e^{2n(p+m)\pi i/\beta_r}$ appears, which is identically zero. Poincaré showed how one example automatically generates infinitely many more, but gave no characterisation of the H for which θ vanishes.

The bulk of the paper was given over to establishing the theorem that every theta-Fuchsian can be written as $\left(\dfrac{dx}{dz}\right)^m F(x, y)$, where F is a rational function and x and y are two Fuchsian functions in terms of which every other Fuchsian function can be written rationally and between which there exists an algebraic relation $\Psi(x, y) = 0$. Conversely, every function of the form $\left(\dfrac{dx}{dz}\right)^m F(x, y)$ can be written as a theta-Fuchsian function provided it vanishes whenever z is a vertex of R_0 which lies on the fundamental circle. The proof distinguished between finite and infinite R_0 and the cases where the genus does and does not equal zero. It follows from the theorem that every Fuchsian function can be written infinitely many ways as a quotient of two theta-Fuchsian functions. Furthermore, the theta-

Fuchsian functions which do not have poles in the fundamental circle form a finite dimensional space, and Poincaré showed how various linear relationships between such functions might be found. The dimension is $q = (2m - 1)(n - 1)$, where m is as above and $2n$ is a number of sides of R_0.

6.6 Poincaré's papers of 1883 and 1884

Mention should be made of Poincaré's other three big papers in *Acta Mathematica*: [1883] on Kleinian groups, [1884a] on the groups associated to linear differential equations, and [1884b] on Zeta-Fuchsian functions. The difficulties involved in studying Kleinian groups and Kleinian functions are much greater than those of the Fuchsian case, and are still not properly resolved today. Poincaré does little more than indicate how much of the analogy goes over, and how three-dimensional non-Euclidean geometry might help. The papers on differential equations and Zeta-Fuchsians are more substantial. Poincaré raised two questions: given a linear equation with algebraic coefficients, find its monodromy group; and, given a second-order linear equation containing accessory parameters, choose them in such a way that the group is Fuchsian. His conclusions, based on the method of continuity (which he admitted was not at all obviously true) were that every equation

$$\frac{d^2v}{dx^2} = \phi(x, y)v.$$

where $\theta(x, y) = 0$, ϕ and θ are rational in x and y, and the exponent differences of the solutions at the singular points are zero or aliquot parts of unit is such that, if z is a quotient of two independent solutions, then $x := x(z)$

(i) will be a Fuchsian function existing only inside a circle for exactly one choice of the accessory parameters;

(ii) will be a Fuchsian or Kleinian function existing for all z for exactly one choice of the accessory parameters;

(iii) will be a Kleinian function existing only in a subregion of $\mathbf{C}$ for infinitely many choices of the accessory parameters.

To give a little more detail, in his [1884a], Poincaré set himself two tasks: given a linear ordinary differential equation, to determine its monodromy group; and given a 2nd order linear ordinary differential equation depending on certain arbitrary parameters, to determine the parameters so that the group is Fuchsian. To address the first problem, he took

$$\frac{d^p v}{dx^p} + \phi_{p-1}(x, y)\frac{d^{p-1}v}{dx^{p-1}} + \cdots + \phi_0(x, y)v = 0 \tag{59}$$

a pth order linear ordinary differential equation for v as a function of x, whose coefficients ϕ are rational functions of x and y, where x and y satisfy an algebraic equation ψ of genus q. The monodromy group G, which he called the group of the equation, is represented by $p \times p$ matrices, and since, without loss of generality $\phi_{p-1} = 0$, Poincaré took these matrices to be of determinant 1. Each monodromy matrix depends on a choice of basis for the solutions of equation (1), so they are determined only up to conjugacy (Poincaré called $g^{-1}Gg$ a *transformé* of the group G); and he considered a group and a conjugate of it to be the same.

If the number of fundamental transformations or generators of G is n, the corresponding matrices have np^2 coefficients, of which $p^2 - 1$ in each matrix are independent. A further $p^2 - 1$ must be subtracted to allow for conjugacy, so one is left with $(n - 1)(p^2 - 1)$ independent coefficients. Poincaré noted that numerical methods for finding the invariants of the monodromy group (essentially, in language he did not use, the eigenvalues and related quantities of the monodromy matrices) are due to Fuchs and Hamburger. He proposed to find properties of the invariants as functions of the coefficients. He wrote the differential equation as

$$\frac{d^p y}{dx^p} = \sum_{k=0}^{p-2} \sum_{i-1}^{n} \sum_h \frac{A_{hki}}{(x - a_i)^h} \frac{d^k v}{dx^k}.$$

He first showed that the invariants are entire functions of the A_{hki}, when the singular points a_i are fixed. His proof uses Kovalevskaya's Theorem, which he had extended in his thesis. It follows that the invariants can be expanded as power series in the A's. Any coefficient in such an expansion will be a function of the a's, and Poincaré next showed that these functions are obtainable by simple quadratures.

Poincaré now turned to the second of his problems and asked conversely about the coefficients of the differential equation as functions of the invariants. He found it necessary to distinguish real and apparent singular points; let there be n real and r merely apparent singular points. In the differential equation there are $\frac{1}{2}n(p(p + 1)) + r - \frac{1}{2}p(p - 1)$ independent parameters, but one can always assume that $a_1 = 0$ and $a_2 = 1$. Moreover, a singular point is merely apparent if $\frac{1}{2}(p+2)(p+1)$ conditions hold. So if there are no apparent singular points, then there are $\frac{1}{2}n(p(p+ 1) - \frac{1}{2}p(p - 1) - 2$ parameters, whereas the group of the equation depends on $(n - 1)(p^2 - 1)$ parameters.

If $p = 2$ these numbers are equal (to $3n - 3$). In this case given a group, there is a differential equation without apparent singular points and having the given group as its monodromy group. But if $p > 2, n > 1$, then

$$\frac{1}{2}n(p(p + 1)) - \frac{1}{2}p(p - 1) - 2 < (n - 1)(p^2 - 1).$$

So apparent singular points will generally be needed in any differential equation having a given group as a monodromy group. To avoid going into too much detail, Poincaré announced that he would therefore restrict his attention to the 2nd order case. He therefore stated the problem in these terms: Given a differential equation $\frac{d^2 v}{dx^2} = \phi(x, y)v$, where as before x and y satisfy an algebraic equation $\psi(x, y) =$

0 of genus q, to determine the coefficients of ϕ so that x is a Fuchsian function of a quotient of two solutions of the differential equation (and the equation, said Poincaré, may be called Fuchsian for brevity).

To solve this problem, Poincaré insisted that at the singular points of the differential equation, the indicial equation has roots that are either 0 or of the form $1/k$ for integers k. When this was done, Poincaré called the differential equation normal. It might be that at a singular point either the function y was no longer holomorphic, or x or y became infinite, but this could be avoided by a suitable rational change of variable. More sweepingly, Poincaré proposed to regard a differential equation and any of its forms under a birational transformation as equivalent, provided the indicial equations at the corresponding singular points were the same, and said that such equations belonged to the same type. He then distinguished between Fuchsian, elliptic, and rational types by the angle sum of corresponding polygon. Next, he observed that by the earlier argument, the function ϕ depends on as many parameters as are needed for the differential equation to be Fuchsian; but the dependence is transcendental and an argument by counting constants is not legitimate: "This proof is at the heart of the problem we are concerned with" (§5). But if it could be shown that every Fuchsian type contains a Fuchsian equation, then it would follow that given any linear ordinary differential equation with algebraic coefficients, the variable and the solutions can be expressed as single-valued functions of the same auxiliary variable; and any algebraic curve can be uniformised.

Poincaré proceeded by defining the subordination of types. The differential equation

$$(1) \quad \frac{d^2v}{dx^2} + v\phi_0(x, y) = 0,$$

with p singular points as above (at a_i exponent difference is $1/k_i$) has the equation

$$(1') \frac{d^2v}{dx^2} + v\phi_1(x, y) = 0,$$

as a subordinate if $(1')$ has extra singular points with suitable exponent differences: at $(a_i, b_i)\alpha_i = 1/N_i$ where $k_i | N_i$, and if $k_i = \infty$ then $N_i = \infty$.

The sought-after results will follow if among the types subordinate to a Fuchsian type there is always one that contains a Fuchsian equation.

It was easy enough for Poincaré to show (§7) that there is at most one Fuchsian equation in each Fuchsian type. As he said, this result had been claimed earlier, by Poincaré himself, *Comptes Rendus* 93, 17 Oct 1881, p. 582, [1881], and by Klein [1881 p. 209] (sic, 1883 is correct), who extended it to domains bounded by infinitely many circles. But to prove existence was much harder, and to accomplish this Poincaré invoked what he called the Method of Continuity.

He said that types of differential equation belong to the same class when the number of singular points and the roots of indicial equation are the same, and they are defined over algebraic curves of the same genus. Similarly, two Fuchsian groups (in the same family) were said to be of the same class when their generating polygons have same number of sides and the edges are grouped in the same cycles. He then fixed a class of differential equations and a class of Fuchsian groups. To each type of differential equation, he associated a point of a manifold S'. To each Fuchsian group of the given class, he associated a point of a manifold S. To each group s there is plainly a corresponding type of equation s', and distinct groups give rise to equations of different types. The problem is to show that each point of S' corresponds to a point of S. This would be immediate if the manifold S was closed — "but this is not at all evident a priori", indeed "It is a difficulty that cannot be overcome in a few lines".

To overcome this profound difficulty, Poincaré proposed to take a case where there was a Fuchsian differential equation and to try and deform its polygon until the corresponding differential equation was equal to a given one. He analysed what happened under this deformation process, and tried to determine all the ways in which it could fail. Thus, given a polygon R_0 whose data depend continuously on a parameter t, there is a Fuchsian equation with respect to R_0 and it is of some type, say T. The parameters of the type are the $3p - 3$ moduli of the Riemann surface and the singular points of the differential equation. However, the functions formed by means of R_0 are continuous in the parameter t, so the corresponding Fuchsian functions depend continuously on t; therefore so do the parameters of type T. It may be that as the parameter t varies, vertices of R_0 go the boundary of the unit disc, and sides vanish. Indeed this is why the idea of subordinate types was introduced.

To see what can happen as a polygon degenerates (from R_0 to R_0'), Poincaré supposed that just two sides shrink to zero (they are necessarily paired by the action of the corresponding Fuchsian group). Now either the differential equation is defined on a curve of genus one less than before but the number of singular points has gone up by two, or it has one fewer singular point. In the latter case, the algebraic curves have the same moduli, and the singular points of the original differential equation are moved onto those of the new one with exactly two coinciding.

Poincaré proved this by looking at the corresponding theta series, but he then observed that when the polygon is symmetrical about an axis, there is a direct proof. He began with a special case, where the differential equation is $\dfrac{d^2 v}{dx^2} = \phi(x)v$, where $\phi(x)$ is a rational function and all the singular points are real, corresponding to the indicial equation having repeated roots. (The corresponding polygon is necessarily symmetric). He now claimed that each symmetric type contains a Fuchsian equation. If there are just three singular points the result is classical, and the differential equation is Legendre's equation for k^2.

If there are 4 singular points, $\{0, 1, a, \infty\}$, the continuity argument is clear. Consider 4 points on the Unit Circle, $\alpha, \beta, \gamma, \delta$, and the Fuchsian function $x = f(z)$ for which $f(\alpha) = 0$, $f(\beta) = 1$, and $f(\delta) = \infty$. Suppose $f(\gamma) = c$. Then the inverse function $z = z(x)$ will be a quotient of two solutions of the Fuchsian equation $\dfrac{d^2 v}{dx^2} = \phi(x)v$, where the singular points are $0, 1, c, \infty$ and at each of these the indicial equation has a double root. Now, as γ goes from β to δ, the value of c goes from near 1 (its value when $\gamma = \beta$) to ∞, its value when $\gamma = \delta$. It cannot take the same value twice, so it takes every value between 1 and ∞ once, in particular the value a.

When there are 5 singular points, $\{0, 1, a, b, \infty\}$, the continuity argument is less clear, but, said Poincaré, it can be pushed through, and so every symmetric Fuchsian type contains a Fuchsian equation.

When there is no symmetry, the situation is much more complicated, and the reader is advised to consult Stillwell's commentary at this point. However, Poincaré now claimed that in all rigour the method of continuity showed that it was possible to deform the polygons within a given type to get from a Fuchsian case to any prescribed differential equation because anything that could go wrong required at least two real parameters to have suitable values, and such conditions could not separate regions of a space of dimension greater than 3. Hence every Fuchsian type contains a Fuchsian differential equation.

Poincaré next considered the possibilities for subgroups of Fuchsian groups. There is a close relation between Fuchsian functions with respect to a given group and Fuchsian functions with respect to a normal subgroup of finite index. The finite index condition ensures that the subgroup is again a Fuchsian group (its fundamental region is a union of copies of a finite polygon). If it is a normal subgroup, then the differential equation is solvable algebraically in terms of the Fuchsian functions associated to the larger group.

In §§17–19, Poincaré calculated, to arbitrary precision, the coefficients of the differential equation in the symmetric case. This involves summing series akin to theta-series; the convergence is delicate and was to be discussed by later writers.

In conclusion, Poincaré observed that he had shown that the solutions of a linear ordinary differential equation are single-valued functions of a uniformising parameter z, and promised that in his next paper he would study the properties of these single-valued functions and expand them in series.

In this next paper, [1884b] the final one of this remarkable series, Poincaré took a pth order linear ordinary differential equation for v as a function of x, whose coefficients ϕ are rational functions of x and y, where x and y satisfy an algebraic equation ψ. He wrote the equation in the form

$$(1) \quad u'' = \phi(x, y)v,$$

where ϕ is a rational function of x and y, which are connected by an algebraic equation

$$(2) \quad \psi(x, y) = 0 \text{ of genus } q.$$

He proposed to compare it with an auxiliary equation

$$(3) \quad w'' = \phi(x, y)w,$$

where ϕ is a rational function of x and y and the singular points of equation (1) (call them $S(1)$) are a subset of those of equation (3) (call them $S(3)$). The conditions on an auxiliary equation involve the indicial equations.

At a point a in $S(1)$ there is an indicial equation, E, of degree p, and because the point a is also in $S(3)$ there is a second indicial equation, E' of degree 2. Let difference of roots of E' be δ. Poincaré required:

that if ϕ is such that $\delta = 0$ or $\delta = 1/k$, for some k, then all roots of E' are multiples of δ;

that if in a neighbourhood of a the solutions of (1) are irregular, then $\delta = 0$, and

that for the points in $S(3) - S(1)$, either $\delta = 0$ or $\delta = 1/k$, for some k, and all solutions of (1) are regular.

This information does not determine θ. It is possible to choose $S(3) - S(1)$ arbitrarily, and the indicial equation for all points of $S(3)$, and θ will still have some arbitrary parameters. Poincaré now supposed that (3) is such that these parameters can be chosen uniquely such that x and y are Fuchsian functions of z defined only within unit circle (it follows that the solutions are single-valued functions of z, and the polygon belongs to Families 1, 2, or 6). Then the solutions of (1) are quotients of two convergent series in z defined for all $|z| < 1$.

When $\delta = 0$ for all points in $S(3)$, and therefore x, y are Fuchsian functions of the 2nd family, the solutions of (1), like x and y, are holomorphic functions of z in the unit disc, and the coefficients can be found by recurrence. But, they satisfy no simple law. So, said Poincaré, it is better to find solutions as series with terms obeying a simple rule, as was done with the theta-Fuchsian functions. This is done here, under the assumption that (1) has all solutions regular in the sense of Fuchs, Frobenius, and Thomé.

Poincaré now classified differential equations. Two pth order linear ordinary differential equations

$$\frac{d^p v}{dx^p} + \sum_k \phi_k(x, y)\frac{d^k v}{dx^k} = 0$$

and

$$\frac{d^p u}{dx^p} + \sum_k \phi'_k(x, y)\frac{d^k u}{dx^k} = 0$$

for v and u as a functions of x, whose coefficients ϕ and ϕ' are rational functions of x and y, (where x and y satisfy an algebraic equation (2) $\psi(x, y) = 0$) were, he said, of the same family if: solutions of (1) are of the form $\Lambda \cdot \left(F_0 v + F_1\frac{dv}{dx} + \cdots + F_{p-1}\frac{d^{p-1}y}{dx^{p-1}} \right)$, where the F's are rational functions of x and y and Λ is an arbitrary function of x and y. They are of the same type (*espèce*) if $\Lambda = 1$.

If (1) and (1′) are of the same type they have the same monodromy group.
If (1) and (1′) are of the same family they do not have the same monodromy group, but the corresponding monodromy matrices differ by the same factor. So the ratios of the solutions are the same. (We could say they have the same projective monodromy group).

Poincaré now considered the function $u' = u/\Lambda$. It satisfies a differential equation (1″) of the same form as (1), with coefficients ϕ''. The coefficient ϕ''_{p-1} is $p\Lambda'/\Lambda = \phi''_{p-1} - \phi'_{p-1}$, so the logarithmic derivative of Λ is a rational function in x and y, which makes it analogous to doubly periodic functions of the 2nd kind. In particular, when the point (x, y) goes on a closed curve on the Riemann surface corresponding to (2) around a singular point, Λ is simply multiplied by a constant factor.

In general, a point $x = a$ can give rise to various kinds of singular behaviour in the solutions. They might be irregular, a case Poincaré proposed to ignore. A case-by-case examination of the implications for the roots of the indicial equations led Poincaré to conclude that one can always pass from one 2nd order differential equation to another in the same family by a series of operations of this form:

(a) multiplying the unknown function by a suitable factor;

(b) differentiating it;

(c) multiplying again by a suitable factor.

The difference of the roots of the indicial equation is preserved by (a) and (c), but (b) alters it, so apparent singular points can appear and disappear (unless the roots are 0, 1). But the parity of such points is unaltered.

In §3, Poincaré sketched the implications for the reduction of linear ordinary differential equations. The aim is to find the simplest in a family. In the second order case, when the differential equation is $\dfrac{d^2v}{dx^2} = \phi(x)v$, he let the singular points be $a_1, \ldots, a_k$, and the apparent singular point be $b_1 \ldots, b_h$ where the roots of the indicial equation are $-1/2$ and $3/2$. Two such differential equations were said to be of the same class if they have the same singular points a, including ∞, same indicial equations there, and the same parity of h.

A class contains infinitely many families, but one can pick out equations in the same family by some invariant functions of the coefficients.

The monodromy group of the differential equation depends on $3k$ parameters, from which it follows that the invariants depend on at most $2k - 4$ parameters. So the reduced (i.e., simplest) differential equation in a family depends on $2k - 4$ parameters. The least number of apparent singular points it can have is $k - 2$ if h and k are of the same parity, and $k - 1$ if not. In the former case there will be only finitely many such differential equations, and any can be chosen as the simplest; in the second case there will be infinitely many, but one can select one by insisting that 0 is an apparent singular point.

In §4 Poincaré took up the theme of zeta-Fuchsian functions and their role in solving every linear differential equation (at least for the Fuchsian families 1, 2, and 6). To motivate this work, he observed that just as not all elliptic integrals arise from inversion, which only gives the first kind, so one cannot solve all differential equation by Fuchsian functions, but only the Fuchsian ones, and in general zeta-Fuchsian functions are needed. The analogy Poincaré had in mind is with Jacobi's function, not, of course, with the Riemann zeta function. The elliptic integral of the first kind,

$$u = \int_0^x \frac{dt}{\sqrt{(1-t^2)(1-k^2t^2)}},$$ defines, by inversion, the elliptic function $x = sn(u)$ with periods $4K$ and $2K'i$ (see page 12 above). The elliptic functions $cn(u)$ and $dn(u)$ are defined by equations $cn(u) = \sqrt{1-x^2}$, $dn(u) = \sqrt{1-k^2x^2}$. The elliptic integral of the second kind $E(u) = \int_0^u dn^2(t)dt$, and since $u = K$ when $x = 1$, the complete elliptic integral of the second kind is $E = E(K) = \int_0^K dn^2(t)dt$. Jacobi (*Fundamenta Nova*, §47) defined the zeta function $Z(u)$ by the equation $Z(u) = -\frac{E}{K}u + E(u)$. From its expression as a power series, it follows that $Z(u)$, and therefore $E(u)$ are single-valued functions of u. Now, $Z(u+2K) = Z(u)$, and $Z(u+2K'i) = Z(u) - \frac{\pi i}{K}$, because $E(u+2K) = E(u) + 2E$, and $E(u+2K'i) = E(u) + \frac{1}{K}(2EK'i - i\pi)$. Introducing the other complete elliptic integral of the second kind, E', which is the same function of k' that E is of k, and recalling the Legendre relation $2(E'K + EK' - KK') = \pi$, we deduce that $E(u+2K'i) = E(u) + 2i(K' - E')$. So $E(u)$ is well-behaved as a function of u, although as a function of x it does not have a single-valued inverse function. Since the elliptic integrals of the second kind satisfy a linear ordinary differential equation, this illustrates how the zeta function and its analogues will necessarily arise in the programme of solving every type of linear differential equation.

Poincaré next supposed that the Fuchsian group g acted linearly on a set of p functions $Z_1, \ldots, Z_p$ defined only in the unit circle. This makes them, by definition, a set of zeta-Fuchsian functions. It also gives rise to a representation of g as G, a zeta-Fuchsian group. He let $x = f(z)$, $y = f_1(z)$ be Fuchsian functions, and $Z_1, \ldots Z_p$ be zeta-Fuchsian functions (with respect to g and G). Then given $q+1$ Fuchsian functions $F_0, F_1, \ldots F_q$ admitting g, the system

$$F_0 Z_i + F_1 \frac{dZ_i}{dz} + \cdots + F_q \frac{d^q Z_i}{dz^q}$$

is a zeta-Fuchsian system.

Conversely, the most general zeta-Fuchsian system (for g) satisfies a linear ordinary differential equation with rational coefficients in x and y of the same type. Then, in §5, for the first family, Poincaré showed that every zeta-Fuchsian function of the form

$$\sum_i [H_\mu(zS_i)] S_i^{-1} \left[\frac{d(zS_i)}{dz} \right]^m$$

where S_i runs through the Fuchsian group, can be written a theta-Fuchsian series. The absolute convergence of such a series was only assured for values of m sufficiently large. The remaining cases were more complicated and I omit them. Note, however, that at this point Poincaré restated the conclusion of two notes which he had published the year before (Poincaré [1883a,b]). Because any finite group can be regarded as a Fuchsian group of the first family, the convergence properties of the Fuchsian functions he was studying were established whenever the monodromy group of the differential equation was finite. This amounted to a solution of the

Riemann problem in this case, and indeed for any group for which the convergence was assured. This was a large class of groups.

At the end of the paper, Poincaré observed that one could extend the theory to zeta-Fuchsian functions of the remaining families, and indeed embark on a study of zeta-Kleinian functions. But he contented himself with the reasonable hope that "This will suffice to make it clear that, in the five memoirs of *Acta Mathematica* which I have dedicated to the study of Fuchsian and Kleinian transcendents, I have only skimmed a vast subject which without doubt will furnish geometers with the occasion for numerous important discoveries".

In fact, subsequent developments were by no means as rapid as Poincaré's own. A number of authors showed that, at least in certain cases, the convergence arguments could be extended to the case where $m = 2$ (Weber [1886], Schottky [1887], Burnside [1891, 1892], Ritter [1892], Lindemann [1899]). This gave those parts of the theory a closer analogy to the theory of Abelian functions and abelian integrals of the usual three kinds. Perhaps the most thorough response, even though it added nothing new, was that provided by Forsyth in the course of his 6-volume treatise on Differential Equations. In view of the real difficulties in Poincaré's account, it may not be out of place to summarise Forsyth's response here.

In his *Theory of Differential Equations* (Part III, Vol. IV, Ch X 1902) Forsyth formulated the problem of uniformisation this way: given an algebraic function $\psi(x, y) = 0$, represent x and y as single-valued functions of a complex variable z. He began by observing that a complex function $x = f(z)$ may always be regarded as providing a conformal representation of one region upon another. He took as an example the function $x = f(z)$ regarded as a conformal map of the upper half z-plane onto a circular-arc polygon where the angle at vertex a_i is $a_i\pi$, and supposed for simplicity that all singular points (denoted a_i) were real. To allow for harmless transformations of the z-sphere by Möbius transformations, he invoked the theory of the Schwarzian derivative $\{z, x\}$ to obtain the differential equation

$$\{z, x\} = 2F(x),$$

where

$$2F(x) = 1/2 \sum_i \frac{1 - \alpha_i^2}{(x - \alpha_i)^2} + \sum_i \frac{A_i}{x - a_i}$$

and all the A_i are real. The A_i must satisfy certain equations if ∞ is to be an ordinary point, and other equations if ∞ is a singular point, but in either case the number of coefficients is sufficient. In the case of an ordinary point, there are $3m - 6$ coefficients in the differential equation and $3m - 6$ for the polygon. When ∞ is to be a singular point, there are $3m - 3$ coefficients in each case.

The function provides a circular-arc polygon, but it may not be the fundamental domain of an automorphic function. For simplicity, Forsyth considered a polygon of the first family, with $2n$ edges in n conjugate pairs and q cycles. The genus of the corresponding group is then p, where $2p = n + 1 - q$. The birational equivalence class of the surface therefore depends on $3p - 3 + q$ complex parameters.

To proceed further, Forsyth then considered the approach via a differential equation of the form $v'' + Rv = 0$, where $R = R(x, y)$ is a rational function of x and y so chosen that z is a quotient of two linearly independent solutions of the differential equation. When this is done, it may be that x and y are single-valued functions of z, but not automorphic functions. For that, extra conditions on the coefficients of the rational function R are needed. These can be obtained by considering what happens at the singular points of the differential equation.

If $(x, y) = (a, b)$ is a singular point of the differential equation, and $\psi(a, b) = 0$, suppose that the indicial equation at the point a is $n(n - 1) + \rho = 0$, with roots m and n. If $m \neq n$, then the corresponding angle of the polygon is $\alpha\pi$, where $\alpha = n - m$. So $\rho = \frac{1}{4}(1 - \alpha^2)$ and near the point $a R$ can be expanded as

$1/4\dfrac{1 - \alpha^2}{(x - a)^2}$ + higher powers of $x - a$. When $\alpha = 0$ (and so $\rho = \frac{1}{4}$) the expansion

of begins with $1/4\dfrac{1 - \alpha^2}{(x - a)^2}$, so there is a condition on R for each singular point. A more complicated analysis is needed if (a, b) is a singular point for the curve $\psi(x, y) = 0$, especially if the solutions are to be irregular there. This can force $z = \infty$.

What holds for a second-order differential equation holds in general. If all solutions are regular, and the exponent differences are commensurable and therefore all multiples of $1/k$, then x, y and the solutions of the differential equation are single-valued functions of z near $z = 0$. If the indicial equation has roots differing by an integer and the solutions involve logarithmic terms, then exponentials arise and $z = \infty$ is an essential singular point of the function $x = f(z)$. Finally, if some solutions are irregular, then $z = \infty$ is again an essential singular point of the function $x = f(z)$.

In each case this information yields conditions on the leading coefficient in some expansion of R, but the remaining coefficients are still undetermined and must be adjusted to make x a Fuchsian function. For example, suppose $\psi(x, y) = 0$ is the elliptic curve

$$y^2 = x(1 - x)(1 - cx), \quad c = 1/a < 1.$$

The singular points are $\{0, 1, a, \infty\}$. At each of them, $\alpha = \frac{1}{2}$. So $R(x, y)$, which is to be a rational function of x alone, is of the form

$$\frac{3}{16}\left[\frac{1}{x^2} + \frac{1}{(x - 1)^2} + \frac{1}{(x - a)^2}\right] + \frac{A}{x} + \frac{B}{x - 1} + \frac{C}{x - a}.$$

The conditions at ∞ are

$$A + B + C = 0$$

and

$$\frac{9}{16} + B + Ca = \frac{3}{16}.$$

So

$$R(x) = \frac{3}{16}\left[\frac{1}{x^2} + \frac{1}{(x - 1)^2} + \frac{1}{(x - a)^2}\right] - \frac{\frac{3}{8}x + \lambda}{x(x - 1)(x - a)}$$

which has only 1 undetermined coefficient, λ. To make x a Fuchsian function of z, note that the condition $\alpha = \frac{1}{2}$ means that the image of the upper half plane is a four-sided figure with angle sum 2π, so the figure is a rectangle. The edges are conjugate in pairs, the vertices form a single cycle, and the familiar diagram of a torus appears before us (but not before Forsyth, who checked that $2p = 2 + 1 - 1$, so $p = 1$). Moreover, the sides cannot be curvilinear, because that would force the angle sum to be less than 2π, so the figure is a Euclidean rectangle, and the Fuchsian functions are elliptic functions.

Taking $x = sn^2z$, $y = snzcnzdnz$, Forsyth obtained the explicit value for λ of $-\frac{1}{8}(a + 1)$, and computed everything explicitly.

In the same thorough-going way, Forsyth then treated the cases where the differential equation has 1, 2, or 3 singular points. In the last case, he found that the hypergeometric equation gave rise to Fuchsian functions when it was the equation for the quarter-periods in elliptic functions:

$$x(1 - x)\frac{d^2w}{dx^2} + (1 - 2x)\frac{dw}{dx} - \frac{1}{4}w = 0.$$

Forsyth then showed how to let the singular values of x be complex, and moved towards a statement of the main existence theorems without proof, noting, for example, that even when the relevant equations were algebraic it was not enough to count constants. Thus he concluded: For a differential equation for which x, y are Fuchsian functions of z, either all roots of indicial equation are commensurable, and the circular-arc polygon is finite, or the roots must be equal and the vertices go off to ∞ (as with the modular equation). The uniformising function can be built out of Fuchsian functions and zeta-Fuchsian functions, but the convergence properties of these are poor.

The person who followed Poincaré's approach to differential equations most directly was Fuchs' pupil, Ludwig Schlesinger, whose monumental *Handbuch* continues to inform mathematicians. He observed that Poincaré's solution to the Riemann problem was valid only under conditions on the exponents that can arise, and sought to establish the result in general by a continuity argument. This was at least plausible, because Poincaré had shown that the coefficients of its monodromy group are determined as transcendental functions of the coefficients of the equation. However, at the end of a series of papers giving a new defense of the method of continuity, a rather acrimonious dispute blew up between Plemelj and Schlesinger.

In a footnote in his [1908] on the solution of the Riemann–Hilbert problem, Plemelj observed that Schlesinger's many papers solving the Riemann–Hilbert problem by appealing to the method of continuity were "entirely unsuccessful". He gave no reason for this judgement, but addressed the point directly in his [1909]. He described Schlesinger's approach in these terms: given an nth order linear ordinary differential equation, the Riemann–Hilbert problem asks for the coefficients of the equation as functions of the coefficients of its monodromy group. However, in Plemelj's opinion, this could only be done when further restrictions were imposed on the monodromy group. For, the coefficients of the differential equation involve

N parameters as do the coefficients of the monodromy, but, and here is the nub of the difficulty, the coefficients in each case are algebraic functions of the parameters. There is no 1–1 correspondence between the coefficients and the parameters, and the continuity method is too simple to establish the required map between them.

He argued that the monodromy group only determines the roots of the indicial equation up to an integer, but that it can be impossible to find a system of first-order linear ordinary differential equations with particular assignments of values to the exponents. The example he gave was obtained from the hypergeometric equation by means of the substitution $z = \zeta^2$. It has four singular points (at $\zeta = 0, -1, 1$, and ∞) and the exponents satisfy the Fuchsian relation (that their sum is zero). But, he showed, there is no system which has the same exponents as this one except at the point $\zeta = 0$, where one exponent has been increased by 1 and the other lowered by 1.

Schlesinger's reply, which took the form of a letter to the editor published immediately after Plemelj's note, was rather lofty. He did allow that his Main Theorem could have been more precisely stated, and he proceeded to offer such an improvement, but he did not accept Plemelj's example. Rather, he said, this belonged to the class of singular cases that posed no obstacle to the method of continuity because this class formed a sub-manifold of cases of codimension at least 2.

Schlesinger's confidence was not universally shared. In his EMW article, concluded in December 1913, Hilb found that the continuity method had come up against difficulties that it had not always been able to overcome, and he put his trust instead in Plemelj's paper of 1908, which he said gave a complete proof. This assessment was influential, and later authors suggested that Plemelj was in the right: the methods of Poincaré and Schlesinger were not rigorous, but those of the new man were. Ironically, it has been put to me that the modern theory of moduli spaces should permit the continuity proof to be put on rigorous terms. It may even be that Schlesinger himself came to share that assessment. In the 3rd edition of his book on ordinary differential equations, [1922, p. 302 n.3], he referred to four solutions of the Riemann–Hilbert problem in the general case: his own [1906], and those due to Hilbert [1905], Plemelj [1908] and Birkhoff [1913a, b]. In fact, with certain notable exceptions, the study of automorphic functions and related differential equations after Poincaré left the field emphasised much more the theory of functions and much less the theory of differential equations. This may reflect the continuing involvement of Klein, manifested through the work of Fricke, and the more tenuous French connection to Poincaré himself. Noteworthy in this endeavour was the work of Ernst Ritter, who sought to extend Klein's work on automorphic functions in the 1890s until it recaptured Poincaré's ideas in the altered conceptual setting of geometric function theory. He wrote a series of profound papers on the subject before his early death at Ellis Island from typhus which he had contracted on the journey across the Atlantic. These were later taken up and simplified by Fricke in the two volumes of Fricke–Klein. In the same way, Fubini's book [1908], which set the subject before an Italian audience, dealt with Fuchsian and Kleinan groups and their automorphic functions, but endorsed the contemporary preference for (complex) function theory over the theory of differential equations.[17]

Conclusion

With Poincaré's papers a certain process was completed. Gauss' insights into the
hypergeometric equation, the theory of modular transformations, the centrality of
the theory of functions of a complex variable, are here combined and in that way
raised to a new level. It would be possible to insist on the neatness of this synthesis,
to stress the unity underlying the apparently varied techniques of non-Euclidean geometry, group theory, and ordinary differential equations. But it is surely preferable
to close by stressing the fortuitous nature of these developments; it was after all the
generously inventive Poincaré and not the more learned Klein who achieved this
consummation. Tradition, even historical momentum, cannot guarantee progress.
The unities of mathematics naturally reflect the unities of a mathematician's training, the shared frameworks within which questions are posed. But they are main-
tained and reformed by the original perceptions of fortunate mathematicians. They
are real, but fragile like all living things. They contain questions as well as answers, and so one task for historians is to capture this aspect of mathematical work.
Poincaré drew together questions asked by different mathematicians, and his answers showed that these concerns belonged together, but his strangely patchy education makes this achievement the more remarkable. Confronted with this fact, the
historian may also feel, and feel pleased, that there are things that cannot be fully
explained.

Appendix 1

Riemann, Schottky, and Schwarz on Conformal Representation

In his Thesis [1851], Riemann sought to prove that if a function u satisfies certain conditions on the boundary of a surface T, then it is the real part of a unique complex function which can be defined on the whole of T. He was then able to show that any two simply connected surfaces (other than the complex plane itself — Riemann was considering bounded surfaces) can be mapped conformally onto one another, and that the map is unique once the images of one boundary point and one interior point are specified; he claimed analogous results for any two surfaces of the same connectivity [§19]. To prove that any two simply connected surfaces are conformally equivalent, he observed that it is enough to take for one surface the unit disc $K = \{z : |z| \leq 1\}$ and he gave [§21] an account of how this result could be proved by means of what he later [1857c, 103] called Dirichlet's principle. He considered [§16]:

(i) the class of functions, λ, defined on a surface T and vanishing on the boundary of T, which are continuous, except at some isolated points of T, for which the integral

$$L = \int_T \left(\left(\frac{\partial \lambda}{\partial x} \right)^2 + \left(\frac{\partial \lambda}{\partial y} \right)^2 \right) dt$$

is finite; and

(ii) functions $\alpha + \lambda = \omega$, say, satisfying

$$\int_T \left(\frac{\partial \omega}{\partial x} - \frac{\partial \beta}{\partial y} \right)^2 + \left(\frac{\partial \omega}{\partial y} + \frac{\partial \beta}{\partial x} \right)^2 dt = \Omega < \infty$$

for fixed but arbitrary continuous functions α and β.

He claimed that Ω and L vary continuously with varying λ but cannot be zero, and so Ω takes a minimum value for some ω. The claim that this value is attained for some λ in the first class of functions had earlier been made by Green [1883, first pub. 1835] and Gauss [1839/40] but it was to be questioned in another context by Weierstrass [1870], who showed by means of a counter-example that there is no general theorem of the kind: "a set of functions bounded below attains its bound". The use of Dirichlet's principle became contentious, and several mathematicians, notably Schwarz, sought to avoid it (Schwarz's work is described below). It was persuasively defended under slightly restricted conditions, by Hilbert [1900a] in a beautifully simple paper.

In his next major paper on complex function theory, *Theorie der Abel'schen Functionen* [1857c], published just after his paper on the hypergeometric equation, Riemann recapitulated most of the above analysis. A complex function ω of $z = x + iy$ is one which satisfies

$$i\frac{\partial \omega}{\partial x} = \frac{\partial \omega}{\partial y},$$

and hence can be written uniquely as a power series in $z-a$ where a is any point near which ω is continuous and single-valued. If ω is defined on a subset of the complex plane it may be continued analytically along strips of finite width. The continuation is unique at each stage, but if the path crosses itself the function may take different values on the overlap. If it does not, it is said to be single-valued or monodromic (*einwerthig*), otherwise multivalued (*mehrwerthig*). Multivalued functions possess branch points; e.g., $\log(z - a)$ has a branch point at a. The different determinations of the function according to the path of the continuation he now called its branches or leaves [*Zweige, Blättern*].

Riemann extended his analysis of surfaces to include those without boundary by the simple trick of calling an arbitrary point of an unbounded surface its boundary. He concluded the elementary part of this paper by repeating his argument, based on what he now explicitly called Dirichlet's principle, that is, that a complex function can be defined on a connected surface T, which, on the simply connected surface T' obtainable from T by boundary cuts,

(i) has arbitrary singularities, in the sense of becoming infinite at finitely many points in the manner of a rational function; and

(ii) has a real part which takes arbitrary values on the boundary.

The conformal representation of multiply connected regions was studied by Schottky [1877]. He considered the integrals of rational functions defined on such a region, A, and showed (§4) by looking at their periods that every single-valued function on A, which behaves like a rational function on the interior of A and is real and finite on the boundary of A, can be written as a rational function of functions like

$$u = \frac{1}{x - a} + u_0 + \sum_{i=1}^{\infty} u_i (x - a)^i$$

$$v = \frac{i}{x-a} + v_0 + \sum_{i=1}^{\infty} v_i(x-a)^i.$$

He defined the genus of A (§16) as Weierstrass had done (Schottky referred to Weierstrass' lectures of 1873–74), noted that it agreed with Riemann's definition and that genus was preserved by conformal maps. He concluded the paper (§16) by taking the special case when A is bounded by a circle L_0 which encloses $n - 1$ circles that are disjoint and do not enclose one another. This region has genus $n - 1$ and there is a group of conformal self-transformations of the disc bounded by L_0 obtained by inverting A in each $L_1, \ldots, L_{n-1}$ and then in the circle which bounds the images, and so on indefinitely. (The limit point set can be quite dramatic; Fricke in Fricke–Klein *Automorphe Functionen*, I, 104, gave an example when A is bounded by three circles and the limit set is a Cantor set). Such regions had already been considered by Riemann, and the fragmentary notes he left on the question edited into a brief coherent text by Weber (Riemann [1953f], and published in 1876). Like Schottky, Riemann showed that a conformal map of the region can always be found which maps the boundary circles onto the real axis and takes every value in the upper half plane, H, n times. Repeated inversion then produces a function, invariant under the group, and this function is the inverse of the quotient of a second-order differential equation with algebraic coefficients, that is, the differential equation is defined on the above covering of H. Conversely, a conformal representation of this covering on A is obtained by solving such a differential equation, provided the coefficients take conjugate imaginary values at conjugate points. Schottky's work is essentially his thesis of 1875, and consequently was done independently of Riemann.

Schwarz wrote several papers on the conformal representation of one region upon another between 1868–1870 which should be mentioned, although they do not immediately relate to differential equations.

In the first of them, [1869a], Schwarz remarked that Mertens, a fellow student of his in Weierstrass' 1863–64 class on analytic function theory, had observed to him that Riemann's conformal representation of a rectilineal triangle on a circle raised a problem: the precise determination of such a function seemed to be beyond one's power to analyse because of the discontinuities at the corners. Discontinuity at that time meant any form of singularity; in this case the map is not analytic. That had set Schwarz thinking, for he knew no examples of the conformal representation of prescribed regions, and in this paper be proposed to map a square onto a circle. He said [1869a, in *Abhandlungen*, II, 66]

> "For this and many other representation problems this fruitful theorem leads to the solution:

> "If, to a continuous succession of real values of a complex argument of an analytic function, there corresponds a continuous succession of real values of the function, then to any two conjugate values of the argument correspond conjugate values of the function".

This result is now called the Schwarz reflection principle. In symbols, it asserts that if f is analytic and $f(x)$ is real for real x, then $f(x + yi) = \overline{f(x - yi)}$. To represent the square on a circle by a function $t = f(u)$ Schwarz said it was simpler to replace the circle by a half-plane, which can be done by inversion ("transformation by reciprocal radii"). The singular points would be $t = \infty, -1, 0$, and 1 and the upper half plane is to correspond to the inside of the square. The boundary of the square will be mapped by t onto the real axis. As the variable u crosses a side of the square $t(u)$ will cross into the lower half plane, and by the reflection principle t is thus defined on the four squares adjacent to the original one; so on iterating this process t is seen to be a doubly periodic function defined on a square lattice, thus a lemniscatic function.

Indeed, Schwarz went on, the function $v = u^2$ converts the wedge-shaped region within an angle of $\pi/2$ to a half plane, and any analytic function with non-vanishing derivative at $v = 0$ will then produce a conformal copy of the wedge. One might speak of such a function straightening out the corner. But even so, this u is not sufficiently general, for an everywhere analytic map sending lines to lines, namely $u' = C_1 u + C_2$, where C_1 and C_2 are constants, is surely equivalent to it. To eliminate the constants Schwarz put forward the equation

$$\frac{d}{dt} \log \frac{du'}{dt} = \frac{d}{dt} \log \frac{du}{dt},$$

which he said was an important step, for

$$\frac{d}{dt} \log \frac{du}{dt}$$

is infinite whenever $\dfrac{du}{dt}$ is zero or infinite, which happens at each singular point. So

$$\frac{d}{dt} \log \frac{du}{dt} = -\frac{1}{2} t^{-1} + d_1 + d_2 t + \cdots,$$

and by considering the vertices $t = -1, 0$ and 1, Schwarz showed that the expression

$$\frac{d}{dt} \log \frac{du}{dt} + \frac{1}{2} \left(\frac{1}{t+1} + \frac{1}{t} + \frac{1}{t-1} \right) = *$$

is to be analytic for all finite t. The transformation $t' = \dfrac{1}{t}$ enabled him to consider t infinite, and it turned out

$$\frac{d}{dt} \log \frac{du}{dt}$$

was zero at infinity, so $*$ is constant and indeed zero, which implies that

$$\frac{d}{dt} \log \frac{du}{dt} = -\frac{1}{2} \left(\frac{1}{t+1} + \frac{1}{t} + \frac{1}{t-1} \right).$$

This led Schwarz to his result:

$$u = C_1 \int_0^t \frac{dt}{(4t(1-t^2))^{1/2}} + C_2.$$

To represent a regular n-gon on a circle, a simple generalization showed that

$$u = \int_0^s \frac{ds}{(1 - s^n)^{n/2}}$$

would suffice, a result which Schwarz said he had presented for his promotion at Berlin University in 1864. A continuity argument which Schwarz attributed to Weierstrass enabled him also to deal with an irregular n-gon.

To consider figures bounded by "the simplest curved lines", circles, Schwarz applied an inversion

$$u' = \frac{C_1 u + C_2}{C_3 u + C_4}$$

where C_1, C_2, C_3 and C_4 are arbitrary constants, to obtain the fullest generality. He eliminated the arbitrary constants and obtained

$$\frac{d^2}{dt^2} \log \frac{du}{dt} - \frac{1}{2} \left(\frac{d}{dt} \log \frac{du}{dt} \right)^2,$$

which he denoted $\Psi(u, t)$. It is equal to

$$\frac{\dfrac{d^3 u}{dt^3} \dfrac{du}{dt} - \dfrac{3}{2} \left(\dfrac{d^2 u}{dt^2} \right)^2}{\left(\dfrac{du}{dt} \right)^2}.$$

If t has no winding point inside the circular-arc polygon, then $\Psi(u, t)$ is analytic inside the polygon, and so, if the polygon is mapped onto the upper half plane, $\Psi(u, t)$ will be a rational function $F(t)$. Furthermore, the general solution of the differential equation $\Psi(u, t) = F(t)$ will be the quotient of two solutions of a second order linear differential equation with rational coefficients. Schwarz thanked Weierstrass for drawing this observation to his attention. We have seen that this observation was a standard move in the theory of elliptic functions.

Schwarz then considered the function

$$u = \int_{t_0}^t (t - a)^{\alpha - 1} (t - b)^{\beta - 1} (t - c)^{\gamma - 1} dt$$

for real constants, a, b, c, α, β, and γ, where α, β, and γ are positive and $\alpha + \beta + \gamma = 1$. It maps the upper and lower half planes onto two rectilinear triangles with angles $\pi\alpha, \pi\beta$, and $\pi\gamma$. By reflection an infinitely many valued function is obtained unless α, β, γ take one of these four sets of values

$$\frac{1}{2}, \frac{1}{4}, \frac{1}{4}; \quad \frac{1}{2}, \frac{1}{3}, \frac{1}{6}; \quad \frac{1}{3}, \frac{1}{3}, \frac{1}{3}; \quad \frac{2}{3}, \frac{1}{6}, \frac{1}{6},$$

in which case the triangles tessellate the plane, as Christoffel [1867] and Briot and Bouquet had shown [1856a, 306–308].

During 1869 Schwarz's views on the Dirichlet principle hardened, one supposes in discussions with Weierstrass, who indeed published his counter-example to a related Dirichlet-type argument the next year. In his paper [1869c] Schwarz observed that Dirichlet's principle still lacked a proof (it was becoming Dirichlet's problem), and he supplied one for simply-connected polygonal regions, or more generally, for convex regions. The proof consisted of showing that one could deform a conformal map of a small convex region onto a circle analytically so that the domain was a slightly larger convex region, which was also mapped conformally into a circle. The deformation of the domain went from one convex boundary curve to another through a series of (non-convex) rectangular approximations. One started trivially with a circle inside the given region and finished with a map of the whole region onto a circle.

By the next year Dirichlet's principle had become quite problematic. In his [1870a] Schwarz brought forward an approach based, he said, on discussion he had had with Kronecker and other mathematicians in November 1869 on the partial differential equation $\Delta u = 0$. This was his "alternating method" (*alternirendes Verfahren*), and it applied to regions bounded by curves with a finite radius of curvature everywhere, and only finitely many vertices, although cusps were not excluded. The idea is an attractive one. He considered two overlapping regions T_1 and T_2 of the kind for which he could already solve Dirichlet's problem, called the boundary of $T_1 L_0$ outside T_2 and L_2 inside, and likewise the boundary of $T_2 L_3$ outside T_1 and L_1 inside, and called the overlap T^*. He sought a map u such that $\Delta u = 0$ and u took prescribed values on L_0 and L_3.

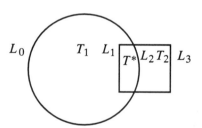

Figure A1.1

Invoking the metaphor of an air pump, he regarded $T_1 - T^*$ and $T_2 - T^*$ as air cylinders, and L_1 and L_2 as valves. The first stroke of the first cylinder maps T_1 by u_1 onto a half-plane, taking arbitrary prescribed values on L_0 and a fixed value on L_2 equal to the least value to be taken by u on L_0 and L_3. The first stroke of the second cylinder did the same for T_2, with a function u_2 which always took the maximum value of u on L_1. This is possible because of his assumptions about T_1 and T_2. The second stroke of the first cylinder converted u_1 into u_3, a function agreeing with u on L_0 and u_2 on L_2. The second stroke of the second cylinder likewise turned u_2 into u_4, which agreed with u on L_3 and u_3 on L_1. The pump works smoothly, and Schwarz found the sequences

$$u' = u_1 + (u_3 - u_1) + (u_5 - u_3) + \cdots + (u_{2n+1} - u_{2n-1}) + \cdots$$

$$u'' = u_2 + (u_4 - u_2) + (u_6 - u_4) + \cdots + (u_{2n+2} - u_{2n}) + \cdots$$

converge to the same, sought-after function u.

He then gave examples of what could be done, remarking at one point [*Abhandlungen*. II, 142]:

"Let a circular arc triangle be given. The conformal representation of the surface of a circle onto the inside or the outside of a circular arc triangle can be carried out without difficulty by means of hypergeometric series".

An extended version of this paper was published in [1870b]. Schwarz took the opportunity to point out that the analytic functions he was constructing might have a natural boundary beyond which they could not be analytically continued, a circumstance of importance for function theory which, he said, Weierstrass had remarked upon in general some years ago. But he gave only a rather contrived example. Finally he again presented his proof of Dirichlet's principle for simply connected regions with, so to speak, "rectifiable" boundaries.[1]

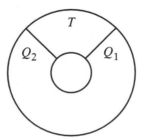

Figure A1.2

Twelve years later, on February 1, 1882, Schwarz wrote a letter to Klein, [1882] in which he showed how to extend the alternating method to multiply-connected regions. A two-fold connected region with boundary, such as an annulus, T, becomes a simply-connected region once a boundary cut is drawn. Let the cut Q produce the region T_1, and Q_2 produce T_2. The alternating method enables one to equalize the two solutions to Dirichlet's Problem for T_1 and T_2, and find a function which jumped by a prescribed constant amount upon crossing a cut. The argument readily extended by induction to regions of higher connectivity, but Schwarz admitted he had found it much harder to deal with unbounded, closed, Riemann surfaces. The difficulty, he said, was overcome by removing two concentric circular patches R_2 containing R_1 from the surface, which lay in the same leaf and enclosed no singular point. The problem could be solved for the Riemann surface without the smaller patch, as it had a boundary, and one could indeed assume the solution function, u_1, took a constant value on the boundary of R_1 and prescribed moduli of periodicity along the cuts. The function did not extend into R_1, but one could solve the Dirichlet problem for the surface patch R_2 by a function u_2 which took the values of u_1 on the boundary of R_2. One could then solve the problem outside R_2 and find a function which agreed with u_2 on R_1, and so on. In this way one could solve the problem.

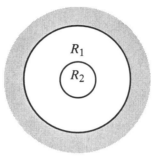

Figure A1.3

One supposes this letter was a reply to one of Klein's about the validity of the Dirichlet principle, perhaps to the publication of Klein's essay on Riemann, to which Schwarz referred. Schwarz still regarded the Riemann surface as spread out above the complex plane. When Klein shortly came to a different view he wrote again to Schwarz, as is described in Chapter VI. I have examined the collection of letters from Schwarz to Klein in the Klein Archive in Göttingen, [N.S.U. Bib. Göttingen, Klein XI, 934–938] but they add nothing to our knowledge of Schwarz's idea. 937 (8 April) looks forward to Klein's visit, 938 (2 July) talks about different matters — models of Kummer's and other surfaces.

Appendix 2

Riemann's Lectures and the Riemann-Hilbert Problem

Much of Riemann's work was not published until 1876 and some extracts from his notebooks and lectures were not published until 1902 in the *Nachträge* added to the second edition of his *Werke*. By then much of what he had found had been independently rediscovered, and these ideas of Riemann's will only be briefly described here. The publication of 1902 excited much interest and Schlesinger [1904] and Wirtinger [1904] reported on it to the Heidelberg International Mathematical Congress.

Riemann had lectured on differential equations at Göttingen in the Winter semester 1856/57, the year before he published his paper on the hypergeometric equation. In those lectures [*Nachträge* 67–68] he analysed the behaviour of two independent solutions of a second order linear differential equation near a branch point, b. He showed that there are two different cases which can arise. The two independent solutions, y_1 and y_2, are each transformed under analytic continuation around b, into $ty_1 + uy_2$ and $ry_1 + sy_2$, respectively. If there is a combination $y_1 + \epsilon y_2$ which returns as $(y_1 + \epsilon y_2)$ constant, then ϵ must satisfy

$$\epsilon(t + \epsilon r) = u + \epsilon s. \tag{A.1}$$

Let ϵ satisfy this equation, and choose α so that $t + \epsilon r = e^{2\pi i \alpha}$. Then $(y_1 + \epsilon y_2) \cdot (z - b)^{-\alpha}$ is unaltered on analytic continuation around b, and so Riemann wrote

$$y_1 + \epsilon y_2 = (z - b)^\alpha \sum_{n=-\infty}^{\infty} a_n(z - b)^n.$$

The two cases arise according as (A.1) has one root or two. If it has two roots, say ϵ and ϵ', then the corresponding α and α' are different and two independent solutions can be found which are invariant (up to multiplication by a scalar) upon analytic continuation around b. But if, and this is the second case, (A.1) has repeated root, ϵ, then the second solution must have the form y,

$$y = (z - b)^{\alpha} \log(z - b) \sum_{-\infty}^{\infty} n(z - b)^n + (z - b)^{\alpha} \sum_{-\infty}^{\infty} b_n (z - b)^n.$$

That is, the second solution contains one term which is $\log(z - b)$ times the first solution, and one term having the form of the first solution.

These two cases correspond to the cases where the monodromy matrix

$$\begin{pmatrix} t & u \\ r & s \end{pmatrix}$$

can be diagonalized and has distinct eigenvalues and where it cannot. They can be read off from the quadratic equation for the eigenvalues of the matrix, so it is quite possible to pass directly from that algebraic equation at the branch points to the analytic form of the solutions, without the intervening geometry being apparent. This has been the method usually adopted subsequently, perhaps at a cost in intelligibility. The connection between the occurrence of matrices in non-diagonal form and the presence of logarithmic terms in the solution, as described by Fuchs, Jordan, and Hamburger, is discussed in Chapter II. From von Bezold's summary of these lectures [*Nachträge*, 108] and the extract [*Werke*, 379–385] it appears that Riemann, after deriving what is essentially the Fuchsian form of a differential equation, considered the nth order case only when the monodromy matrices can be put in diagonal form [1953a, 381] but gave a complete analysis of the 2×2 case, neglecting the trivial diagonal case $u = r = 0, t = s$.

Riemann lectured on the hypergeometric equation [*Nachträge*, 67–94] again in the Winter semester 1858–59. In Section A(67–75) he gave a treatment of his P-function in terms of definite integrals of the form

$$\int s^a (1 - s)^b (1 - xs)^c dx.$$

The theme of differential equations emerges more strongly in Section B(76–94), which teems with ideas that were not to be developed by other mathematicians until much later, and then without direct knowledge of Riemann's pioneering insights.

Riemann began by considering the equation $a_0 y'' + a_1 y' + a_2 y = 0$, where a_0, a_1, a_2 are functions of z, and

$$y' = \frac{dy}{dz}, y'' = \frac{d^2 y}{dz^2}.$$

If Y_1 and Y_2 are independent solutions, and $Y = Y_1/Y_2$, then under analytic continuation around a closed path in the z domain, Y is transformed into

$$\frac{\alpha Y + \beta}{\gamma Y + \delta},$$

where α, β, γ and δ are constants. The inverse function to Y, say $z = f(Y)$, has the attractive property that

$$f\left(\frac{\alpha Y + \beta}{\gamma Y + \delta}\right) = f(Y) = z,$$

so it is invariant under all the substitutions corresponding to circuits in z.

Conversely, Riemann claimed that the inverse of any function f of Y invariant under a set of substitutions

$$Y \to \frac{\alpha Y + \beta}{\gamma Y + \delta} = Y'$$

satisfies a second order linear ordinary differential equation. Since $\dfrac{dY}{dz}$ is transformed by the substitution

$$Y \to \frac{\alpha Y + \beta}{\gamma Y + \delta}$$

into

$$\frac{dY'}{dz} = \frac{d}{dz}\left(\frac{\alpha Y + \beta}{\gamma Y + \delta}\right) = \frac{\alpha\delta - \beta\gamma}{(\gamma Y + \delta)^2}\frac{dY}{dz} = \frac{1}{(\gamma Y + \delta)^2}\frac{dY}{dz},$$

assuming, as Riemann did, that $\alpha\delta - \beta\gamma = 1$, so

$$\left(\frac{dY'}{dz}\right)^{-1/2} = \left(\frac{dY}{dz}\right)^{-1/2}(\gamma Y + \delta)$$

and

$$Y'\left(\frac{dY'}{dz}\right)^{-1/2} = \left(\frac{dY}{dz}\right)^{-1/2}(\alpha Y + \beta).$$

Thus, setting

$$Y_1 = \left(\frac{dY}{dz}\right)^{-1/2}, \quad Y_2 = Y\left(\frac{dY}{dz}\right)^{-1/2},$$

Y_1 and Y_2 are two particular solutions of the equation $y'' + a_2 y = 0$, where

$$a_2 = -\left(\frac{dY}{dz}\right)^{-1/2}\frac{d^2}{dz^2}\left(\frac{dY}{dz}\right)^{-1/2}$$

is, moreover, an algebraic function (it is

$$\frac{d^2 Y_2}{dz}\cdot\frac{dY_1}{dz} - \frac{d^2 Y_1}{dz^2}\cdot\frac{dY_2}{dz}$$

and so a constant multiple of the Schwarzian of Y). Riemann observed that this analysis is very important for the study of algebraic solutions of a differential equation. I cannot find that the idea of specifying the monodromy group in advance of the differential equation was taken up by anyone before Poincaré in 1880, who undoubtedly came to it independently (see Chapter VI). It is the origin of Hilbert's 21st problem, see Hilbert [1900b]. Riemann then turned to the hypergeometric equation and took the constants α, α', β, β', γ, γ' to be real. Then the real axis is mapped by Y, the quotient $P^\alpha/P^{\alpha'}$, onto a triangle with sides the real interval $[0, Y(1)]$, a straight line from 0, and circular arc from $Y(1)$. The angles of this triangle are $\pi(\alpha - \alpha')$ at 0, $\pi(\gamma - \gamma')$ at $Y(1)$, and $\pi(\beta - \beta')$ where the two other sides meet. The function takes every value inside this triangle exactly once. Conversely, the inverse function $Z = Z(Y)$ is single-valued inside the triangle, since $\dfrac{dY}{dz}$ never vanishes, and maps it onto the upper half plane.

This triangle will be a spherical triangle if the angles $\lambda\pi = (\alpha - \alpha')\pi$ at $Z = 0$, $\mu\pi = (\beta - \beta')\pi$ at $Z = \infty$, $\nu\pi = (\gamma - \gamma')\pi$ at $Z = 1$ are such that $\lambda + \mu + \nu > 1$. Riemann studied the particular cases which are connected to the quadratic and cubic transformations that can sometimes be made to a P-function, so this is a convenient place to record what he had earlier found in [1857a, §5]. Let $P(z_1)$ and $\tilde{P}(z_2)$ denote P-functions of their respective arguments, possibly with different exponents. A quadratic transformation replaces z by $\sqrt{z}$, and one wants to define $\tilde{P}(\sqrt{z}) := P(z)$. This clearly forces $\tilde{P}(z) = \tilde{P}(-z)(= P(z^2))$, so the exponent difference at 0 in the $\sqrt{z}$-domain must be -1. There are singular points at $\sqrt{z} = +1$ and $\sqrt{z} = -1$ which must have the same exponent difference, say $\gamma - \gamma'$, and the exponents at $\sqrt{z} = \infty$ are, say 2β and $2\beta'$. The map $\sqrt{z} \to z$ therefore maps the quadrilateral

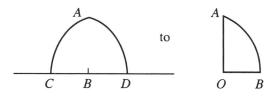

Figure A2.1

and one has the P-function

$$P\left\{\begin{matrix} 0 & \infty & 1 \\ 0 & \beta & \gamma \\ \frac{1}{2} & \beta' & \gamma' \end{matrix}\; z\right\} = \tilde{P}\left\{\begin{matrix} -1 & \infty & 1 \\ \gamma & 2\beta & \gamma \\ \gamma' & 2\beta & \gamma' \end{matrix}\; \sqrt{z}\right\}.$$

Entirely similar considerations of the transformation $\sqrt[3]{z} \mapsto z$ show that

$$P = \left\{\begin{matrix} 0 & \infty & 1 \\ 0 & 0 & \gamma \\ \frac{1}{3} & \frac{1}{3} & \gamma' \end{matrix}\; z\right\} = \tilde{P}\left\{\begin{matrix} 1 & \rho & \rho^2 \\ \gamma & \gamma & \gamma \\ \gamma' & \gamma' & \gamma' \end{matrix}\; \sqrt[3]{z}\right\}, \qquad \rho^3 = 1$$

is also possible. So quadratic and cubic transformations can be made if the exponents are suitable.

Riemann wrote $P(\mu, \nu, \frac{1}{2}, x)$, $P(\mu, 2\nu, \mu, x_1)$, and $P(\nu, 2\mu, \nu, x_2)$ when he wanted to display the exponent differences, where $x = 4x_1(1-x_1) = 1/4x_2(1-x_2)$. He discussed these transformations in the *Nachträge* extract. If x lies in triangle AOB, then x_1 lies in ADB and x_2 in ACB. Conversely, it was straight-forward for him to show that the rational function of x which transforms AOB into ADB is necessarily of the form $x = cx_1(1 - x_1)$ for some constant c, which can be normalized to $c = 4$. This gave him a geometric interpretation of the quadratic transformations of the P-function. He also knew that the regular solids provide further examples in which the function mapping the triangle onto the half-plane is algebraic. There is no suggestion, however, that Riemann knew these are essentially the only example of such functions. The recognition of that fact had to wait for Klein (see Chapter III). Riemann did not consider the problem of analytically continuing Y around several circuits of the branch points, when the image triangles may overlap (unless $\lambda\pi$, $\mu\pi$, and $\nu\pi$ are the angles of a regular solid with triangular faces), an idea Schwarz was the first to develop (see Chapter III).

Finally, Riemann considered the periods K and iK' of an elliptic integral as functions of the modulus k^2, and considered a branch of the function $\frac{K'}{K}$. As k^2 goes from 0 to 1, $\frac{K'}{K}$ is real and goes from ∞ to 0. The part of the real axis $(-\infty, 0)$ is mapped onto the right half line through $-i$ parallel to the real axis, the part $(1, \infty)$ is mapped onto a semicircle centre $\frac{-i}{2}$ radius $\frac{1}{2}$ (see Figure A2.2). So one branch of $\frac{K'}{K}$ maps the values of k^2 on the real axis onto the figure made up of two horizontal straight lines and a semicircular arc. As k^2 ranges over the upper half plane $\frac{K'}{K}$ takes every value exactly once within its domain. When k^2 was allowed to vary over the plane, Riemann found that each branch of $\frac{K'}{K}$ mapped a half plane into a similar circle arc triangle in the right-hand half-plane, and $\frac{K'}{K}$ takes every value exactly once in that space, whatever circuit k^2 performs about 0, 1, ∞. So, given an arbitrary function, $Y(z')$, with branch points at 0, 1, and ∞, setting $z' = \phi(z)$, where $k^2 = \phi\left(\frac{K'}{K}\right)$, produces $Y(z)$ which is single-valued in z.

Riemann has shown that a multi-valued function Y defined on C with branch points at 0, 1, ∞ can be lifted to a single-valued function on H, a half plane, by means of the modular function k^2.

Thus Y has been globally parameterized by k^2, for which the term is 'uniformized'.

Interestingly, the following appears in the first edition of volume III of Gauss' *Werke* [1866, 477]:

"If the imaginary part of t, $\frac{1}{t}$ lies between $-i$, $+i$ then the Real Part of $\left(\frac{Qt}{Pt}\right)^2$ is positive

"The equation $\left(\frac{Qt}{Pt}\right)^2 = A$ has exactly 1 solution in the

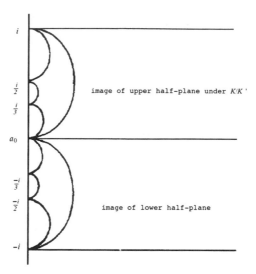

Figure A2.2

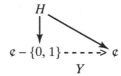

Figure A2.3

space $\alpha\delta - \beta\gamma = 1, \alpha \equiv \delta \equiv 1 \bmod 4, \beta, \gamma$ even

$$t' = i\left(\frac{\alpha t + \beta i}{\gamma t + \delta i}\right) \quad \text{then} \quad \left(\frac{Qt'}{Pt'}\right)^2 = i^\gamma \left(\frac{Qt}{Pt}\right)^2 \text{''}.$$

P and Q play the role of theta-functions in Gauss' theory of elliptic functions, and Gauss here described the fundamental domain for the modular function t but the editor (Schering) missed the point and rendered the arcs as mere doodles. In the second edition the diagrams have been fattened up to represent semicircles, as they should. Had Riemann seen that part of the *Nachlass*? Certainly he would not have failed to see the import of Gauss' drawings.

The question of the boundary values of $\frac{K'}{K}$ was also considered by Riemann and the fragment he left [*Werke* 455–465] was discussed by Dedekind [Riemann *Werke* 466–479]. The matter is a subtle one with number-theoretic and topological implications, and Dedekind was able to develop it to advantage when Fuchs, on Hermite's instigation, raised it again in 1876. This was discussed in Chapter IV.

Before turning to the Riemann–Hilbert problem, let me note that Riemann also showed how the hypergeometric equation plays a role in the study of minimal sur-

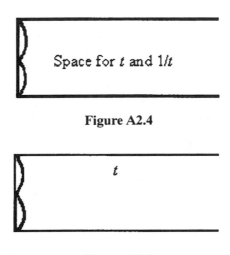

Figure A2.4

Figure A2.5

faces (Riemann [1867]) later taken up and examined at length because the monodromy considerations are so delicate — by Garnier, [1928].

The Riemann–Hilbert problem

Riemann's name was attached to a problem in the theory of differential equations by David Hilbert because of his general theory of linear ordinary differential equations with algebraic coefficients sketched in the fragmentary text mentioned above. In it, Riemann showed that the solution to an nth order equation with a certain number of fixed singular points $(a, \ldots g)$ can be thought of as an n-tuple of functions $(y) = (y_1, \ldots, y_n)$. To obtain a single-valued set of functions, he imagined a simple closed curve drawn through the singular points, dividing the x-domain into two regions.. The effect on such an n-tuple of a circuit of a singular point is described by a monodromy matrix (here denoted $A, \ldots G$ respectively — Riemann wrote (A) etc), and $G.F. \ldots B.A = (0)$, where (0) was Riemann's notation for the identity matrix.

Riemann denoted such a system $Q \begin{pmatrix} abc & g & \\ & & x \\ ABC & G & \end{pmatrix}$. He put into the same class

systems of functions with the same branch points and monodromy matrices, and observed that there would be infinitely many equations in a class. He then said that in general, as Jacobi had shown, each matrix could be diagonalised and the diagonal entries were the roots of an equation determined by the matrix; he proposed to ignore for the moment the special case when the equation had repeated roots and the matrix could not be diagonalised. He then explained the connection between the diagonal entries and the leading exponent of the power series of the corresponding solution. Moreover, given any $n+1$ systems of functions of the same class, it is easy to write down the linear ordinary differential equation they satisfy. In particular, by considering the derivatives of a function y, it follows that the functions in a system y satisfy an nth order differential equation with polynomial coefficients. Finally he showed that the coefficients of such an equation were not completely determined by the information that determined its class, so each class would contain infinitely many members (although one could normalise the solutions to avoid this problem).

In the second part of the fragment, Riemann sought to determine the simplest form of the differential equation in a given class, and found that it was

$$\omega^n X_0 y^{(n)} + \cdots + \omega X_{n-1} y' + X_n y = 0$$

where

$$\omega = (x - a)(x - b) \ldots (x - g),$$

the function X_i are polynomials of degree $r + (m - 1)i$, and $r = \frac{1}{2}(m - 2)n(n - 1) - s$, where s denotes the sum of the characteristic exponents and there are m singular points. This guarantees that the solutions have at worst poles.

To ensure that the solutions are branched only at the branch points $a, b, \ldots g$, a detailed examination of the power series representation of the solutions was carried out to ensure that they contained n arbitrary constants. Riemann's analysis said nothing about what happened at ∞, and apparently Riemann only hinted at how the enumeration of constants could be carried out. Weber's account in the first edition had to be corrected, as Hilbert pointed out. The upshot was that if the solutions of the differential equation have at worst poles and are branched only at the branch points of the coefficients, then the differential equation has

$$r + n + \frac{1}{2}(m - 2)n(n + 1) = (m - 2)n^2 + n - s$$

free constants, and an arbitrary system of n particular integrals (with n^2 constants of integration) has $(m - 1)n^2 + n - s$ free constants. On the other hand, the monodromy matrices contain mn^2 constants and satisfy one equation, so they involve $(m - 1)n^2$ free constants, and once a system of solutions is specified, there are $n - s$ constants left over. So, for at least one differential equation to exist with the given branch points and monodromy, it is necessary that $s < n$.

Riemann's analysis of the problem ends here. Klein's comments on the Riemannian fragment at this point are trenchant [1894, p. 247]. He observed that transcendental questions about the existence of functions cannot be solved by counting constants, and went on: "The central problem, which must now be disposed of, but which Riemann nowhere tackled and which has still not been resolved, is the proof of existence, a proof that for arbitrary given [branch points] $a, b, \ldots, n$ and [monodromy matrices] $A, B, \ldots, N$, corresponding functions really exist, and then to study how many such functions exist".

In 1900, Hilbert chose to present a variant of this problem as the 21st of his 23 problems at the International Congress of Mathematicians in Paris. He stated it in these terms: to show that there always exists a linear ordinary differential equation of the Fuchsian class, with given singular points and monodromic group. This is not identical with the problem introduced by Riemann. As authors from Schlesinger to Yoshida have observed, Riemann's problem starts with local data and seeks a global object, whereas Hilbert's problem start with global data and seeks a global object (Schlesinger [1897], §162, Yoshida [1987, 28]). Even with this distinction made clear, as Anosov and Bolibruch have pointed out, Hilbert's remark is still not as easy to interpret as it might seem. A linear differential equation of the Fuchsian class (defined over the Riemann sphere) cannot be determined if one specifies the singular points, and at each point a monodromy matrix. Simply by counting constants, one sees that there are more coefficients in the matrices than there are in the differential equation. Indeed, an nth order Fuchsian differential equation with p singular points has $N_e = 1/2((p - 2)n^2 + np)$ constants, while the set of conjugacy classes of representations (and so the monodromy group) depends on $N_r = n^2(p - 2) + 1$ constants. The difference is

$$N_r - N_e = \frac{1}{2}(p - 2)n(n - 1) + 1 - n,$$

which is greater than 0 if $p > 3$ and $n > 1$, and if $p = 3$ and $n > 2$. One solution would be to increase the number of constants in the differential equation by introducing apparent singularities — points at which the coefficients have poles but the solutions remain unbranched.

Another possibility would be that Hilbert proposed the study a system of first-order differential equations. Indeed, he indicated that a count of the constants indicated that this was probable, and that Schlesingerhad given a proof of a special case using Poincaré's theory of zeta-Fuchsian functions, but that the problem remained to be tackled by some perfectly general method.

In 1904, in a short paper presented to the International Congress of Mathematicians at Heidelberg, Hilbert showed how questions about the existence of analytic functions satisfying certain monodromy conditions can be tackled using Fredholm's newly-discovered method of integral equations. This insight was then taken up as the basis of Plemelj's more thorough treatment of the Riemann–Hilbert problem, which was to be regarded as definitive for 81 years.

In what became his famous paper of [1908], Plemelj expressed the Riemann–Hilbert problem in this fashion. As Riemann had done, he supposed the complex sphere was divided into two regions (called the inside and the outside) by a simple closed curve C, and asked: are there n regular analytic functions $f_i^-(z)$ defined on the outside and n such functions $f_i^+(z)$ on the inside, where the outer functions $f_i^-(z)$ take constant values c_i at ∞, such that on the curve C, they are linearly related by a prescribed matrix $A(x)$. In matrix notation, $F^-(z) = A(z)F^+(z)$, where the n-tuple of the $f_i^-(z)$ is denoted by $F^-(z)$ and of the $f_i^+(z)$ by $F^+(z)$. Strictly speaking, for the Riemann–Hilbert problem the coefficients of $A(z)$ were piecewise constants; following Hilbert, Plemelj massaged the problem so that the entries in $A(z)$ were at least once differentiable functions and the curve C had a normal everywhere.

Plemelj defined $A(z, \zeta) = A(z)A^{-1}(\zeta)$, so the Cauchy Integral Theorem implies that if the sought-after function s exists then

$$c = F^-(z) - \frac{1}{2\pi i} \int_C [A(z, \zeta) - I]F^-(\zeta)\frac{d\zeta}{\zeta - z}, \text{ where } \mathbf{c} = (c_1, \ldots, c_n).$$

Conversely, if this system of equations can be solved, then the Riemann–Hilbert problem is solved.

To solve these equations Plemelj, using a technique pioneered by Fredholm, turned them into an integral equation of the form $c(z) = \phi(z) + \int K(z, \zeta)\phi(\zeta)\,d\zeta$. By the 'Fredholm alternative' this has a solution if and only if the homogeneous equation $\phi(z) + \int K(z, \zeta)\phi(\zeta)\,d\zeta = 0$ does not, unless $c(z)$ satisfies certain conditions. I shall not discuss this part of Plemelj's paper. It turns out that the conditions on $c(z)$ give rise to functions that are regular at ∞, and the solutions that vanish at ∞ correspond to solutions of the homogeneous equation. Since there is only a finite number of linearly independent solutions of that equation, the order of the zeros at ∞ is bounded.

To connect this problem more directly to the Riemann problem, Plemelj supposed that the curve C enclosed m singular points $a_1, \ldots, a_m$. There were now n functions $y_1(z), \ldots, y_n(z)$ which were transformed linearly among themselves as z circled the singular points. They defined single-valued functions $y^+(z)$ inside a curve C that passed through the singular points, and one could choose single-valued functions $y^-(z)$ outside the curve C. Consequently the inner and outer functions were related along C by equations of the form $\mathbf{y}^-(z) = K\mathbf{y}^+(z)$, where the constant matrices K depend on the piece of the curve C under consideration. Plemelj formed the functions $Y_k^+(z) = K\mathbf{y}^+(z)$. He showed, in a lengthy analysis, how to replace the matrices K with a matrix of continuous functions, so that Fredholm's theory could be applied. The argument extends automatically to deal with n-tuples of functions, (such as specify a solution to an nth order linear ordinary differential equation).

Only at the end of his paper did Plemelj consider the Riemann–Hilbert problem as providing the solution to a system of differential equations. He showed that the functions he had shown to exist were regular, and the corresponding differential equation had rational coefficients. Moreover, it had the right number of constants in its coefficients. Significantly, he did not investigate if the differential equation was in fact of the Fuchsian class. Plemelj concluded his paper by observing that the same arguments could be used to show that if a Galois group with m generators is represented as a permutation group acting on n objects, then n functions can be found on the Riemann sphere which are permuted at m branch points in ways described by the generators. In this way a Riemann surface can be associated with any given Galois group.

Some five years later, but apparently independently, Plemelj's methods received independent confirmation when the young American mathematician G.D. Birkhoff published an article [1913a] in which he almost without realizing it solved the Riemann–Hilbert problem. He set himself the task of generalizing the following fact about functions of a single variable: a function $f(z)$ which is single-valued and analytic for large z but not analytic at ∞, and which does not vanish at ∞, may be written as a product, $f(z) = a(z)e(z)z^k$, where the function $a(z)$ is analytic and non-zero at ∞, the function $e(z)$ is entire and nowhere zero, and k is an integer. He established the following theorem: If $(l_{ij}(z))$ is a matrix of functions which are single-valued and analytic for large z but not analytic at ∞, and if $\det(l_{ij}(z))$ does not vanish in a neighbourhood of ∞, then $(l_{ij}(z))$ may be written as a product, $f(z) = (a_{ij}(z))(e_{ij}(z)z^{k(j)})$, where the functions $a_{ij}(z)$ are analytic at ∞ and the matrix $a_{ij}(z)$ reduces to the identity matrix at ∞, the function $e_{ij}(z)$ are entire, the determinant is $(e_{ij}(z))$ nowhere zero, and the $k(j)$ are integers.

His method of proof was first to establish that the matrix equation $L.E = B$, as an equation for E, has a solution, where $L = (l_{ij}(z))$ is as above and the matrix B is composed of functions that are either analytic or have poles at ∞. The matrix E is to be composed of entire functions. This problem lead Birkhoff to the equations $L^{-1}.B = E$ and $L(z)^{-1}. L(t) = I + K(z, t)$.

The entries in K are such that $\dfrac{k_{ij}(z, t)}{t - z}$ is analytic for large z and t, since $k_{ij}(z, t)$ vanishes identically when $z = t$. Moreover, $L(z).K(z, t) = L(t) - L(z)$. The Cauchy Integral Theorem suggested the equations for unknown functions $f(z)$ and polynomials $p(z)$,

$$\frac{1}{2\pi i} \int_C \frac{k_{ij}(z, t) f_j(t)}{t - z} \, dt = f_j(t) - (l_{ij}(z))^2 (p_1(z)),$$

where C is the circle $|z| = R$, for some large R. Birkhoff rewrote this equation after a switch to polar coordinates, in the language of Fredholm integrals as

$$\int_0^{2n\pi} K(\theta, \phi) F(\phi) d\phi = F(\theta) - P(\theta),$$

where the functions F and P are continuous except where they have finite jumps. Fredholm's theory permitted Birkhoff to deduce that the integral equation had a solution, at least when the polynomials $p(z)$ satisfy mild conditions. So the functions $f_i(t)$ exist for $|t| > R$. The initial observation about functions of a single variable therefore reduced the matrix problem to checking on the conditions on the polynomials $p(x)$, and these were easily satisfied.

The solution to the equation $L.E = B$ was massaged so that the $\det(E)$ was nowhere zero, then so that $B = (b_{ij}(z)) = (a_{ij}(z)z^{k(j)})$, for some $k(j)$ and finally so that by a suitable rearrangement of these k's every $e_{ij}(z)$ vanishes at $z = 0$ to the right order. I omit these arguments. In a note at the end of volume 74 of the *Mathematische Annalen*, Birkhoff observed that he had now noticed, and Toeplitz had also pointed out to him, that his proof not only amounted to a proof of the Riemann–Hilbert problem, previously solved by Hilbert and Plemelj, but that his was indeed "not markedly different from theirs".

In a second paper the same year [1913b] Birkhoff returned to the problem, with a view to simplifying the treatment of Hilbert and its "elegant completion" by Plemelj by eliminating the use of Fredholm theory, as well as adding certain extensions of his own (to irregular singular points and difference equations). In this paper, following Plemelj, Birkhoff introduced the variable $\alpha = \exp(2\pi i s/l)$, where the variable s measures arc length along a simple closed curve C of length l. He then defined a function f^+ by the equation

$$\frac{1}{2\pi i} \int_C \frac{\tau^P(t)g^+(t)}{t-z} a(t)\, dt,$$

where $g^+(t)$ is function analytic within C and continuous along it (what Plemelj had called a regular inner function) and $a(t)$ is an infinitely differentiable function along the curve C, save perhaps for a finite number of points. A function f^- was defined similarly, using a regular outer function. A simple argument showed that along C,

$$f^+(z) - f^-(z) = \tau^P(t)g^+(z)a(z).$$

Birkhoff then showed that given any ϵ, a value of p can be found so that along C

$$\text{Max}\{f^-(z)\} < \epsilon\, \text{Max}\{g^+(z)\},$$

and

$$\text{Max}\{f^+(z)\} < \epsilon\, \text{Max}\{g^-(z)\}$$

Birkhoff now proposed the following matrix equations for matrices F^+ and F^-, G^+ and G^-:

$$F^+(z) - F^-(z) = \tau^P(z)G^+(z)A(z)$$

$$G^-(z) - G^+(z) = \tau^{-P}(z)F^-(z)A^{-1}(z) - I.$$

He solved them by the method of successive approximations, under the assumption that p was large enough. This gave him control over the matrix equation

$$F^+(z) = \tau^P(z)[I + G^-(z)]A(z),$$

which he showed could be further simplified to an equation between a matrix of regular inner functions and a matrix of functions that were regular outer functions except for poles of order p at ∞:

$$\Phi(z) = \Psi(z)A(z) \text{ along } C.$$

Moreover, $|\Phi(z)|$ does not vanish inside or on C, and $|\Psi(z)|$ does not vanish outside or on C. The solution to this equation gave him the solution to the matrix problem of the previous paper and with that a solution to the Riemann–Hilbert problem, as he proceeded to explain.

Birkhoff stated the problem in this form. Given a system of differential equations

$$\frac{dY(x)}{dx} = R(x)Y(x),$$

where all the elements of $R(x)$ are regular and is not a singular point of the system, a set of monodromy matrices is obtained. Conversely, given such a set of matrices, it is required to find a set of functions $Y(x)$ with that monodromy. Let the singular points be $a_1, \ldots, a_i$ and suppose them to be surrounded by separate closed curves C_i and joined by a closed analytic curve D. Let T_i be the monodromy matrix at the singular point a_i and choose matrices $A_1, \ldots, A_m$ such that $A_{i+1}^{-1} A_i = T_i$. Define a matrix $A(x)$ along D to be A_i between C_i and C_i and a suitable linear combination of A_{i-1} and A_i on that part of D within C_i. Now smooth out $A(x)$ so that it is in fact analytic, except for finitely many points of D. Then the theorem above ensures that there is a matrix satisfying

$$\lim_{x \mapsto x_1^+} \Phi(x) = [\lim_{x \mapsto x_1^-} \Phi(x)] A(x_1) .$$

Let the matrix $U(x)$ be obtained from $\Phi(x)$ by analytic continuation. It is analytic outside the C_i and is transformed by T_i on a positive circuit of a_i. Finally, Birkhoff called a matrix of functions $Z_i(x - a_i)$ a Cauchy matrix if it is transformed to $Z_i(x - a_i) T_i$ as x makes a positive circuit of a_i. The matrix $U(x)$ has suitable regularity properties, so the Riemann–Hilbert problem is solved by exhibiting the matrix $Y(x) = Z(x) U(x)$.

For many years it was agreed that the problem was solved in the affirmative, and new proofs were given indicating how the question and its answer fitted into one or another more general setting. For example, Röhrl [1957] reworked the problem in the context of the newly-emerging theory of sheaves. It therefore came as a great surprise when two Russian mathematicians, Anosov and Bolibruch, announced in 1989 that the problem had not been solved, and indeed that the Riemann–Hilbert problem is to be answered correctly in the negative: it is possible to specify singular points and monodromy data in such a way that there is no corresponding differential equation of the Fuchsian class. To wrap the matter up, they exhibited a whole family of counterexamples.[1]

The matter is presented most accessibly in their book (Anosov and Bolibruch [1994]) which briefly indicates how the gap in Plemelj's proof was noticed and exploited by a number of Soviet mathematicians (see pp. 8–9). In the book they make a number of useful distinctions to exploit the distinction between an nth order differential equation and a system of n first order differential equations. It is straightforward to write an nth order differential equation as a system of equations for y and its first $n - 1$ derivatives, $y', y'', \ldots, y^{(n-1)}$. Given the equation

$$y^{(n)} + q_1 y^{(n-1)} + q_2 y^{(n-2)} + \cdots + q_n y = 0 ,$$

define $y_1 = y, y_2 = y_1', y_{n-1} = y_{n-2}', y_n' = -(q_1 y_n + q_2 y_{n-1} + \cdots + q_n y_1)$. This is then read as a system of first-order differential equations. A differential equation or system of first-order differential equations is said to be *regular,* if at each singular point there is a sector such that solutions approaching within the sector have at most polynomial growth. It is the distinction between being Fuchsian and being regular that is crucial.

A system of first-order differential equations is Fuchsian if the coefficients have at most simple poles. For a single differential equation of any order, the concepts of regularity and Fuchsian coincide; a differential equation is regular if and only if it is Fuchsian. For systems, a differential equation which is Fuchsian is regular, but the converse is false. As an example of a system of differential equations which is regular but not Fuchsian, Anosov and Bolibruch gave the example of the system obtained from the Fuchsian differential equation $y'' + (1/x)y' + (1/x^2)y = 0$. The corresponding system is not Fuchsian because of the second order pole.

Their simplest counterexample to the Riemann–Hilbert problem was a 3rd order system with given singular points and monodromy which is regular but not Fuchsian, and to which, as they showed, there is no Fuchsian system with the same singular points and monodromy. The matter is delicate, because, as they also showed, any regular system can be modified locally to become a Fuchsian one, although this can only be done globally by introducing apparent singularities. This, however, violates the terms of the problem, which specify the singular points in advance.

Anosov and Bolibruch gave this as their simplest counter-example. Define the following four matrices of coefficients:

$$A_1 = \begin{pmatrix} 0 & 1 & 0 \\ 0 & x & 0 \\ 0 & 0 & -x \end{pmatrix}; \qquad A_2 = \begin{pmatrix} 0 & 6 & 0 \\ 0 & -1 & 1 \\ 0 & -1 & 1 \end{pmatrix};$$

$$A_3 = \begin{pmatrix} 0 & 0 & 2 \\ 0 & -1 & -1 \\ 0 & 1 & 1 \end{pmatrix}; \qquad A_4 = \begin{pmatrix} 0 & -3 & -3 \\ 0 & -1 & 1 \\ 0 & -1 & 1 \end{pmatrix}.$$

Then the system

$$A(x) = \frac{1}{x^2}A_1 + \frac{1}{6(x+1)}A_2 + \frac{1}{2(x-1)}A_3 + \frac{1}{3(x-1/2)}A_4, \quad \text{(N1)}$$

is singular at the points $a_0 = 0$, $a_1 = -1$, $a_2 = 1$, and $a_3 = \frac{1}{2}$. It is not singular at ∞. Of the singular points, a_1, a_2, and a_3 are Fuchsian, but a_0 is not (the pole at 0 is or order 2). So the system is not Fuchsian, but it is regular.

The matrix $A(x)$ can be written in the form

$$\begin{pmatrix} 0 & a_{12}(x) & a_{13}(x) \\ 0 & & \\ & & B(x) \\ 0 & & \end{pmatrix}$$

where

$$a_{12}(x) = \frac{1}{x^2} + \frac{1}{(x+1)} - \frac{1}{(x-1/2)}; \qquad a_{13}(x) = \frac{1}{2(x-1)} - \frac{1}{(x-1/2)},$$

and

$$B(x) = A(x) + \frac{1}{x} B_1 + \frac{1}{6(x+1)} B_2 + \frac{1}{2(x-1)} B_3 + \frac{1}{3(x-1/2)} B_4,$$

where

$$B_1 = \begin{pmatrix} 1 & 0 \\ 0 & -1 \end{pmatrix}; B_2 = \begin{pmatrix} -1 & 1 \\ -1 & 1 \end{pmatrix}; B_3 = \begin{pmatrix} -1 & -1 \\ 1 & 1 \end{pmatrix}; \text{ and } B_4 = \begin{pmatrix} -1 & 1 \\ -1 & 1 \end{pmatrix}.$$

This gives the sytem of differential equations

$$\frac{dy^1}{dx} = a_{12}(x)y^2 + a_{13}(x)y^3,$$

where y_2 and y_3 satisfy

$$\frac{dy}{dx} = B(x)y. \tag{N2}$$

Notice that this last equation, for a system of two unknown functions, is Fuchsian.

Equation N1 has four singular points and some monodromy. To show that no Fuchsian system exists with the same singular points and monodromy, Anosov and Bolibruch constructed a detailed theory (the theory of Fuchsian weights) which I do not describe here. The crucial point is that the monodromy representation is reducible. This enabled Anosov and Bolibruch to show that any Fuschian system having the same monodromy as equation N1 necessarily had properties that the 2-dimensional system involving only the B's did not. On the other hand, if the monodromy representation is irreducible, they could show that the Hilbert Problem was indeed to be answered in the affirmative.

Appendix 3

Fuchs' Analysis of the nth Order Equation

It has been seen that an equation such as (2.1.1) always has n linearly independent solutions in the neighbourhood of any non-singular point in the domain of definition of the coefficients $p_1, \ldots, p_n$; Fuchs took this domain to be

$$\mathbf{C} - \{a_1, a_2, \ldots, a_{\rho+1} = \infty\} = T',$$

Suppose $y_1, \ldots, y_n$ is a system of solutions which, upon analytic continuation around a_k, return as $\tilde{y}_1, \ldots, \tilde{y}_n$, respectively. Then there are linear relations between each $\tilde{y}_i$ and the $y_1, \ldots, y_n$:

$$\tilde{y}_i = \sum_j \alpha_{ij} y_j,$$

and the matrix $A = (a_{ij})$ has eigenvalues which determine the properties of the fundamental system near a_k. (Of course, the matrix depends upon which a_k has been chosen.) The equation for the eigenvalues

$$|A - wI| = 0$$

was called by Fuchs the fundamental equation (*Fundamentalgleichung*) at the singular point a_k [1866,§3].

A change of basis permits one solution, u, to be found near a_k for which $u = w.u$, where w is a solution of the fundamental equation. If an r is chosen so that $w = e^{2\pi i r}$ (Fuchs called r an exponent at a_k), then $u.(x - a_k)^{-r}$ is single-valued near a_k.

If all the eigenvalues $w_1, \ldots, w_n$ are distinct, then a fundamental system of solutions can be made up of n elements, each with the property that $\tilde{y}_i = w_{ik} y_i$ and $(x - a_k)^{r_{ik}} y_i$ is single-valued near $x = a_k$, and $w_{ik} = e^{2\pi i r_{ik}}$. All solutions near a_k are therefore sums of the form

$$\sum c_i (x - a_k)^{r_{ik}} \phi_{ik}(x - a_k)$$

for single-valued functions ϕ_{ik}. If, however, the eigenvalues are not all distinct, and w is an l-times repeated root, then Fuchs showed [1866, 136 = 1904, 175] a fundamental system of solutions exists in blocks of the form $u_1, u_2, \ldots, u_l$ which transform under analytic continuation around a_k as

$$u_1 = wu_1$$

$$u_2 = w_{21}u_1 + wu_2$$

$$\ldots \quad \ldots$$

$$u_l = w_{l1} + w_{l2}u_2 + \cdots + wu_l$$

where the w_{ij} are constants. So Fuchs presented the solutions in the form

$$u_1 = (x - a_k)^r \phi_{11}$$

$$u_2 = (x - a_k)^r \phi_{21} + (x - a_k)^r \phi_{22} \log(x - a_k)$$

$$\ldots \ldots \ldots$$

$$u_l = (x - a_k)^r \phi_{l1} + (x - a_k)^r \phi_{l2} \log(x - a_k) + \cdots$$
$$+ (x - a_k)^r \phi_{ll}[\log(x - a_k)]^{l-1}$$

where r again satisfies $w = e^{2\pi i r}$, and each ϕ_{ij} is a linear combination $\phi_{11}, \phi_{21}, \ldots, \phi_{l1}$. This is a mildly confusing presentation, but blocks of solutions were found for each repeated eigenvalue w, and the resulting solution then involved logarithmic terms. Fuchs did not write his solutions in the simplest possible way, which would have amounted to putting the monodromy matrix in its Jordan canonical form; this was first done by Jordan [1871] and Hamburger [1873] (see Hawkins [1977]), as was discussed in Chapter II. On the other hand, Fuchs' presentation is two years earlier than Weierstrass' theory of elementary divisors, which gives a canonical presentation of a matrix equivalent to Jordan's but one couched in a more forbidding study of the minors of the matrix. Weierstrass' theory was designed to explain the simultaneous diagonalization of two matrices and Fuchs' simpler approach benefits from needing to consider only one. Fuchs' minor modifications of his presentation in the solution in his [1868, §1], but it is possible that Fuchs' paper joined with others to reawaken Weierstrass' interest in the problem of canonical forms.

To relate the solution valid near a_1 to those valid near a_k, Fuchs extended them analytically in T''. He found relations of the form

$$y^1 = B_k y^k,$$

and $\tilde{y}^k = B_k R_1 B_k^{-1} y^k = R_k y^k$, extending the earlier notation in the obvious way, and as before

$$\Pi_i \det R_i = \Pi_i(w_{i1} \ldots w_{in})\Pi_i e^{2ki} \sum_{k^r} ik = 1, \quad \text{so} \quad \sum_{i=1}^{\rho+1} \sum_{k=1}^{n} r_{ik} = K,$$

an integer, i.e., the sum of the exponents is an integer. To find K, Fuchs formed the determinants

$$\Delta_0^K = |(d_{pq})| = \left| \frac{d^{n-q}}{dx^{n-q}} y_p^k \right|,$$

and Δ_k^i, which is obtained from Δ_0^i by replacing the kth row of Δ_0^i with the transpose of

$$\left(\frac{d^n y_1^i}{dx^n}, \ldots, \frac{d^n y_n^i}{dx^n} \right), k = 1, \ldots, n.$$

So $\Delta_0^i p_k = -\Delta_k^i$, and

$$\Delta_k^1 = (\det B_i) \Delta_k^i \text{ for } k = 1, \ldots, n.$$

Again, $\tilde{\Delta}_k^i = \det R_i \cdot \det \Delta_k^i$, so $\Delta_k^i = (x - a_i)^\epsilon \psi$, where ψ is a single-valued function of x near a_i, and ϵ satisfies $e^{2\pi i \epsilon} = \det R_i$. Therefore $\Delta_0^i (x - a_i)^{-\Gamma_i}$ is finite, single-valued, and continuous near a_i, where $\Gamma_i = \sum_k r_{ik} - \dfrac{n(n-1)}{2}$,

for Δ_0^i has a pole of order $\sum_k r_{ik} - \dfrac{n(n-1)}{2}$, being made up of products of y_{ik} and its derivatives and $y_{ik}(x - a_i)^{-r_{ik}}$ has only a logarithmic infinity at a_i, as does $\dfrac{d^s y}{dx^s} ik(x - a_i)^{-r_{ik}+s}$. Similarly $\Delta_0^{\rho+1}$ has a pole of order $\sum r_{\rho+1,k} + \frac{1}{2}n(n-1)$ at ∞.

The expression $K_0 = \Delta_0^1 (x - a_1)^{-\Gamma_1} \cdots (x - a_\rho)^{-\Gamma_\rho}$ is finite, continuous, and single-valued everywhere in **C**. It is infinite at ∞ in a way that can be determined by looking at the solutions near ∞, and the behaviour of $\Delta_0^{\rho+1}$. Indeed K_0 has a pole at ∞ of order at most

$$-\sum_{i=1}^{\rho} \sum_{k=1}^{n} r_{ik} + (\rho - 1)\frac{n(n-1)}{2} = -K + (\rho - 1)\frac{n(n-1)}{2}.$$

Fuchs took as a basis of solutions

$$y_1^i, y_2^i = y_1^i \int z_1 \, dx, y_3^i = y_1^i \int dx z_1 \int u_1 \, dx, \ldots y_n^i = y_{n-1}^i \int w_1 \, dx,$$

so that

$$\Delta_0^i = C(y_1^i)^n z_1^{n-1} \cdots w_1.$$

$$\Delta_0^i (x - a_i)^{-\Gamma} i \text{ does not vanish at } a_i, \text{ and}$$

$$\Delta_0^i = C \cdot e^{\int p_1 \, dx}, \text{ so}$$

$e^{\int p_1 \, dx}(x - a_i)^{-\Gamma_i}$ is regular a_i, $\lim\limits_{x \to a_i} (x - a_i)p_1$ is finite, and Γ_1 is the coefficient of $(x - a_i)^{-1}$ in the power series for p_1 near a_i. Likewise near $a_{\rho+1} = \infty$, $\Gamma_{\rho+1} =$

$$\sum_k r_{\rho+1,k} + \frac{n(n-1)}{2} \text{ is the coefficient of } x^{-1} \text{ in the power series for } p_1 \text{ near } \infty.$$

So

$$\sum_{i=1}^{\rho+1} \Gamma_i = K - (\rho-1)\frac{n(n-1)}{2}.$$

But $\lim_{x\to\infty} xp_1$ being finite entails first that p_1 is rational, second, that the degree of the denominator of p_1 exceeds that of the numerator by at most 1, so $\sum_{i=1}^{\rho+1} \Gamma_i = 0$ and Fuchs has obtained the following equation for K, the sum of the exponents:

$$K = (\rho-1)\frac{n(n-1)}{2}$$

1865, §4, equation 8, 1866, §4 equation 10].

This reduces K_0 to a constant, which cannot be zero because Δ_0^1 cannot vanish.

$$K_k := \Delta_k^1 (x-a_1)^{-\Gamma_1} \cdots (x-a_\rho)^{-\Gamma_\rho} \text{ for } k = 1, \ldots, n$$

is likewise related to p_k, and turns out to have a degree at most

$$-K + (\rho-1)\frac{n(n-1)}{2} + (\rho-1)k = (\rho-1)k,$$

so it is of the form

$$F_{(\rho-1)k}(x)/[(x-a_1)\ldots(x-a_\rho)]^k,$$

where $F_s(x)$ is a polynomial in x of degree at most s.

The general linear ordinary differential equation of the Fuchsian class is therefore of the form:

$$\frac{d^n y}{dx^n} + \frac{F_{\rho-1}(x)}{\psi}\frac{d^{n-1}y}{dx^{n-1}} + \frac{F_{(\rho-1).2}(x)}{\psi^2}\frac{d^{n-2}y}{dx^{n-2}} + \cdots + \frac{F_{(\rho-1)n}(x)y}{\psi^n} = 0.$$

It can be put in various alternative forms. Near $x = a_i$ for example, setting $F_{k(\rho-1)}(x)(x-a_i)^k \psi^{-k} = P_{ik}(x)$, it takes the form

$$\frac{d^n y}{dx^n} + \frac{P_{i1}(x)}{x-a_i}\frac{d^{n-1}y}{dx^{n-1}} + \frac{P_{i2}(x)}{(x-a_i)^2}\frac{d^{n-2}y}{dx^{n-2}} \cdots + \frac{P_{in}(x)}{(x-a_i)^n}y = 0$$

Upon further setting $y = (x-a_i)^r u$ it becomes

$$\frac{Q_{i0}(x)}{x-a_i}\cdot\frac{d^n u}{dx^n} + \frac{Q_{i1}(x)}{x-a_i}\frac{d^{n-1}u}{dx^{n-1}} + \cdots + \frac{Q_{in}(x).u}{(x-a_i)^n} = 0,$$

where $Q_{ij}(x)$ is a linear combination of the P_{ik}'s, $k = 1, \ldots, j$ and r is one of the exponents $(r_{i1}, \ldots, r_{in})$ at a_i and so satisfies

$$r(r-1)\cdots(r-n+1) - r(r-1)\cdots(r-n+2)p_{i1}(a_i)$$

$$-r(r-1)\cdots(r-n+3)P_{i2}(a_i) - \cdots - P_{in}(a_i) = 0.$$

The advantage of this form is that it can easily be shown to have solutions of the form

$$\sum_0^\infty C_k(x-a_i)^k.$$

To show this Fuchs cleared the denominator $(x-a_i)^n$ from the equation and rearranged it in a form suitable for calculation of the constants C_k. A comparison with the more familiar method of Frobenius is given in Section 2.3 above.

Appendix 4

On the History of Non-Euclidean Geometry

Throughout the nineteenth century geometries were developed which differed from Euclid's. The most prominent of these was projective geometry, whether real or complex; n-dimensional geometries were also increasingly introduced. However, these geometries were considered to be mathematical constructs, generalizing the idea of real existing Space which could still be regarded as *a priori* Euclidean. The overthrow of Euclid is due to the successful development of what may strictly be called non-Euclidean geometry: a system of ideas having as much claim as Euclid's to be a valid description of Space, but presenting a different theory of parallels. In non-Euclidean geometry, given any line l and point P not on l there are infinitely many lines through P coplanar with l but not meeting it. The consequences of this new postulate (together with the other Euclidean postulates, which are taken over unchanged) include: the angle sum of a triangle is always less than π, by an amount proportional to the area — so all triangles have finite area; and there is an absolute measure of length. Moreover, a distinction must be made between parallel lines (which are not equidistant) and the curve equidistant to a straight line (which is not itself straight).

As is well-known, non-Euclidean geometry was first successfully described by Lobachevskii [1829] and J. Bolyai [1831] independently, although Gauss had earlier come to most of the same ideas. The story is well told in Bonola [1912, 1955], and Klein [1972, Chs. 36, 38]. I have argued elsewhere [Gray 1979a, b, and 1989] that the crucial step taken by Lobachevskii and Bolyai was the use of hyperbolic trigonometry. Their descriptions of non-Euclidean geometry are in this sense analytic; it is the power and wealth of their formulae which convince the reader that non-Euclidean geometry must exist, for its existence is in fact taken for granted. Hyperbolic trigonometry, which in turn, successful because it allowed for a covert use of the differential geometric concepts of length and angle in a more flexible way than earlier formulations of the problem would permit. Moreover, the problem of parallels was not taken to be one of the mere logical consistency of a set of postulates. Saccheri, Lambert, and Taurinus, three of its most vigorous investigators, knew that spherical geometry differed from Euclid's, but regarded it as irrelevant. For them the problem had to do with the nature of the straight line and the plane (which they could not really define), and so broke into two parts: could there be a system of geometry with many parallels? and, if so, could it describe Space? The Lobachevskii–Bolyai system seemed to answer both questions affirmatively, and thus to reduce the nature of Space to an empirical question, one Lobachevskii sought unsuccessfully to resolve by measuring the parallax of stars (which is bounded away from zero in non-Euclidean geometry). However, not everyone was convinced, and for a generation little progress was made. It was not until the hyperbolic trigonometry was given an explicit grounding in differential geometry and non-Euclidean geometry based explicitly on the new, intrinsic, metrical ideas that it could be said to be rigorously established. The central figures in this development are Riemann [1854, 1867] and Beltrami [1868]. From 1868 on, non-Euclidean geometry met with increasing acceptance from mathematicians, if not, predictably, from among philosophers.

The growth of non-Euclidean geometry from, say, 1854 to 1880, has seldom been described. I shall concentrate on the more purely mathematical developments, chiefly its formulation in projective terms, for that line leads to Klein and Poincaré, and leaves aside the more philosophical aspects,[1] which are well discussed in Radner [1979], Scholz [1980], and Torretti [1978].

In his *Habilitationsvortrag* [1854] "On the hypotheses which lie at the foundations of geometry", Riemann does not explicitly mention non-Euclidean geometry, and he never mentions the names Bolyai and Lobachevskii. He refers only to Legendre when describing the darkness which he says has covered the foundations since the time of Euclid, and later (Sections II, 5 and III, I) he describes 2 homogeneous geometries in which the angle sum of all triangles is determined once it is known for one triangle, as occurring on surfaces of constant non-zero curvature. This oblique glance at the characterisation of 3 homogeneous geometries is typical of the formulation of the "problem of parallels" since Saccheri, it can be found in many editions of Legendre's *Eléments de Géometrié*, e.g. [1823], and would have alerted any mathematician to the implications for non-Euclidean geometry without its being mentioned by name. On the other hand, not naming it explicitly would avoid philosophical misapprehensions about what he had to say, so one may perhaps ascribe the omission to a Gaussian prudence. It is less certain, however, that Riemann had read Lobachevskii, and very unlikely he had read Bolyai. Only Gauss appreciated them in Göttingen at that time, and it is not known if he discussed these matters with Riemann. Whatever might be the resolution of that small question of Riemann's sources, the crucial idea in Riemann's paper in his presentation of geometry as intrinsic, grounded in the free mobility of infinitesimal measuring rods, and to be expressed mathematically in terms of curvature. This is an immense generalization of Gauss' idea of the intrinsic curvature of a surface [1827], itself a profound novelty. It has the effect of basing any geometry on specific metrical considerations, and so removes Euclid's geometry from its paramount position as the geometry of space and the source of geometrical concepts which are induced onto embedded surfaces. Following Minding [1839], Riemann gave an informal local description of a surface of constant negative curvature as a piece of a curved narrowing cylinder but left the global existence question unresolved, and he did not discuss the connection with non-Euclidean geometry.

Riemann did not seek to publish these ideas. They were somewhat further developed in his Paris Prize entry of 1861, also unpublished until 1876, and first appeared[2] in 1867, after his death. By then Beltrami had independently discovered the import of Gauss' ideas for non-Euclidean geometry. Beltrami's famous *Saggio* [1868] (translated into English in Stillwell [1996]) begins with an account of geometry as based upon the notion of superposability of figures and thus existing on a surface of constant curvature. Beltrami modified the metric on the sphere to obtain a metric with negative curvature $K = -1/R^2$ and a coordinate system in which straight lines had linear equations. This enabled him to map the geometry, which Beltrami said resided on the surface of a pseudosphere (p. 290) onto the interior of a disc with radius a; in this map a point with coordinates (r, θ) is, in the non-Euclidean metric, $\rho - \frac{a}{2} \log \frac{a+r}{a-r}$ away from the (Euclidean) centre of the disc. This metric allows him to derive the formulae of non-Euclidean trigonometry, and he observed that they agreed with the trigonometric formulae of Minding [1839] and Codazzi [1857] for the surface of constant negative curvature, as well as those of Lobachevskii. Beltrami also sketched an account of non-Euclidean three-dimensional geometry — the original starting point of Lobachevskii and Bolyai's descriptions — but his grasp of higher dimensional differential geometry was not yet sufficiently certain.

Beltrami described the relationship between non-Euclidean geometry and its image in the disc as one in which approximate or qualitative Euclidean pictures are given of non-Euclidean figures. He discussed the way in which such pictures, necessarily distorted, may mislead the mind if features proper to the one geometry (such as points at infinity) are illicitly transposed to the other. He did not state that the non-metrical projective properties of non-Euclidean geometry are represented exactly in the disc — a point made soon afterwards by Klein — and he did not indicate that this gave a projective proof of the existence of the new geometry — for him the new geometry was grounded in differential geometric considerations.[3] The crucial point, clearly grasped by Beltrami, is that his description is the first global description of the space of non-Euclidean geometry. By contrast, Riemann's model has unacceptable features globally: self-intersecting geodesics and a boundary of singular points.

In a paper published soon afterwards, [1868b], Beltrami was able to extend his reasoning to n dimensions and give a thorough account of three-dimensional non-Euclidean space. This paper, but not the earlier one, makes extensive reference to Riemann's recently published *Habilitationsvortrag*, so it seems that Beltrami only learned of them after his first paper was finished. Scholz [1980, 107] confirms that this is so, but that a delay in the publication of the [1868a] enabled Beltrami to read Riemann's work before his own was published, and to add a note linking it so the forthcoming [1868b]. His source for Lobachevskii's work was Houël's translation of some of it [1866], but he does not seem to have known of Houël's French translation of Bolyai's *Tentamen* [1867].

The Italians had become very interested in non-Euclidean geometry. Battaglini and Forti published reports on it in 1867 and Battaglini also translated Bolyai into Italian in 1868. The French, led by Houël, were discovering the new geometry too, and in 1866 German interest, awakened by the publication of the Gauss-Schumacher correspondence[4] [Gauss 1860], was decisively quickened by Helmholtz's essays [1866] and, more important, [1868]. Christoffel's first works on differential geometry [1868, 1869a,b] also appeared at this time, on the subject of transformations of differential expressions like $\sum g_{ij} dx_i dx_j$ (which represent Riemannian metrics). The English, notably Cayley, were anomalies and require separate discussion.

The generally accepted view of these developments is that only ignorant philosophers and elderly mathematicians refused to accept the new geometry, but this view has been claimed to be historically inexact by Toth [1977, 144]. However, Toth's criticism of philosophers seems to be that they were not ignorant (and so had no excuse), and the mathematicians he assembles are a mixed bunch. In Russia, Ostrogradskii and Buniakovskii; in England Dodgson (= Lewis Carroll), de Morgan, and Cayley; and in Hungary the astonishing case of Wolfgang Bolyai, the father of Janos. The Russians' furious criticism of Lobachevskii dates from the 1830s and 1840s (when Gauss taught himself Russian so as to be able to follow it) and so predates the acceptance of differential geometric concepts as suitable foundation for geometry. Without it Lobachevskii's work is indeed open to criticism, and in other works [1829, 1837] Lobachevskii made unsuccessful attempts to ground his geometry in something more basic than his intuitive ideas of lines and surfaces. As for Wolfgang Bolyai, Toth cites no evidence for his claim, and I can find none to support it in the two volumes written by Stäckel on the Bolyai's [Stäckel, 1913], so it must be set aside.

Dodgson's and de Morgan's importance should not be greatly stressed, so Toth's argument rests on only one major mathematician, Cayley, and cannot really be accepted. Once a grave weakness in the Bolyai–Lobachevskii approach had been repaired, non-Euclidean geometry commanded widespread acceptance among mathematicians, as can be seen from the annual reviews in *Fortschritte*, if depressingly little among philosophers[5] (Frege among them). The case of Cayley is interesting, however, and has recently been ably treated by Richards [1979]. She argues that Cayley accepted non-Euclidean geometry as a piece of mathematics, but refused to accept differential geometry as the foundations of geometry because of its empiricist philosophical undertones. Cayley preferred to regard the new geometry as an intellectual construct based ultimately on projective or even Euclidean geometry given *a priori*. Richards also cites Jevons [1871] who gave a related argument, endorsed by Cayley, that the geometry on the tangent space to the surface of constant negative curvature is Euclidean. Their absolutism was opposed by Clifford precisely because Clifford wanted to entertain empiricism elsewhere, in religion, morality, and politics, whereas conservatives saw mathematics as an example of *a priori* reasoning which endorsed such reasoning on those other issues. So the English debate on non-Euclidean geometry became embroiled in other contemporary debates which gradually advanced science at the expense of religious orthodoxy.

Cayley's purely mathematical papers made a more secure contribution. In "A Sixth Memoir upon Quantics" [1859a] he described how a metrical geometry may be defined with respect to a fixed conic which he called the "Absolute" in (complex) projective space. He first developed the projective geometry of such a space, taking as the points of the geometry all the points of projective space. He then introduced a metric by considering two points (x, y) and (x', y') as defining a line which met the Absolute in two points, and defining equidistance in terms of cross-ratio (making it a projective invariant). The additivity of distance led him to define the distance between the given points as

$$\frac{\cos^{-1}(a, b, c)(x, y)(x', y')}{\sqrt{(a, b, c)(x, y)^2}\sqrt{(a, b, c)(x', y')^2}}$$

(§211)

His notation is:

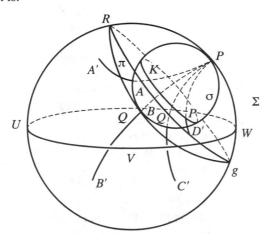

Non-Euclidean 3-space in the Poincaré model. The 2-sphere Σ, PRU-VWS, bounds a ball in Euclidean 3-space but its interior can be made to carry a metric which makes it a model of non-Euclidean 3-space. The non-Euclidean lines $PAA', \dots PDD'$ are represented by arcs of circles perpendicular to Σ. The smaller sphere, σ, $PABCD$ touches Σ internally at P, so it represents what Lobachevskii called a horosphere. The spherical shell, π, RQS, meets Σ at right angles, so it represents a non-Euclidean plane. It touches π at Q. The non-Euclidean metric inside Σ induces, remarkably, the Euclidean metric on σ. Lines like PDD' which meet π only at points of Σ are asymptotic to π. They cut σ in a circle which represents Cayley's absolute conic, κ, which bounds a model of non-Euclidean 2-space in σ as Σ (Cayley's absolute quadric) bounds a model in 3-space:

$$(a, b, c)(x, y)(x', y') := axx' + b(xy' + x'y) + cyy'$$

$$(a, b, c)(x, y)^2 := ax^2 + 2bxy + cy^2$$

and the equation of the absolute is $(a, b, c)(x, y)^2$. In the simplest general case (§225) the absolute is $x^2 + y^2 + z^2 = 0$ (in homogeneous form), and the distance between (x, y, z) and (x', y', z') is

$$\frac{\cos^{-1} xx' + yy' + zz'}{\sqrt{(x^2 + y^2 + z^2)}\sqrt{(x'^2 + y'^2 + z'^2)}} .$$

Cayley defined angles dually. What he considered most important was the distinction between a proper conic as the Absolute and a point-pair as Absolute. The choice of proper conic led to spherical geometry; the choice of the point-pair to Euclidean geometry (§225).

Cayley did not consider non-Euclidean geometry at all in this paper. Subsequently, in [1865], he quoted Lobachevskii's observation that the formulae of spherical trigonometry become those of hyperbolic or non-Euclidean trigonometry on replacing the sides a, b, c of the spherical triangle by ai, bi, and ci, respectively [Lobachevskii 1837], and went on to say, "I do not understand this; but it would be very interesting to find a *real* geometrical interpretation [of the equations of hyperbolic trigonometry]" (Cayley's italics). Such an interpretation was soon to be given by Felix Klein.[6]

In early 1870 Felix Klein was in Berlin with his new friend Sophus Lie, attending the seminars of Kummer and Weierstrass on geometry. Weierstrass was interested in the Cayley metrics, and Klein raised the question of encompassing non-Euclidean geometry within this framework, but Weierstrass was skeptical, preferring to base geometry on the idea of distance and to define a line as the curve of shortest length between its points.[7] Klein was later to claim that during seminars he and Lie had had to shout their view from the back, and he was not deterred by Weierstrass' opinion. Spurred on by his friends, notably Stolz, who had read Lobachevskii, Bolyai, and von Staudt (which Klein admitted he could never master [1921, 51–52]) he continued to work on his view of geometry. It became crucial to his unification of mathematics behind the group concept, and his *Erlanger Program* [1872] reflects his satisfaction at succeeding. His first paper on non-Euclidean geometry [1871] presents the new theory. The space for the geometry is the interior a fixed conic (possibly complex) in the real projective plane. Klein called the geometries he obtained hyperbolic, elliptic, or parabolic, according as the line joining any two points inside the conic meets the conic in real, imaginary, or coincident points (§2). Non-Euclidean geometry coincides with hyperbolic geometry. Euclidean geometry is Cayley's special case of a highly degenerate conic. In the hyperbolic case the distance between two points is defined as the log of the cross-ratio of the points and the two points obtained by extending the line between them to meet the conic (divided by a suitable real constant c(§3)).[8] This function is additive, since cross-ratio is multiplicative, and puts the conic at infinity. Angles were again defined dually. The Cayley metric yields a geometry for which the infinitesimal element of length can be found, and in this way Klein showed that the geometry on the interior of the conic was in fact non-Euclidean (§7). Klein showed that his definition of distance gave Cayley's formulae when $c = i/2$ and the conic is purely imaginary. Klein then considered the group of projective transformations which map the conic to itself, showed that these motions were isometries, and interpreted them as translations or rotations (§9). He also considered elliptic and parabolic geometry, and extended the analysis to geometries of three dimensions. In a subsequent paper [1873] he developed non-Euclidean geometry in n dimensions.

Klein's view of non-Euclidean geometry was that it was a species of projective geometry, and this was a conceptual gain in that it brought harmony to the proliferating spread of new geometries which Klein felt was in danger of breaking geometry into several separate disciplines (see the *Erlanger Program*, p. 4). In particular it made clear what the group of non-Euclidean transformations was, and Klein regarded the group idea as the key to classifying geometries. But this view was perhaps less well adapted to understanding the new geometry in its own right than was the more traditional standpoint of differential geometry.

The most brilliant exponent of the traditional point of view was Poincaré. It would be interesting to know how he learned of non-Euclidean geometry, but we are as ignorant of this as we are of the details of most of his early career. He might have read Beltrami's [1868], perhaps in Houël's translation [Beltrami 1869] published in the Annals of the Ecole Normale Supérieure, or any of Houël's articles, or the papers [J. Tannery 1876, 1877] or [J.M. de Tilly 1877]. He might have learned of it from Hermite, or might have been introduced to the works of Helmholtz and Klein (he could read German, and Helmholtz had been translated into French), but in any case did not say. The description he gave of non-Euclidean geometry is a novel one. In it geodesics are represented as arcs of circles perpendicular to the boundary of a circle or half plane. The group of non-Euclidean proper motions is then obtained as all matrices $\begin{pmatrix} a & b \\ \bar{b} & \bar{a} \end{pmatrix}$, $a\bar{a} - b\bar{b}$ if the circle is $|z| = 1$, or as all matrices $\begin{pmatrix} a & b \\ c & d \end{pmatrix}$, $ad - bc = 1$, $a, b, c, d, \in \mathbf{R}$, in the half-plane case. In each case a single reflection ($z \to \bar{z}$ or $z \to -\bar{z}$, respectively) will then generate the improper motions. Poincaré's representation of non-Euclidean geometry is conformal, and may be obtained from Beltrami's as follows. Place a sphere, having the same radius as the Beltrami disc, with its South pole at the disc and project vertically from the disc to the Southern hemisphere. Now project that image stereographically from the North pole back onto the plane tangent to the sphere at the south pole. The first projection maps lines onto arcs of circles perpendicular to the equator, and since stereographic projection is conformal, these arcs are then mapped onto circular arcs perpendicular to the image of the equator. To see that the groups correspond, it is enough to show that they are triply transitive on boundary points and that the map sending any given triple to any other is unique, which it is. The metric may be pulled back from the one representation to the other, or calculated directly from the group. It is particularly simple in the half-plane model:

$$ds^2 = \frac{dx^2 + dy^2}{y^2}.$$

Since the elements of the group are all compositions of Euclidean inversions in non-Euclidean lines, they are automatically conformal.

The main text of this chapter describes how Poincaré published his interpretation of non-Euclidean geometry. To my knowledge, there is no account of how he came to discover it. I have argued elsewhere [1982] that Poincaré did not know of Klein's Erlanger Program in 1880 and seems only to have known of Beltrami's work on non-Euclidean geometry. This belief is confirmed by a letter Poincaré wrote to Lie in 1890 in which he explicitly disclaimed any knowledge of Klein's tract, which was surely little known in 1880. I am indebted to T. Hawkins for informing me of this letter, which he discusses in passing in his convincing paper arguing that the *Erlanger Program* had much less influence than is generally believed [Hawkins 1984]. D. Rowe [1983] has also pointed out that the written Programm is quite different from the address Klein gave at his inauguration, which was on mathematics and education.

Appendix 5

The Uniformisation Theorem

The uniformisation theorem is one of the central theorems in the theory of Riemann surfaces.[1] Its original proclamation was based on little more than counting constants, but with surprising rapidity in 1883 Poincaré stated a remarkable generalisation of the theorem, and gave it the outlines of a proof (Poincaré [1883]). He let y be any many-valued analytic function of x, and claimed that one can always express x and y as single-valued functions of a complex variable z. To prove this, Poincaré considered m functions $y_1, y_2, \ldots, y_m$ of x. The value of each y_i would be known at a point x when its value was known at some initial point x_0 and a path from x_0 to x was specified. He supposed that the point x was one coordinate of a point moving on a Riemann surface having infinitely many leaves. To study this surface he proposed to construct what today is called its universal cover, which topologically is a disc. His method was to construct a new Riemann surface by opening out the loops on the surface which corresponded to non-trivial analytic continuations of any of the m functions $y_1, y_2, \ldots, y_m$. To obtain this surface took the arbitrary point x_0 as the starting point for all loops drawn on the surface. As one then traced a loop starting and finishing at x_0, the values of the m functions $y_1, y_2, \ldots, y_m$ varied. For each loop he asked whether the values of any of the functions were different at the start and the end of the loop. If any were, he said the loop was of the first sort, otherwise it was of the second sort. Among the loops of the second sort, some could be continuously shrunk to a point without ever losing their defining property; they were said to be of the first type. Those that could not be shrunk in this manner he said were of the second type. Poincaré then said that the initial and final points of a Riemann surface corresponding to these functions had a point for each loop of the first sort and for each loop of the second type and the second sort. Each loop of the first type and the second sort yields only one point on the new Riemann surface. Because all the non-trivial loops have been opened out, and have a common base point, O, the new Riemann surface is simply connected, and is topologically a disc, as Poincaré remarked.

As later mathematicians commented, this stage amounts to a topological construction of the universal covering surface. It remains to show that the covering map is itself analytic. To this end Poincaré used potential theory to obtain a suitable harmonic function. He considered a sequence of nested discs spanned by non-intersecting circles C_n such that every point of the surface is contained in such a disc. The first disc is obtained by finding a small disc, centered on O, within which all the functions y_i are holomorphic. This boundary of this disc is then surrounded by circles within which the functions y_i remain holomorphic; the boundary of the region so formed is the first circle, C_1. The construction is then repeated indefinitely. Poincaré then looked for a suitable Green function. He used the elliptic modular function ϕ which is holomorphic except at 0, 1, and ∞ to define a function ψ which is holomorphic and maps the upper half plane to the interior of the Unit Disc. Without loss of generality, he set $y_m = \psi$. Then defining the function t to be

$$\log \left| \frac{1}{\psi} \right|,$$

Poincaré obtained a function t which was essentially positive, logarithmically infinite at certain points, and harmonic.

He then introduced a sequence of functions u_n which are harmonic inside C_n and vanish on C_n, except at the common point O where they became logarithmically infinite. It followed that each function u_n was positive everywhere inside its boundary contour C_n. Consequently, so were

$$u_{n+1} - u_n, \quad \text{and} \quad t - u_n,$$

and so he deduced that the series

$$u = u_1 + (u_2 - u_1) + \cdots + (u_{n+1} - u_n) + \cdots$$

converges, because the u_n's increase with n and are bounded above by the function t. The demonstration fails when the function t is infinite, and Poincaré gave a separate argument to show how it can be modified.

Poincaré then showed that the function u is continuous, and that the series defining it converges uniformly. Moreover, it is harmonic away from the point O (where it is logarithmically infinite). To prove this, Poincaré assumed that there was a function U which solved the Dirichlet problem for any contour c in a small region of the Riemann surface, in that U agreed with the function u on the contour c and was harmonic inside c. But U was the limit of the functions u_n, so $U = u$, and therefore the function u is harmonic. A separate argument ensured that u was harmonic at other points where t became infinite but the functions u_n remained finite. So Poincaré assumed that the Dirichlet problem was always solvable, i.e., that there always exists harmonic function on a simply connected domain having prescribed continuously varying values on an arbitrary boundary. It is remarkable that this part of his proof was not to be questioned.

Because the function u is harmonic, it has a harmonic conjugate v making $u+iv$ a holomorphic function. The functions z and z_n defined by the equations

$$z = e^{-(u+iv)}, z_n = e^{-(u_n+v_n)}$$

are then well-defined and Poincaré showed that z_n is one-to-one inside C_n and that z is one-to-one everywhere. It followed that the functions y_k are uniformised by z, because the surface S was constructed in such a way that they have a unique value at each point of S.

Poincaré added two notes which show that he was aware of the central weakness in his approach. He excluded singular points from the Riemann surface, banishing them to its boundary. In particular, he noted that the points where the modular function is not holomorphic are singular. They spawn infinitely many points on the boundary of the Riemann surface S. For this reason his proof was to be criticised by Hilbert, in the course of presenting the topic as the 22nd of his 23 mathematical problems. Hilbert stressed that it was extremely desirable to check that the uniformising map was indeed surjective. Poincaré was to endorse this criticism when he took up the question again in 1907, and gave two ways of dealing with it. He added that his original method left it uncertain whether the conformal map of S mapped it onto the unit circle or merely onto a part of it. "The problem", he said, "is none other than the Dirichlet problem applied to a Riemann surface with infinitely many leaves".

In 1907 Poincaré and Koebe independently proved the uniformisation theorem by rigorous methods. Poincaré began his [1907], published in *Acta Mathematica*, by reviewing his earlier paper and the attempts by others to overcome or else avoid the problems which it raised. Then he outlined his new approach. He characterised the problem, as he had done earlier, as a Dirichlet problem for a Riemann surface with infinitely many leaves. Then he set out, first to make more precise and more supple the concept of a Riemann surface, by enlarging Weierstrass' notion of an analytic element to include ramification points. Then he constructed a Green function for his surface, using his *méthode de balayage* and simplifying it using Harnack's theorem (Harnack [1887]). From the Green's function he deduced the existence of a function that mapped the Riemann surface conformally into the unit disc. He then compared the different functions that could be used to this end, and showed that they are all linearly related. He could then show, following Osgood [1900], that the conformal representation was indeed onto the interior of the disc.

Poincaré's definition of a Riemann surface, D, implied that it was covered by a countable number of open discs, D_n, each the domain of a convergent power series. For such a surface he proceeded as follows. He defined a function that was positive at all but one point, O, of the first disc, D_0, where it was logarithmically infinite, and which was zero on the boundary of D_0. He extended this function to the rest of D by defining it to be zero outside D_0, so it is continuous away from O, but its derivatives are not continuous on the boundary of D_0. He then defined a function u_{n+1} inductively, using his *méthode de balayage*. In this way he obtained at stage $n+1$ a function that was generally positive on the domain consisting of all the discs contiguous with the discs encountered at stage n. The points of discontinuity in the derivatives are the boundary points of the domain of definition of u_{n+1}. The function was harmonic everywhere it was positive except at O. To apply Harnack's theorem it was therefore enough to ensure that there was a point, P, of the domain D at which the sequence of function values $u_n(P)$ did not increase indefinitely. The sought-after convergence was ensured by introducing a suitable majorising function, which Poincaré found among the classes of Fuchsian functions he had studied at the start of his career. It followed that the sequence (u_n) tended to a Green function on D.

From the Green function, which is harmonic inside D except at the point O, Poincaré could obtain by repeated use of Harnack's theorem the function called v in his memoir of 1883, and thence the function $z = e^{-(u+iv)}$ that mapped the domain D into the unit disc. He also showed that if instead of the arbitrarily chosen point O he had begun with another, O' say, and had been led to a function z', then the functions z and z' would be connected by a Möbius transformation.

Poincaré gave two proofs of what the memoir of 1883 had left obscure: the function z mapped the domain D onto the interior of the unit disc, one following Osgood's and one of his own. The memoir then proceeded to analyse the cases when a given simply connected region is to be represented conformally not on a circle but on the whole plane, and showed how these cases could be distinguished.

Poincaré's paper came out at the same time that Koebe began his work. Koebe was a student of Schwarz's in Berlin, and wrote his doctoral dissertation in 1905. He then embarked on a lengthy series of papers which quite deliberately and successfully brought him to the attention of the leading German mathematicians. He published at length in the *Göttingen Nachrichten*, making sure his papers were presented by Hilbert and Klein, in the *Journal für Mathematik, Mathematische Annalen*, and the *Comptes Rendus*. In these papers he solved the problem of uniformising algebraic and analytic curves, the Riemann mapping theorem and its generalisation to non-simply connected domains, and then turned to rescue the old continuity method of Klein and Poincaré. As a result of all this activity, he was invited to speak at the International Congress of Mathematicians in Rome on the subject, quite an honour for a young man.

In the second of these papers, [1907b] he observed that Poincaré's original method had the defect that "certain points of the domain must be excluded which cannot necessarily be excluded in the nature of the problem". He then outlined his own approach to the uniformisation theorem for analytic curves. The main result he established is that the interior of any Riemann surface over a simply connected domain in the plane may be mapped one-to-one and conformally onto one of three regions on the sphere: the entire sphere, the sphere minus a point, or a disc. So his proof of the uniformisation theorem includes the Riemann mapping theorem as a special case. The paper concludes with the interesting comment that

> The idea of an analytic domain (*analytische Gebilde*) with an independent variable is connected by Weierstrass with the representation of infinitely many uniformly convergent series with rational terms . . . .

The fundamental problem is to find a selection of these uniformly convergent series of rational functions which represents the whole domain. Here it is shown that this can always be done. In this way a problem which one might say belongs to the Weierstrassian mode of analysis is solved by principles which belong to the Riemannian circle of ideas. [p. 210]

Koebe also took note of every other entrant in what he plainly saw as a competition to establish the uniformisation theorem, and in this spirit Koebe compared his approach with that of Poincaré in his [1907c]. He noted that he avoided the use of modular functions entirely, unlike Poincaré and others such as Osgood, which he felt had led to a notable simplification of the argument. Like Poincaré, he had relied on Schwarz's methods, and he had made a modest use of Harnack's theorem, upon which Poincaré had relied heavily. The comparison inspired Koebe to give a new proof of his theorems the next year, completely avoiding Harnack's theorem. Indeed, Koebe's ambition drove him to give several proofs of his results. When Hilbert revived Dirichlet's principle by establishing a somewhat different minimising principle, Koebe responded with an explanation of how these ideas lay close to his own. When others gave simpler proofs using techniques drawn only from complex function theory, Koebe showed that he too could operate in that way. Knowing that Fricke was at work on improving the original approach to the uniformisation theorem due to Poincaré and Klein (the continuity method) Koebe showed how that too could be rigorised, thus entering Brouwer's territory and extending Brouwer's proof of the invariance of dimension.

Faced with so much to read, no-one disputed the rigour of Koebe's methods. His division of the uniformisation theorem into two parts was also accepted. The first part is topological and asserts the existence of a simply connected covering surface for any Riemann surface. The second part is analytic and asserts that the map from this covering surface to exactly one of three surfaces (the sphere, plane, or disc) is analytic. Indeed, his proofs of the uniformisation theorem were usually taken as definitive; Fricke acknowledged "with the greatest thanks" not only Koebe's papers but "the many hours of conversations on a whole series of points" that had helped him write his (and Klein's) [1912]. The chief response of several authors despite, or perhaps because of Koebe's work, was the desire to give short direct proofs of what seemed buried under the torrent of his papers, many of them very long.

Koebe's work caused such a stir that a special meeting of the *Deutsche Mathematiker Vereinigung* was held (in Karlsruhe, 27 September 1911) to discuss recent work on automorphic functions, and a report was published the next year (*Jahresbericht*, **21**, 153–166). Brouwer spoke on his proof of the invariance of dimension under a one-to-one, continuous map, with reference to the Fuchsian case. Koebe replied that he had been able to extend this to other cases, and gave a Schottky-type example, where, he said, Poincaré's methods could not work. He then gave his own report on the uniformisation theorem for both analytic and algebraic curves. Bieberbach reported briefly on single-valued automorphic functions, and Hilb on many-valued ones. Klein summed up, expressing the view that "one must learn to calculate with single-valued automorphic functions as easily as one can calculate with elliptic functions", and referred in this spirit to his old paper on *Primformen* [1889]. Later Osgood (like Fubini [1908]) adopted Koebe's new argument about Green's functions. He summarised Koebe's papers ("a task", he dryly noted, "of some labour"), in the second edition of his *Funktionentheorie*, vol. 1, and in a paper [1913]. Koebe's proofs, like Poincaré's, made use of a sequence argument in which the right sort of function is defined on a steadily larger domain, composed of copies of the cut-up, simply-connected version of the original Riemann surface. Koebe proved uniform convergence directly; Osgood simply regarded this as a consequence of Montel's theorem (Montel, [1907]) By Montel's Theorem, the limiting function is analytic, uniquely defined at each point of the limit domain and maps it onto a single-leaved region of the t-plane with a discrete boundary.

The matter did not end there. Courant [1912] eliminated potential theory with an astute use of conformal mappings. Weyl gave his own account, based on Koebe's work, in his influential *Ideé* [1913, reprinted 1923]. Bieberbach drew on his study of the distortion theorem to write his [1918], where he gave a proof, using Weyl's idea of triangulating the Riemann surface, which its author claimed placed the uniformisation theorem firmly in the spirit of Weierstrassian function theory and avoided not only potential theory but also considerations of topology. All was to be rewritten by Teichmüller in the late 1930s, from whom the modern approach derives.

Appendix 6

Picard–Vessiot Theory

Picard began his work on the Galois theory of differential equations with a paper in 1883, which he amplified in his paper [1887]. Basing himself on the work of Lie, Picard outlined the general theory as far as what became called the Picard–Vessiot double Theorem, and illustrated it in a discussion of algebraic groups in 2 or 3 parameters that drew heavily on the work of Lie. This work was taken up by Vessiot [1892], who made more explicit the connection to the earlier work of Lie, and gave necessary and sufficient conditions for a linear ordinary differential equation to be solvable by quadratures (the analogous case to being solvable by radicals). This work was conveniently republished by Picard in the third volume of his *Traité d'analyse* published in 1895 (see Ch. XVII, section II), and I follow that account here before comparing it with Klein's response and the account given by Schlesinger in his *Handbuch* [1897], noting only that the study of algebraic groups for their own sake was not resume in the *Traité*.

Picard considered the situation where $\mathbf{Y}$ and $\mathbf{y}$ are n-dimensional vectors of functions related by linear equations, so $\mathbf{Y} = \mathbf{Ay}$, where $\mathbf{A}$ is a matrix of constants chosen arbitrarily in a group.[1] He proposed the problem of finding rational functions of the y's and the derivatives y', y'', etc., that are invariant under the group. Mindful of the analogy, he called such functions *symmetric functions*; Schlesinger's name for them was *rational differential functions*, which will be preferred here.

Picard argued that given the y's, one should form the linear differential equation which has them as a basis of solutions:

$$\frac{d^n y}{dx^n} + p_1 \frac{d^{n-1} y}{dx^{n-1}} + \cdots + p_n y.$$

It is clear that the coefficients of this differential equation are rational differential functions. More interestingly, Picard established that any rational differential function is rational in the p's and their derivatives (I omit the proof).

Picard next considered rational differential functions in the y's and their derivatives. In particular he looked for such functions $R(\mathbf{y})$ which are invariant with respect to matrices $\mathbf{A}$, so $R(\mathbf{Y}) = R(\mathbf{y})$. In general, only the identity matrix has this property, but for some functions R a non-trivial homogeneous linear algebraic group of such transformations can be found. He noted that the converse question (given such a group, find the corresponding rational differential functions R) had already been dealt with by Lie. This led him to discuss the reducibility of a differential equation, and to set out the elementary theory of linear algebraic groups.

In Section IV, Picard took up the theme of what has become known as Picard–Vessiot theory. He took a linear ordinary differential equation (A) in x of order m, whose coefficients, the $p(x)$'s, are rational functions of x, and let a set of functions $\{y_1, \ldots, y_m\}$ be a basis of solutions of (A). He defined

$$V = \sum_{i,k} u_i y_k,$$

where the u_i are m arbitrary rational functions of x, and noted that V satisfies a differential equation (E) of order m^2 (called by Schlesinger [1897, 2.1, p. 60] the Picard resolvent).

Solutions are of equation (E) are sums of $u_i y_k$. For arbitrary u's, these $u_i y_k$ are linearly independent (Picard had originally assumed this, Beke proved it in [1895] and Picard recapitulated that proof here). Evidently, a solution of equation (E) would give rise to m solutions of equation (A), but these need not be a basis. They would be a basis if and only if the determinant $D = |y_k^{(i)}|$ of the y's did not vanish. In general, solutions of equation (E) have no solutions in common with a differential equation of lower order (other than the determinant equation), so it is interesting when this happens. When it does, suppose that (F), the other differential equation for V, is of order p. Then G, the group of equation (F) depends algebraically on p parameters. Picard now required that the functions y be solutions of equation (F), and denoted the group of transformations of equation (F) by G; it depends algebraically on p parameters. To distinguish it from the monodromy group of an equation, Picard called this group the group of transformations of the equation. Here, Klein's name for this group, the rationality group, will be preferred (Klein [1984], p. 511).

Picard now proved an important result later called by Schlesinger [1897, 2.1, p. 71] the Picard–Vessiot double Theorem:

Theorem. *Every rational differential function of x, y's and their derivatives, as a rational function of x is invariant as a function of x when acted upon by the rationality group G. Conversely, every rational function of x, the y's and their derivatives which is invariant under G is a rational function of x. (i.e., of x alone.)*

Picard gave examples to illustrate what can happen in low-dimensional cases. The rationality group might contain only one parameter. This can happen in one of only two ways.

$$(1) \qquad \frac{dy_1}{dx} = ay_1, \qquad \frac{dy_2}{dx} = by_2.$$

The solutions to this system of equations is algebraic if and only if a/b is rational. The group is now represented as

$$Y_1 = y_1 \theta^m,$$

$$Y_2 = y_2 \theta^n.$$

The following expressions are invariant for this group:

$$\frac{dy_1}{dx}/y_1, \quad \frac{dy_2}{dx}/y_2, \quad \text{and} \quad \frac{y_2^m}{y_1^n}.$$

It follows that there is a basis of solutions for which these three expressions are rational functions of x. This has implications for Riccati's equation, which may be written either as

$$\frac{dy}{dx} + ay^2 = bx^m,$$

or, on setting

$$y = \frac{1}{a}\frac{u'}{u},$$

as

$$\frac{d^2y}{dx^2} = abx^m u.$$

Precisely because $\dfrac{u'}{u}$ is a solution, it has two rational solutions which can be found "by quadratures."

(2) $\qquad \dfrac{dy_1}{dx} = ay_1, \quad \dfrac{dy_2}{dx} = y_1 + ay_2.$

This leads to the group

$$Y_1 = y_1 e^{at},$$

$$Y_1 = y_1 t e^{at} + y_2 e^{at},$$

which is algebraic if and only if $a = 0$, and the group reduces to

$$Y_1 = y_1,$$

$$Y_1 = y_1 t + y_2.$$

In this form, it is clear that the function y_1 is invariant and therefore rational. A second invariant is

$$y_2 \frac{dy_1}{dx} - y_1 \frac{dy_2}{dx}.$$

It is therefore also rational and so y_2 is obtained from a rational function by quadrature.

Picard could also show that if the rationality group contained two parameters, the solutions of the differential equation could be found by quadrature. Although all the groups depending algebraically on three parameters could also be enumerated, Picard preferred to point out that now that there was no reason to suppose that the corresponding differential equations were solvable by quadrature. Indeed, the simple equation

$$\frac{d^2y}{dx^2} + p_2(x)y = 0,$$

where $p_2(x)$ is an arbitrary rational function of x, has this three parameter group:

$$Y_1 = ay_1 + by_2$$

$$Y_2 = cy_1 + dy_2,$$

with $ad - bc = 1$, so these groups do not imply that the original differential equation is solvable by quadrature.

Instead, vindication of the Galois metaphor was to be found elsewhere. Picard took a 3rd order linear ordinary differential equation with rational coefficients and regular singular points, and supposed that its solutions, y_1, y_2, and y_3, satisfied an algebraic equation $F(y_1, y_2, y_3) = 0$ of genus greater than 1. Such a surface has only a finite group of symmetries, by Hurwitz's Theorem. But this symmetry group is also that of equation (E). So

$$\frac{y_2}{y_1} \quad \text{and} \quad \frac{y_3}{y_1}$$

are rational functions of x, and now indeed y_1, y_2, and y_3 are to be found by quadrature. If moreover the roots of the indicial equation are rational, then all solutions are algebraic.

The problem is harder if the genus is 0 or 1. Picard went back to results of Klein and Lie from 1870 to deduce Fuchs' Theorem: if F of degree greater than 2, then (E) is solvable algebraically, and if F has degree 2 then, as Fuchs had shown, equation (E) has as its solution the square of a solution to a linear ordinary differential equation of order 2.

In the final section of this chapter, Picard summarised, from his own rather different point of view, the results of Vessiot's Thesis on the reduction of the group G of a linear ordinary differential equation which is solvable by quadratures. Picard began with a rational function ϕ of a basis, whose coefficients as functions of x belong to the given domain of rationality. Then replacing the y's by their values as functions of V, ϕ satisfies a differential equation of order

$$p : \phi = \chi \left(V, \frac{dV}{dx}, \ldots, \frac{d^p V}{dx^p} \right).$$

In general, ϕ depends on p parameters and so it satisfies a differential equation of order p. But it can happen that ϕ depends on $p' < p$ parameters, so satisfies a differential equation S of order p', whose coefficients as functions of x belong to the same original domain. In this case, replace V by v, a solution of the differential equation for ϕ in terms of V and its derivatives, and consider the function ϕ. It will be invariant under a group G' of $p - p'$ parameters, which depend on v, which in turn might have subgroup Γ independent of the choice of v. When this happens, Γ leaves invariant all the solutions of equation S.

The reduction process consists of adjoining a function ϕ which is a general solution of S, and invariant under Γ. When this is done, the group of the equation becomes Γ. Moreover, the group Γ is an invariant (i.e., normal) subgroup of the group G. Following Lie, Picard looked for a descending chain of normal subgroups, each successive group depending on one fewer parameter until the last group had no subgroup which depends on arbitrary parameters, and into which no more normal subgroups could be inserted. This Lie had called a *normal decomposition* of a group G, and Picard observed that one could construct a theory analogous to Jordan's (but he gave no details). A group was said to be *integrable* if it had a normal decomposition in which the number of arbitrary parameters went down by 1 each time until the last group depends on a single parameter. The main result could then be given:

Theorem. *A linear ordinary differential equation is solvable by quadratures if and only if its rationality group is integrable.*

Picard proved both halves of the theorem by induction.

Klein's response to the work of Picard and Vessiot was swift. Basing himself on Picard's Toulouse publication [1887] and Vessiot's thesis [1892], he considered its implications for the Schwarzian differential equation towards the end of his lectures on the hypergeometric function (Klein [1894]). It is clear, he said, that a rational differential invariant of the group of all projective transformations

$$\eta' = \frac{\alpha\eta + \beta}{\gamma\eta + \delta}$$

is again a function of x, but this does not mean that this group is the rationality group of the equation, only that it contains it. The rationality group is the smallest group whose differential invariants are rational functions of x. In fact, said Klein, he had already listed the 12 cases that arise when studying the hypergeometric equation in his lectures on higher geometry ([1892], Vol. 2, p. 278). These included the group consisting of the identity transformation alone, for which η is the invariant, the group of transformations of the form $\eta_1 = \epsilon^i \eta$, where ϵ is an nth root of unity, and the corresponding invariant is $\eta^n + \eta^{-n}$, the dihedral group and those associated with the Platonic solids, groups of such transformations as $\eta_1 = \alpha\eta$, for which the invariant is $\dfrac{\eta'}{\eta}$, and so on up to the largest case, where the group consists of all transformations of the form

$$\eta_1 = \frac{\alpha\eta + \beta}{\gamma\eta + \delta}$$

and the invariant is

$$\frac{\eta'''}{\eta'} - \frac{3}{2}\left(\frac{\eta''}{\eta'}\right)^2.$$

Klein discussed some of these cases in detail, and then commented (pp. 517–8): "The advance of the new conception of Picard–Vessiot consists in subsuming the discovery of special cases that earlier authors had discovered under a general principle, the group principle, in particular the principle of the rationality group. The example of the η-function is the first to which the Picard–Vessiot analysis can be applied". He went on to suggest that their approach be carried through fully for higher-dimensional cases.

Schlesinger's summary was in some ways easier to follow (*Handbuch*, II.1, Section 9). He too began with the observation that, given a linear ordinary differential equation, if one takes a rational differential function and forms a differential equation of least order with coefficients in the same ground field as the given equation, then adjoining a solution of the new differential equation reduces the rationality group of the differential equation to a normal subgroup. More generally, suppose one is given a differential equation (A) with rationality group G, and another differential equation (A') with rationality group G'. If adjoining a solution of (A') reduces G to a normal subgroup H depending on s fewer parameters, then adjoining a solution of (A') reduces G' to a normal subgroup H' depending on s fewer parameters. So in particular, if (A') is of order 1, then the solution of the 1st order differential equation is a rational differential function of the basis of (A), and the H involves one fewer parameter. Consequently, the existence of a chain of such 1st order differential equations implies that the group G is integrable, and if the group is integrable, then the differential equation is solvable by quadrature.

Schlesinger then investigated the implications of assuming that the original differential equation was of the Fuchsian class. For example, a differential equation with all its solutions algebraic has a finite rationality group. In this case the rationality group and the monodromy group of the differential equation coincide. For a general differential equation of the Fuchsian class, its rationality group is the smallest algebraic group that contains the monodromy group. This is false in general, as the example of a linear ordinary differential equation with constant coefficients shows. When the roots of the characteristic equation of such a differential equation are distinct, the solutions are exponential functions, $y_i = e^{r_i x}$ and so the monodromy group is trivial, and trivially algebraic. But the rationality group of the differential equation is determined by

$$ r_1 = \frac{1}{y_1} \frac{dy_1}{dx}, $$

which of course belongs to the domain of rationality. So G is generated by the infinitesimal transformations

$$ y_1 \frac{\partial f}{\partial y_1}, \ldots, y_n \frac{\partial f}{\partial y_n}, $$

which makes it an integrable group. Indeed, Schlesinger showed that the Picard–Vessiot double Theorem is valid with G replaced by the monodromy group if and only if the differential equation is of the Fuchsian class. This corrected a remark due to Klein (*Höhere Geometrie*, vol 2, p. 361).

Finally, he observed that a linear ordinary differential equation with rational coefficients is reducible (i.e., has solutions in common with another such linear ordinary differential equation of lower order) if and only if its rationality group is reducible (i.e., can be expressed as a group in fewer variables). So if the differential equation is of the Fuchsian class, it is reducible if and only if its monodromy group is reducible.

For much of the 20th Century, differential algebra was kept alive by Ritt, Kaplansky and Kolchin. The first edition of this book brought me into contact with Michael Singer and the recent revival of work on differential Galois theory in which he has played a prominent role (see also Magid [1994]). The wish to find computer algorithms for solving linear ordinary differential equations has evidently been a further stimulus to reopen questions th' t, as Singer has pointed out in several papers, were opened up and then forgotten in the pre-computer age.

Joseph Liouville was perhaps the first to work systematically on the question of what functions could be obtained from what others by the standard processes of algebraic operations, taking exponentials, and integration. He showed that the elliptic functions could not be obtained in this way from the known (algebraic and trigonometric) functions (see Lützen [1990]). For this reason extensions that are obtained in this way are called Liouvillian (Singer [1981]). A natural question is to ask if every Picard–Vessiot extension of the differential field $C[z]$ (the field of rational functions with complex coefficients) is a Liouvillian extension. This is dealt with by the remarks on integrable groups already mentioned. In his paper [1992] Singer revived the methods of Fuchs (in Chapter III above) to compute the minimal polynomial of algebraic solutions of a differential equation with an primitive differential Galois group. He pointed out that an imperfect algorithm had already been found by Pépin [1881] for the second-order case.

In his paper Singer listed the primitive groups that can arise, and showed how to rederive Fuchs' results (in his [1875]) on a case-by-case basis. It emerges that every solution of the differential equation is a primitive element of a Picard–Vessiot extension, and the degree of an algebraic solution always equals the order of the Galois group, as Pépin and Fuchs had shown. The situation is, however, different for third-order equations.

Finally, I note that the so-called inverse problem of differential Galois theory is solved in the affirmative. Indeed, building on the work of many authors, notably Magid [1994], Mitschi and Singer [1994] establishes in a uniform and constructive way that every connected linear algebraic group is the Galois group of a Picard–Vessiot extension of a field $C(x)$ where $x' = 1$ and C is an algebraic closed field of characteristic zero. In the classical case where $C = \mathbf{C}$, the complex numbers, Tretkoff and Tretkoff [1979] established that every linear algebraic group is the Galois group of a Picard–Vessiot extension. Their work used a weak solution of Hilbert's 21st Problem in which a certain differential equation was required to be Fuchsian; Mitschi and Singer observe that this can be done despite the work of Bolibruch.

In 1992 I was unable to answer Singer's question about Pépin and Boulanger, who also worked in this area, and I take this opportunity to do so. In 1863 P. Th. Pépin published a memoir in the *Annali di Matematica* on Liouville's problem of finding rational integrals of linear differential equations with rational coefficients. Liouville had considered the differential equation in the form $\dfrac{d^2y}{dx^2} = Py$, and looked for solutions in the form of roots of the algebraic equation $f(x, y) = 0$, where he supposed that the degree of y was known in advance. Pépin removed this restriction on y. His solution was largely correct, but it contained errors later pointed out by Fuchs. In his [1881] Pépin returned to the question, which, as we saw in Chapter III, had now embraced the search for algebraic solutions to a differential equation. He simplified and corrected his earlier account by using the theory of complex functions, and then gave explicit and as he hoped effective methods for finding algebraic solutions, when they exist.

Boulanger, in his [1898], observed that despite some work by Jordan and Poincaré (noted in Chapter III) the explicit study of the third order equation had not been much studied, although Painlevé [1887], in one of his earliest publications, had indicated how to extend Klein's method. The method encountered difficulties in estimating upper bounds on the degree of certain rational functions that appeared in it, and in order to make the analysis truly effective Boulanger then devoted the rest of his memoir to that problem. For a modern account, see Singer [1992].

Jean François Théophile Pépin was born in the Haute Savoie on 14 May 1826 and died in Lyons on 3 April 1904. He became a Jesuit in 1846 and a professor at different Jesuit Colleges at various stages between 1850 and 1873, when he became a professor at Canon Law. He went to Rome in 1880. His obituary[2] lists 52 publications. Apart from the two discussed here, there were two lengthy accounts of Gauss' posthumously published theory of elliptic functions; the rest are almost all on number theory. Boulanger was born in Lille in 1866. He studied at the *École Polytechnique* and at the Faculty of Sciences in Lille, where he was much influenced by Painlevé, and in due course he became an adjoint professor of mechanics there. He escaped from Lille shortly after the start of the war, in which he had a distinguished career, and retired as Director of Studies at the *École Polytechnique* in 1921 after a brief reign. He was involved in editing *L'Intermédiaire des Mathématiciens* (chiefly a problem magazine, but his obituary is in (2) **2** 1923, 73–5) and with the French edition of the *Encyklopädie der Mathematischen Wissenschaften*, he was at one time President of the French Mathematical Society, and was awarded its Prix Poncelet. Apart from the work described above, he wrote on mechanics, theoretical and experimental elasticity, and hydraulics. He died in 1923.

Appendix 7

The Hypergeometric Equation in Higher Dimensions; Appell and Picard

In 1770 Euler introduced a partial differential equation which generalised the wave equation, and which was to turn out to have deep connections to the hypergeometric equation and generalisations to functions of two variables. His motivation appears to have been didactic, finding other partial differential equations that can to solved, and not, as Darboux was to suggest (*Surfaces*, **2**, p. 54) to study problems in the propagation of sound. It was studied again by such authors as Laplace and Poisson before being taken up by Riemann [1860] who gave a general method for solving partial differential equations and applied it to a special case of the Euler equation. A very special case of this partial differential equation is the so-called telegraphist's equation. It was introduced by Kirchhoff [1857] and exploited brilliantly by Heaviside [1876], but only caught the interest of mathematicians with Poincaré's paper of [1893]. By then, this work had led a number of French and Italian mathematicians to study the hypergeometric equation in two variables.

The partial differential equations that Euler wrote down in his *Institutiones Calculi Integralis*, **3**, Section 2, Ch 3, are more general than the one that bears his name, and the one that gets closest is a special case. He generally sought power series solutions for them in the form of $f(x) + F(y)$. Darboux, in a remarkably thorough study of this equation (*Surfaces*, **2**,) wrote what he called Euler's equation in the form

$$E_{\beta\beta'} : \frac{\partial^2 z}{\partial x \partial y} - \frac{\beta'}{x-y} \frac{\partial z}{\partial x} + \frac{\beta}{x-y} \frac{\partial z}{\partial y} = 0,$$

where β and β' are constants. The special case $\beta = \beta'$ studied by Euler (§328, p. 217) is also the one Riemann studied. He introduced the adjoint of this equation, and showed that the general solution of the Euler equation can be expressed in terms of simple solutions of the adjoint equation. Darboux pointed out that the special case reduces, on making the substitution $z = f(x - y)^{-\beta}$, to the equation $\dfrac{\partial^2 f}{\partial x \partial y} =$ $\dfrac{\beta(1 - \beta)f}{(x - y)^2}$. Any solution $Z(\beta)$ of this equation is of the form $(x - y)^\beta Z(\beta, \beta)$, where $Z(\beta, \beta')$ denotes a solution of Euler's equation.

Solutions of the general Euler equation which are homogeneous in x and y are found using the substitution $y/x = t$, $z = x^\lambda \phi(t)$. This leads to the ordinary differential equation

$$t(1 - t)\frac{d^2\phi}{dt^2} + (1 - \lambda - \beta - (1 - \lambda - \beta')t)\frac{d\phi}{dx} + \lambda\beta'\phi(t) = 0$$

which is the hypergeometric equation. Darboux showed, following Appell [1882], that if $f(x, y)$ is a solution of Euler's equation, then the general solution of the equation is of the form

$$\varphi\left(\frac{cx + d}{ax + b}, \frac{cy + d}{ay + b}\right)(ax + b)^{-\beta}(ay + b)^{-\beta'},$$

where a, b, c, and d are arbitrary constants. In this way, Euler's equation is completely solved.

Riemann, dealing with the special case, made some ingenious substitutions to reduce it to

$$(1 - z)\frac{d^2 y}{d\log z} - z\frac{dy}{d\log z} + (\lambda + \lambda^2)zy = 0.$$

This equation, as he remarked, has this solution (which takes the value 1 at $z = 0$):

$$P\begin{pmatrix} 0 & -\lambda & 0 \\ & & z \\ 0 & 1+\lambda & 0 \end{pmatrix},$$

and a great many other solutions are also available as hypergeometric series and as integrals.

Riemann's paper is one of the major paper in applied mathematics of the period. In it he gave a rigorous method for solving certain types of hyperbolic partial differential equations, described how shock waves form, and predicted solutions in which pitch varies as the wave propagates. A brief history going from Riemann via Christoffel and Hugoniot to Hadamard and beyond will be found in Hölder [1981]. In his short note on Riemann's paper (in Riemann [1990] 807–810) Lax comments that in the linear case Riemann's solution method was successfully extended by Hadamard to admit any number of space variables, but that much remains to be done in that direction for the non-linear case, despite much successful numerical work.

If in the equation $\dfrac{\partial^2 f}{\partial x \partial y} = \dfrac{\beta(1 - \beta)f}{(x - y)^2}$ one substitutes $\beta - x$ for x and then lets β tend to ∞, one obtains successively $\dfrac{\partial^2 f}{\partial x \partial y} = \dfrac{\beta(\beta - 1)f}{(\beta - x - y)^2}$ and then the

equation $\dfrac{\partial^2 f}{\partial x \partial y} = f$, which is called the telegraphist's equation, first written down
by Kirchhoff in his [1857] and then studied by Heaviside in a series of papers from
1876. It is interesting to explain the physical interpretation of this equation.

The full form of the telegraphist's equation

$$A \frac{\partial^2 U}{\partial t^2} + B \frac{\partial U}{\partial t} = \frac{\partial^2 U}{\partial x^2}$$

describes the propagation of electricity in a long straight wire. It is satisfied by
the current at any point, and by the potential. The constants that appear in the
equation involve the conductance, the self-inductance, the resistance and the leakage
conductance of the wire. In the ideal case of no resistance and no leakage B and
C vanish, and the equation reduces to the wave equation. In the case when the
inductance is negligible by comparison with the resistance, the constant A may be
taken to be zero, and the equation is parabolic. It is in fact the one-dimensional heat
equation, and was treated in this spirit by William Thomson (later Lord Kelvin) in
1855 when he was advising on the laying of Atlantic cable. It follows that the time
necessary to produce the maximum electrical effect at a distance x is proportional
to x^2.

Despite the success of that enterprise, the character of the equation is very
different when neither resistance nor leakage is negligible. In this case neither B
nor C vanish, and the equation can be reduced to the form

$$\frac{\partial^2 U}{\partial t^2} = \frac{\partial^2 U}{\partial x^2} + U$$

(provided there is some resistance and some leakage). The method of separation of
variables combined with the theory of Bessel functions now shows that the general
solution with initial conditions at $t = 0$ of $U = F(x)$, $\dfrac{\partial U}{\partial t} = G(x)$ is the sum
of three terms. One is a wave propagating with velocity proportional to $1/x$, as
before, although now it is exponentially damped. But the other two terms are like
the tail of a wave that never dies. Therefore, when an attempt is made to transmit a
periodic wave down the wire, the velocity and wavelength depend on the frequency
and the waves undergo dispersion, unless, and this was to be Heaviside's remarkable
discovery in 1887, the values of the physical constants can be so adjusted that the
rate of dispersion is zero. This can be done both mathematically and physically;
it merely requires that the leakage be non-zero. Far from being an inconvenience,
this condition is necessary for the production of distortionless telephony. The signal
becomes fainter over distances, but this can be corrected by fitting amplifiers. Long
distance telegraphy had dealt with distortion by accepting a low transmission rate,
so as to separate the pulses. Telephony required much higher frequencies; with
some leakage and a deliberately high self-inductance it became distortionless. Long
distance communication was reborn – although the money for the first successful
patents went to the American electrical engineer Michael Pupin in 1901, and not to

Heaviside. The reader is referred to Yavetz's fascinating book for a detailed account of the whole story, which explains among other things why it took Heaviside from 1881 to 1887 to make the (apparently) simple observation about distortion, and how his work fitted into contemporary ideas of electromagnetic theory.

Solutions of the telegraphist's equation were given using Riemann's method by Picard in 1894, the year after Poincaré had solved it by means of complex Fourier integral methods under the initial conditions that $U(x, 0) = f(x)$ and $\dfrac{dU}{dt}(x, 0) = f_1(x)$ when $t = 0$ (Poincaré [1893]). Poincaré considered that case when the functions $f(x)$ and $f_1(x)$ are given by polynomials inside the interval $b < x < a$ and vanish outside it. He quickly found using the theory of residues that U was zero outside the interval $b - t < x < a + t$, and had four discontinuities, at $a \pm t$ and $b \pm t$, which propagate at the speed of light. Other initial conditions were investigated. For example, if $f(x) = 0$ for all x, and $f_1(x) = \pi/2$ inside $(-\epsilon, \epsilon)$ and 0 outside, then the solution is given by a Bessel function on the interval $(-t, t)$ and is zero outside. In any case, the head of the disturbance moves with a finite speed, as is the case with the transmission of light but not of heat, and the head, once it has passed, leaves behind a disturbance that never vanishes, which does not happen with the wave equation.

In his short paper, Poincaré did not mention the ingenious discoveries of Heaviside. No reason why he should, but the omission of Heaviside's discovery from his *Cours sur les oscillations électriques* is more striking. It is not surprising that French mathematicians were not reading the English Maxwellian. Such a breadth of scholarship would have been remarkable, and it is apparent from his other writings that Poincaré read Maxwell but not all his British successors, but it also shows the originality of Heaviside's insight. Poincaré may have missed it because his simplification of the telegraphist's equation depends on the non-vanishing of $B^2 - 4AC$, whereas the distortionless condition is exactly that $B^2 - 4AC = 0$.

Meanwhile, in 1880 the two variable story had begun, when Paul Appell, one of a new generation of French mathematicians that included Picard and Goursat as well as Poincaré, wrote down a generalisation of the hypergeometric equation as an equation in 2 variables, and found four series generalizing the hypergeometric series [1880]. He took the general term in the product of the series for $F(\alpha, \beta, \gamma; x)$ and $F(\alpha', \beta', \gamma'; y)$, which is

$$\frac{(\alpha, m)(\alpha', n)(\beta, m)(\beta', n)}{(\gamma, m)(\gamma', n)(1, m)(1, n)} x^m y^n, \text{ where } (\lambda, k) = \lambda(\lambda + 1) \ldots (\lambda + k - 1),$$

and replaced one, two, or three of $(\alpha, m)(\alpha', n)$, $(\beta, m)(\beta', n)$, or $(\gamma, m)(\gamma', n)$ by the corresponding expression in $(\alpha, m + n)$, $(\beta, m + n)$, or $(\gamma, m + n)$. This gave him five possibilities. He rejected the double series whose general term is $\dfrac{(\alpha, m + n)(\beta, m + n)}{(\gamma, m + n)(1, m)(1, n)} x^m y^n$, because it only gives $F(\alpha, \beta, \gamma; x + y)$. The remaining four double series he called F_1, F_2, F_3, and F_4. He investigated the domain of convergence of the double series, found the linear relations that exist between contiguous functions (defined by analogy with Gauss' work), and showed how the new functions can degenerate to the usual Gaussian hypergeometric functions. He also showed that each of these new functions satisfies two simultaneous partial differential equations. Full details followed in the much longer [1882].

Very quickly Picard, in his [1880c], obtained the first of these generalised hypergeometric functions in a way modelled on Riemann's theory of P-functions. He showed that there was a (many-valued) function $F(x, y)$ of two complex variables with these properties: any 4 branches satisfy a linear relation with constant coefficients; the function is holomorphic unless one of x or y is 0, 1, or ∞ or $x = y$; there is specified branching behaviour near singular points of the form $(0, 0)$, $(0, 1)$, $(0, \infty)$, $(1, 0)$, etc., and at points (α, α), $\alpha \neq 0, 1, \infty$. More precisely, two branches are holomorphic and a third is a holomorphic function multiplied by a power of x, $(x - 1)$ or $1/x$.

For each fixed value of y, the function thus specified satisfies a third-order linear ordinary differential equation in x, and similarly when x is fixed, the function satisies a differential equation as a function of y. In each case this is Pochhammer's differential equation. When the exponents in the branching conditions are symmetrical in x and y, the differential equations have three linearly independent solutions in common, which, Picard showed, can be expressed as integrals of the form

$$\int_g^h u^{b_1-1}(u-1)^{b_2-1}(u-y)^{b_3-1}(u-x)^{\lambda-1}\, du$$

where h and g are different choices among $\{0, 1, y, x\}$. For suitable values of the exponents, this function agrees with Appell's function F_1. In his [1881] Picard studied the analytic continuation of the new functions by representing them as double integrals, and showed that F_1 satisfies three simultaneous partial differential equations, one of which is the Euler equation. In his thesis of 1881 Goursat used contour integration to rederive the 24 Kummer relations between solutions to the hypergeometric equation, an approach he said (in his *Acta* paper) that had greatly pleased Hermite. He showed that there were 6 solutions represented as integrals of the form

$$\int_g^h u^{\beta-1}(u-1)^{\gamma-\beta-1}(1-ux)^{-\alpha}\, du = \int_g^h V\, du$$

where the limits of integration are $(0, 1)$, $(0, -\infty)$, $(1, \infty)$, $(0, 1/x)$, $(1, 1/x)$, $(1/x, \infty)$. Each can be represented in 4 ways by a hypergeometric series, whence Kummer's 24 solutions. Between any 3 such integrals there is a linear relation (so 20 relations in all), all of which can be derived from comparing $\int_{-\infty}^0 V\, du$, $\int_0^1 V\, du$, and $\int_1^\infty V\, du$ using the Cauchy Integral Theorem.

In two papers in the *Comptes Rendus* [1882a,b] Goursat showed that Kummer's 24 solutions generalise to a family of 60 solutions of the simultaneous partial differential equations, and that F_2 and F_3 can be characterised in the way that Picard had characterised F_1, by presupposing linear relations between any five branches of one of the functions. Then, in a long paper in *Acta Mathematica* [1883], Goursat extended this analysis and drew on work of Pochhammer [1870] to show that F_2, F_3, and F_4 each satisfy a pair of simultaneous partial differential equations (which are of the kind to which Darboux was to show Riemann's methods apply) and that they each satisfy a pair of partial differential equations. In his [1885], Picard showed how to connect this work to his study of what he called hyper-Fuchsian functions, which uniformise certain kinds of algebraic surface, by showing that they satisfy systems of three partial differential equations similar to the Euler equation.

After this profusion of activity, Pincherle [1888] and Lauricella [1893] made a successful attempt to obtain the generalisation to n variables, and then, perhaps daunted by the complexity of the problem, it seems that mathematicians turned aside to other matters. Accounts of the more conceptual modern approach, which leads deep into algebraic geometry, can be found in the book and article by Holzapfel [1986a, b], and in the book by Yoshida.

Notes

Introduction

1. Revised 1998.

Chapter I

1. The most thorough recent treatment of all these topics is Houzel's essay in [Dieudonné, 1978, II]. Gauss' work on elliptic functions and differential equations is treated in the essays by Schlesinger [Schlesinger 1909a,b] and in his *Handbuch* [Schlesinger 1898, Vol. II, 2]. Among many accounts of the mathematics, two which contain valuable historical remarks are Klein's *Vorlesungen Über die hypergeometrisché Function* [1894] and, more recently, Hille's *Ordinary Differential Equations in the Complex Domain* [1976].

2. An integral form of the solution was also to prove of interest to later mathematicians:

$$f(x) = \int_0^1 u^{b-1}(1-u)^{c-b-1}(1-xu)^{-a}\, du$$

where the integral is taken as a function of the parameter. It is assumed that $b > 0, c - b > 0, x < 1$, so that the integral will converge. It is easy to see, by differentiating under the integral, that $f(x)$ satisfies (1.1.1). To obtain f as a power series, expand $(1 - xu)^{-a}$ by the binomial theorem and replace each term in the resulting infinite series of integrals by the Eulerian Beta functions they represent. An Euler Beta function is defined by

$$B(b + r, c - b) = \int_0^1 u^{b+r}(1-u)^{c-b-1}\, du;$$

it satisfies the function equation or recurrence relation

$$B(b + r + 1, c - b) = \frac{b+r}{c+r} B(b + r, c - b),$$

which yields precisely the relationship between successive terms of the hypergeometric series, and so $f(x)$ is, up to a constant factor, represented by (1.1.2). This argument is given in Klein [1894, 11], see also [Whittaker and Watson, 1973, Ch. XII].

3. Biographical accounts of Gauss can be found in Sartorius von Waltershausen, *Gauss zum Gedächtnis* [1856] G.W. Dunnington [1955], and H. Wussing [1979]. Accounts of most aspects of his scientific work are contained in *Materialien für eine wissenschaftlichen Biographie von Gauss*, edited by Brendel, Klein, and Schlesinger 1911–1920 and mostly reprinted in [Gauss, *Werke* X.2, 1922–1933]. The best introductions in English are the article on Gauss in the *Dictionary of Scientific Biography* [May, 1972], which is weak mathematically, Dieudonné [1978], and W. K. Bühler [1981].

4. There are detailed discussions in Schlesinger [1898] and [Gauss, *Werke*, X.2, 1922–1933], [Krazer, 1909], [Geppert, 1927], and [Houzel in Dieudonné 1978 vol. II]. In particular Geppert supplies the interpretation of the fragments in [Gauss, *Werke* III, 1866, 361-490] on which this account is based.

5. Later Gauss calculated $M(1, \sqrt{2})$ to twenty decimal places, his usual level of detail, and found it to be 1.19814 02347 35592 20744 [*Werke* III, 364].

6. This is not contiguous in the sense of Arbogast [1791], which more or less means continuous.

7. The lemniscate has equation $r^2 = \cos 2\theta$ in polar coordinates, and arc-length $\int_0^t \frac{dx}{(1-x^4)^{\frac{1}{2}}}$. Introduced by Jacob Bernoulli in 1694, it had been studied by Fagnano (1716, 1750) who established a formula for the duplication of arc, and by Euler (1752) who gave a formula for increasing the arc n times, n an integer. Euler considered the problem an intriguing one because it showed that the differential equation $\frac{dx}{(1-x^4)^{\frac{1}{2}}} = \frac{dy}{(1-y^4)^{\frac{1}{2}}}$ had algebraic solutions: $x^2 + y^2 = c^2 + 2xy(1-c^4)^{\frac{1}{2}} - c^2x^2y^2$. The comparison with the equation $\frac{dx}{(1-x^2)^{\frac{1}{2}}} = \frac{dy}{(1-y^2)^{\frac{1}{2}}}$, which leads to $x^2 + y^2 = c^2 + 2xy(1-c^2)^{\frac{1}{2}}$, guided his researches. (It provides the algebraic duplication formula for sine and cosine.) Gauss observed that, whereas in the trigonometric case the formula for multiplication by n leads to an equation of degree n, for the lemniscatic case the formula is of degree n^2, and invented double periodicity on March 19, 1797, see Diary entry #60, to cope with the extra roots. He was fond of this discovery, and hints dropped about it in his *Disquisitiones Arithmeticae* (§335) inspired Abel to make his own discovery of elliptic functions. An English translation of Gauss' diary is available in [Gray, 1984b and corrigenda].

8. Gauss had discussed the multiple-valued function $\log x$ in a letter to Bessel the previous year [*Werke* II, 108] and also stated the residue theorem for integrals of complex functions around closed curves. From the discussion in Kline [1972, 632–642], it seems that Gauss was well ahead of the much

younger Cauchy on this topic; see also Freudenthal [1971], and Bottazzini [1981, 133].

9. For a comparison of the work of Abel and Jacobi see [Krazer 1909] or [Houzel in Dieudonné, 1978, II]. It seems that Abel was ahead of Jacobi in discovering elliptic functions; there is no doubt he was the first to study the general problem of inverting integrals of all algebraic functions. Jacobi's work was perhaps more influential because of his efforts as a teacher and an organizer of research, and also because Abel died in 1829 at the age of 26.

10. The expression $\dfrac{\frac{d\lambda}{dk}\frac{d^3\lambda}{dk^3} - \frac{3}{2}\left(\frac{d^2\lambda}{dk^2}\right)^2}{\left(\frac{d\lambda}{dk}\right)^2}$ has come to be known, following Cayley [1883], as the Schwarzian derivative of λ with respect to k. Schwarz's use of it is discussed in detail in Chapter III.

11. See [Edwards, 1977] for a discussion of Kummer's number theory and [Lampe 1892–3] and [Biermann 1973] for further biographical details about Kummer.

12. I am indebted to S. J. Patterson who sent me his copy of Kummer's own summary of his [1836], which he found in the Widener Library, Harvard.

13. Kummer wrote this equation as

$$2\frac{d^3z}{dx\,dx^2} - 3\left(\frac{d^2z}{dz\,dx}\right)^2 = \left(2\frac{dP}{dz} + P^2 - 4Q\right)\left(\frac{dz}{dx}\right)^2 + (2\frac{dp}{dz} + p^2 - 4q)$$

using the then customary notation for higher derivatives.

14. The symmetries in the arrangement of the 24 solutions are analysed in Prosser [1994].

15. In section III Kummer studied the special cases when γ depends linearly on α and β and quadratic changes of variable produce other hypergeometric series. Typical of his results is his equation 53:

$$F(\alpha, \beta, \alpha-\beta+1, x) = (1-x)^{-\alpha} F\left(\frac{\alpha}{2}, \frac{\alpha - 2\beta + 1}{2}, \alpha - \beta + 1, \frac{-4x}{(1 - x)^2}\right).$$

His results were incomplete and were extended by Riemann [1857a, §51].

16. Neuenschwander [1979, 1981b] discusses the limited use Riemann had for the theory of analytic continuation, of which he was certainly aware, see e.g., [1857c, 88–89].

17. A discussion of Riemann's topological ideas and their implications for analysis will be found in Pont [1974 Ch. II]. Pont does not make as much as he should have done of the distinction between homotopy- and homology-theoretic ideas. Better discussions will be found in [Scholz, 1980, Chapter 2], and Laugwitz [1996].

18. For a history of Cauchy's Theorem see Brill and Noether [1892–93, Chapter II], who also discuss Gauss' independent discovery of it.

19. Riemann wrote (A), (B), and (C) for A, B, and C respectively.

20. Riemann wrote (b) for B' and (c) for C'. I have introduced vector notation purely for brevity.

21. E. Scholz tells me there is evidence in the Riemann *Nachlass* to show that Riemann had read Puiseux. See also Neuenschwander, [1979, 7].

Chapter II

1. These biographical details are taken from Biermann, [1973b, 68, 94, 103].

2. See Bottazzini [1986].

3. Weierstrass [1841]. These series are called Laurent series after Laurent's work, reported on in Cauchy, [1843]. Laurent's paper is reprinted in Peiffer [1978].

4. *Singuläre Punkte* = singular points, in this case poles of finite order.

5. In his [1868, §8] Fuchs investigated the singularities more closely, and found that there are exceptional cases when a singular point of a coefficient function does not give rise to a singular point of the solutions. Such a point he called an accidental singular point (*ausserwesentlich singulärer Punkt*) in contradistinction to the other singular points which he called essential (*wesentlich*), a term he attributed to Weierstrass. Since "essential" when applied to singularities now means something different, I shall use the word "actual" for them instead.

A singular point $x = a$ is accidental if the determinant of a fundamental system vanishes, for $p_i = \dfrac{-\Delta_i}{\Delta_0}$; so if $p_i(a) = \infty$ and no solution is infinite at $x = a$, then $\Delta_0(a) = 0$. Fuchs showed that for $x = a$ to be accidental, it is necessary and sufficient:

(i) that the equation have the form

$$(x-a)^n \frac{d^n y}{dx^n} + (x-a)^{n-1} p_1(x) \frac{d^{n-1} y}{dx^{n-1}} + \cdots + p_n(x) y = 0,$$

where the p_i's are analytic near $x = a$;

(ii) that $p_i(a)$ be a negative integer;

(iii) that the roots of the indicial equation at $x = a$ all be different and are positive numbers or zero; and finally

(iv) that none of the solutions contain logarithmic terms.

6. See Hawkins [1977].

7. Fuchs did not call the w *Eigenwerthe* (eigenvalues), and had no special term for them other than "roots of the fundamental equation".

8. I introduce the vector notation purely to abbreviate what Fuchs wrote in co-ordinate form.

9. Fuchs did not specify the sense in which a circuit of a point is to be taken, but it must be taken in an agreed way each time.

10. See Biermann [1973b, 69–70].

11. Strictly, Fuchs has made a mistake: (2.1.9) admits solutions $n = 1$, ρ arbitrary or $n = 2$, $\rho = 2$.

12. If n_λ is the period taken along the λth part, $n_\lambda = \frac{1}{2} \int_{(\lambda)} y \, dx$, then

$$\frac{d^i n_\lambda}{du^i} = \frac{1}{2} \frac{\alpha^i u}{\alpha y^i} \, dx,$$

where (λ) is any appropriate curve. The solutions of (2.2.2) are regular. Indeed, if Π is the product of the difference of the roots of y^2, then for each λ, $\Pi \eta_\lambda$ is everywhere finite and non-zero away from $k_1, \ldots, k_n$:

$$\tilde{\Pi} = \prod_{i<j} (k_i - k_j) = \prod_{i<j} [(k_i - x) - (k_j - x)]$$

$$\sum c_\ell (k_1 - x)^{\ell_1} (k_2 - x)^{\ell_2} \ldots (k_{n-1} - x)^{\ell_n}.$$

where c_ℓ is an integer, the numbers $\ell_1, \ell_2, \ldots, \ell_n$ are all taken from the set $\{0, 1, \ldots, n - 1\}$, and $\ell_1 + \ell_2 + \cdots + \ell_n = \frac{1}{2} n(n - 1)$. The quotient $\tilde{\Pi}/y$ is a sum of terms of the form

$$(-1)^{n/2} c_\ell (k_1 - x)^{\ell_1 - \frac{1}{2}} (k_2 - x)^{\ell_2 - \frac{1}{2}} \ldots (k_{n-1} - x)^{\ell_n - \frac{1}{2}}$$

is finite even at $k_1, \ldots, k_{n-1}$. Near k_i, $\tilde{\Pi}$ has the form $(u - k_i)^{m\pi}$ where m is a positive integer and $\tilde{\Pi}'$ is a finite, continuous, single-valued function of n vanishing at $u = k_i$. This makes the functions η_λ regular.

13. Legendre's relation had been generalised by Weierstrass [1848/49] to a system of $2n^2 - n$ relations between periods of a hyperelliptic integral of order n, and, as a determinant, by Haedenkamp [1841], who, however, considered the

periods only over real paths. Fuchs was easily able to extend Haedenkamp's, work, and he found (Fuchs [1870b]) that if

$$\eta_{ij} = \int_{\gamma_j} \frac{x^i \, dx}{[\phi(x)]^{\frac{1}{2}}},$$

where

$$\phi(x) = (x - u)(x - k_1) \ldots (x - k_{n-1})$$

and γ_j is a path from k_{j-1} to k_j $(1 \le j < n - 1)$ and γ_1 connects u and k_1, then the $(n - 1) \times (n - 1)$ determinant

$$H := \begin{vmatrix} \eta_{01} & \eta_{02} \cdots & \eta_{0,n-1} \\ \eta_{11} & \eta_{12} \cdots & \eta_{1,n-1} \\ \cdots\cdots & \cdots\cdots & \cdots\cdots \\ \eta_{n-2,1} & \eta_{n-2,2} \cdots & \eta_{n-2,n-1} \end{vmatrix}$$

is independent of u, as Haedenkamp had claimed. So it can be evaluated when $\phi(x) = x^n - 1$, and Fuchs found

$$H = \frac{(-1)^{\frac{n-1}{4}} (2\pi)^{\frac{n}{2} - 1}}{(n - 2)(n - 4) \ldots 3.1}, n \text{ odd}.$$

$$H = \frac{(-1)^{\frac{n}{4}} (2\pi)^{\frac{n}{2} - 1} \pi}{(n - 2)(n - 4) \ldots 2}, n \text{ even}. \quad [1870, 135, \text{ in } 1904, 291]$$

Legendre's relation is the special case $n = 2$.

14. The first generalization of the hypergeometric series is due to Heine [1848, 1847], who studied the series

$$\phi(\alpha, \beta, \gamma, q, x) = 1 + \frac{(1 - q^\alpha)(1 - q^\beta)}{(1 - q)(1 - q^\gamma)} x$$
$$+ \frac{(1 - q^\alpha)(1 - q^{\alpha+1})(1 - q^\beta)(1 - q^{\beta+1}) x^2}{(1 - q)(1 - q^2)(1 - q^\gamma)(1 - q^{\gamma+1})} + \cdots$$

for α, β, γ, x real or imaginary, and q real. He showed that, as $\dfrac{1 - q^\epsilon}{1 - q} \to \epsilon$ as $q \to 1$, $\phi(\alpha, \beta, \gamma, 1, x) = F(\alpha, \beta, \gamma, x)$. For particular values of a, b and particular q, Heine's series can represent any of Jacobi's θ-functions, so it stands in the same relation to elliptic functions as Gauss' series to the trigonometric functions. However, Heine's series satisfies not a differential equation but a difference equation with respect to x. This difference equation was later studied by Thomae [1869] from a Riemannian point of view. Thomae also

sought to generalise the hypergeometric equation. Thomae [1870] studied the power series $y = 1 + \dfrac{a_0 \cdot a_1 \cdot a_2}{1 \cdot b_1 b_2} x + \dfrac{a_0(a_0 + 1)a_1(a_1 + 1)a_2(a_2 + 1)x^2}{1.2.b_1(b_1 + 1)b_2(b_2 + 1)} + \cdots$
and showed that many of the properties of the hypergeometric series carried over to the higher hypergeometric series, as he called them. For example, the series converges inside $|x| = 1$; it satisfies an nth order linear ordinary differential equation, it can be written as an $(n - 1)$-fold integral, contiguous functions can be defined in the obvious way, and any $n + 1$ contiguous functions satisfy linear relationships with rational coefficients. Thomae was much influenced by Riemann's [1857a] and presented his solutions as generalised P-functions, which he denoted $F\begin{pmatrix} \alpha & \alpha' \dots \alpha^{(h-1)} \\ \beta & \beta' \dots \beta^{(h-1)} \end{pmatrix} x$. For computational reasons he concentrated in his [1870] on the 3rd order case, and deduced a host of relationships for the monodromy coefficients, his generalizations of Riemann's α_β's. In his [1874], Thomae generalised the P-function to a function $P\begin{pmatrix} \alpha, & \beta, & \gamma, & \delta \\ \alpha', & \beta', & \gamma', & \delta, \end{pmatrix} k, x$ which had branch points at $0, \infty, 1$, and $1/k$ and for which the exponent pairs were α, α', at 0, etc. (and $\alpha + \alpha' + \beta + , \cdots + \delta' = 2$). He showed that much of Riemann's theory went over to the new functions with inessential changes, for example P satisfies a second order differential equation, but that P depended in an essential and complicated way on a parameter t (which, in later terminology, is an accessory parameter). Thomae showed that if $\delta = \delta' = 0$ then t can be chosen arbitrarily without affecting the differential equation satisfied by P, and that if $\delta = 1, \delta' = 0$, the equation reduced to Riemann's form of the hypergeometric equation.

15. New readers can start with Thomé's own summary [1884].

Chapter III

1. There are several alternative forms for the Schwarzian which can be of use. For example

$$\Psi(s, x) = \frac{2s's''' - 3s''^2}{2s'^2} = \frac{s'''}{s'} - \frac{3}{2}\left(\frac{s''}{s'}\right)^2 \text{ (where } s' = \frac{ds}{dx}, \text{ etc.)}$$

and this can be written in terms of the logarithmic derivative as

$$\Psi(s, x) = \frac{d^2}{dx^2}\left(\log \frac{ds}{dx}\right) - \frac{1}{2}\left(\frac{d}{dx}\log \frac{ds}{dx}\right)^2.$$

2. There is a change of variable which can simplify the original differential equation and help to make the argument clearer. Set $y = gu$ where $g = e^{-\frac{1}{2}\int p\,dx}$. Then the differential equation for y becomes the following equation for u:

$$\frac{d^2u}{dx^2} = Pu, \text{ where } P = \frac{1}{4}p^2 + \frac{1}{2}\frac{dp}{dx} - q.$$

P is a rational function whose only singularities are those of the original p and q, and for such equations, if the quotient of two linearly independent solutions is algebraic, then so are all solutions. It was this form of the general second order differential equation that was later studied by Fuchs. Set $\eta = \frac{f}{g}$, the quotient of two independent solutions of $y'' + Py = 0$. Then

$$\frac{\eta'''}{\eta'} - \frac{3}{2}\left(\frac{\eta''}{\eta'}\right)^2 = 2P.$$

3. Kronecker's example is discussed in Neuenschwander, [1977, 7] which is based on Casorati's notes of a conversation with Kronecker, 16 October 1864.

4. See Kiernan [1972], Wussing [1969], Purkert [1971, 1973].

5. $A_m(ju_1)^m + A_{m-1}(ju_1)^{m-1} + \cdots + A_0 = 0$, and (3.2.2) is irreducible, so $\frac{A_{m-i}}{A_m} = \frac{A_{m-i}}{A_m}j^{-i}$, which implies, since $A_m \neq 0$,

$$A_{m-i} = 0 \quad 1 \leq i \leq m, \text{ or } j^{-i} = 1.$$

The irreducibility of (3.2.2) precludes $A_{m-i} = 0$ for all i, so j is a primitive root of unity.

6. The most general expression which is the root of a rational function is $y = (z - a_1)^{\alpha_1}(z - a_2)^{\alpha_2}\ldots(z - a_\rho)^{\alpha_\rho}g(z)$, where $g(z)$ is a polynomial, and $\alpha_1 \ldots, \alpha_n$ are rational numbers which can be assumed not to be positive integers. If y is to satisfy an nth order differential equation, then the singular points of the equation must be $a_1, \ldots, a_\rho$ and the exponents must be $\alpha_1, \ldots, \alpha_n$. When y is substituted into the given differential equation, an equation is obtained for the coefficients of $g(z)$. If this equation has solutions, then y is a solution of the differential equation, and if not, not.

7. Suppose $g_1, g_2, \ldots, g_N$ are the elements of G. Then $\Pi(x) = (x - g_1a)(x - g_2a)\ldots(x - g_Na)$ is a polynomial of order N which vanishes on the orbit of G containing the point a, and any other polynomial with those zeros is a linear multiple of Π. In particular $\Pi(gx)$ has the same zeros, is monic, and so coincides with $\Pi(x)$, and $\Pi(x)$ is said to be G-invariant. Let $\Pi'(x) = (x - g_1a')\ldots(x - g_Na')$ be the monic polynomial defining the orbit containing a' and $Q(x)$ a polynomial defining the orbit of a point b. To show $Q(x) = \kappa\Pi(x) + \kappa'\Pi'(x)$, consider $Q(x) - Q(a)$.

It vanishes when $x = a$ and is G-invariant, so it is some multiple of $\Pi(x)$, say $Q(x) - Q(a) = \beta\Pi(x)$. Likewise $Q(x) - Q(a') = \beta'\Pi'(x)$, so

$$Q(x)[Q(a') - Q(a)] = Q(a')\beta\Pi(x) - Q(a)\beta'\Pi'(x)$$

or $Q(x) = \dfrac{Q(a')\beta\Pi(x) - Q(a)\beta'\Pi'(x)}{(Q(a') - Q(a))}.$

8. Transvectants are a somewhat mysterious collection of invariants of given forms. Let $f(x_1, x_2)$ and $\phi(y_1, y_2)$ be two binary forms, of degrees m and n respectively, and introduce the symbolic operator

$$\Omega = \frac{\partial^2}{\partial x_1 \partial y_2} - \frac{\partial^2}{\partial x_2 \partial y_1}.$$

so, e.g., $\Omega(f.\phi) = \dfrac{\partial f}{\partial x_1} \dfrac{\partial \phi}{\partial y_2} - \dfrac{\partial f}{\partial x_2} \dfrac{\partial \phi}{\partial y_1}$

Then the form

$$\frac{1}{m.n} \Omega(f.\phi)$$

is the *first transvectant* of f, and ϕ.

The rth transvectant of f and ϕ is similarly defined as $\dfrac{(m-r)!}{m!} \dfrac{(n-r)!}{n!}$ $\Omega^r(f.\phi)$.

To obtain the rth transvectant of f one calculates with $f(x_1, x_2)$ and $f(y_1, y_2)$ and then sets $y_1 = x_1$, $y_2 = x_2$ after the differentiation is over. So, for example, the second transvectant of f with itself is obtained by first calculating

$$\left(\frac{1}{m(m-1)}\right)^2 \Omega^2(f.f) = \left(\frac{1}{m(m-1)}\right)^2$$
$$\left\{ \frac{\partial^2 f}{\partial x_1^2} \cdot \frac{\partial^2 f}{\partial y_2^2} - 2\frac{\partial^2 f}{\partial x_1 \partial x_2} \cdot \frac{\partial^2 f}{\partial x_1 \partial x_2} \cdot \frac{\partial^2 f}{\partial y_1 \partial y_2} + \frac{\partial^2 f \partial^2 f}{\partial x_2^2 \partial y_2^2} \right\}$$

and then substituting x_1 and x_2 for y_1 and y_2 respectively, obtaining

$$\left(\frac{1}{m(m-1)}\right)^2 .2. \left\{ \frac{\partial^2 f}{\partial x_1^2} \frac{\partial^2 f}{\partial x_2^2} - \left(\frac{\partial^2 f}{\partial x_1 \partial x_2}\right)^2 \right\}.$$

This, up to a constant factor, is the Hessian of f. The odd order transvectants of a form with itself all vanish. A modern account of transvectants is Dieudonné and Carrell [1971], and light is also shed on them by the aptly-named umbral calculus of Kung and Rota [1984].

9. This problem has been studied in great generality in Baldassarri and Dwork [1979] and Baldassarri [1980]. They regard Klein's method as getting close to solving the recognition problem, although insufficient to deal with the di-hedral case, but as not well adapted for the construction of all second order linear differential equations with algebraic solutions and prescribed singular parts.

10. $H(f)$ is the Hessian up to a constant factor. For $f_6 = y_1^5 y_2 + y_1 y_2^5$, $H(f_6) = y_1^8 - 14 y_1^4 y_2^4 + y_2^8$. For $f_{12} = y_1^{11} y_2 + 11 y_1^6 y_2^6 + y_1 y_2$, $H(f_{12}) = y_1^{20} - 228 y_1^{15} y_2^5 + 494 y_1^{10} y_2^{10} + 228 y_1^5 y_2^{15} + y_2^{20}$.

11. This paragraph draws on Slodowy's account [1986, p. 91].

12. Sylow presented his discovery that subgroups of order p^r exist in a group of order n, whenever p is a prime and p^r divides n, in [1872]. footJordan was much impressed with this result, and wrote to Sylow about it; Sylow's letters are preserved in the collection of Jordan's correspondence at the Ecole Polytechnique (catalogue numbers I, 19–22). Sylow's proofs were permutation theoretic, and the history of their reformulation in abstract terms is well traced in Waterhouse [1980]. Jordan himself never published a proof of Sylow's theorems.

13. Jordan called the linear transformations 'substitutions' and the name attached itself to the elements of groups as his theory of groups became progressively more abstract. In matrix form this transformation would be written

$$\begin{pmatrix} u_1 \\ u_2 \end{pmatrix} \to \begin{pmatrix} \alpha & \beta \\ \gamma & \delta \end{pmatrix} \begin{pmatrix} u_1 \\ u_2 \end{pmatrix}$$

14. They are as follows, where Jordan's notation for a typical element has been replaced by matrix notation:

  1. Groups generated by

$$\begin{pmatrix} a & 0 & 0 \\ 0 & b & 0 \\ 0 & 0 & c \end{pmatrix} \text{ and } \begin{pmatrix} 0 & a' & 0 \\ 0 & 0 & b' \\ c' & 0 & 0 \end{pmatrix}, a, b, c, a', b', c$$

     all roots of unity, and their subgroups [no. 198].

  2. Extensions of such groups by the group generated by

$$\begin{pmatrix} 0 & a'' & 0 \\ b'' & 0 & 0 \\ 0 & 0 & c'' \end{pmatrix},$$

     a'', b'', c'' roots of unity [no. 204].

  3. A group generated by mI, $\begin{pmatrix} \tau & 0 & 0 \\ 0 & \tau^{-1} & 0 \\ 0 & 0 & 1 \end{pmatrix}$, $\begin{pmatrix} 0 & 1 & 0 \\ 1 & 0 & 0 \\ 0 & 0 & -1 \end{pmatrix}$, and

$$C = \begin{pmatrix} a & -(1+a) & -2a^2 \\ -(1+a) & a & 2a^2 \\ 1 & -1 & -(1+2a) \end{pmatrix},$$

     where m is a root of unity, $\tau^5 = 1$, and a is defined by $a(\tau + \tau^{-1} - 2) = 1$. If $m = 1$ this group reduces to the group of proper motions of an icosahedron [no. 208].

4. A group generated by mI,

$$A = \begin{pmatrix} 1 & 0 & 0 \\ 0 & \theta & 0 \\ 0 & 0 & \theta^2 \end{pmatrix}, B = \begin{pmatrix} 0 & 1 & 0 \\ 0 & 0 & 1 \\ 1 & 0 & 0 \end{pmatrix},$$

$$rD = \begin{pmatrix} rj & 0 & 0 \\ 0 & rj\theta^2 & 0 \\ 0 & 0 & rj \end{pmatrix}, \text{ and } r^2E = r^2a \begin{pmatrix} 1 & 1 & 0 \\ 1 & \theta & 1 \\ 1 & \theta^2 & \theta^2 \end{pmatrix},$$

where $\theta^3 - 1$, $j^3 = \theta$, $a^3 - \dfrac{1}{3(1-\theta^2)}$, $m^{3\rho} = 1$, $r^3 = m^\mu$,

(p arbitrary, $\mu = 0$, 1, or 2). If $r = 1 = \rho$ this group has order 27.24 [nos. 201–3, 209].

5. A subgroup of the previous one of order 27.2.2 generated by mI, A, B, and sDE, where $s^4 = 1$, m, m^2, or m^3.

6. A subgroup of order 27.2.4 generated by mI, A, B, sDE, and tED where $t^2 = s^2$ or ms^2, which Jordan called Hesse's group.

Hesse's group is a subgroup of the Galois group of the equation of the 9 inflection points on a general cubic. Jordan had first considered the larger group in his *Traité*, 302–305, where he showed it had order 432 and was the symmetry group of the configuration of the lines joining the inflection points, by treating it as the symmetries of the affine plane over the field of 3 elements. Hesse's group consists of those matrices having determinant 1. The group in (5) above, of order $27.2.2 = 108$ is the symmetry group of the Pappus configuration. The group is discussed in Miller, Blichfeldt, and Dickson, *Theory and Applications of Finite Groups*, Chapter XVIII.

15. In a letter from Jordan to Klein, 11 October 1878, [N.S.U. Bibliothek zu Göttingen, Klein, cod Ms. F. Klein, 10, number 11] Jordan wrote:

"Mon cher ami "Vous avez parfaitement raison. En énumérant les groupes linéaires à trois variables, j'ai laissé échapper celui d'ordre 168 que vous me signalez".

Jordan corrected his mistake, and went on to discuss the group of the modular equation at the prime 11 (PSL (2; 11), of order 660) which Klein had presumably raised in his letter to Jordan. Jordan observed that the group has an element, say A', of order 11, and one, B', of order 5. If it is to be a linear group in three variables, which he doubted, then, he wrote:

"En supposant qu'il n'y ait que 3[?] variables, il faudra trouver une substitution C' telle que A', B', C' combines entre elles ne donnent que 12.11.5 substitutions. Il ne serait pas difficile sans doute de vérifier si la chose est possible, mais je suis trop occupé en ce moment pour pouvoir exécuter ce calcul."

Jordan went on to discuss the Hessian group of order 216, and then congratulated Klein on his good fortune in working with Gordan.

> "Je ne me sens pas aucune en état de le suivre sur le terrain des formes ternaires; mais j'ai longuement approfondi sa belle démonstration de l'Endlichkeit der Grundformen des formes binaires..."

Jordan's intuition about PSL (2; 11) was correct, it cannot be represented as a group in three variables. However, both he and Klein missed a group of order 360 which can be represented by ternary collineations. This is Valentiner's group (Valentiner, [1889]) which is discussed in Wiman [1896], where it is shown to be abstractly isomorphic to A_6, the group of even permutations of 6 objects, and, in Klein, [1905 = 1922, 481-502.].

16. See Hermite's letter to P.du Bois-Reymond, 3 September 1877, in Hermite [1916], referred to in Hawkins, [1975, 93n].

17. Jordan was somewhat imprecise. Poincaré showed [1881x = *Oeuvres* III, 95–97] that to each finite monodromy group in three or more variables there correspond infinitely many differential equations having rational coefficients and algebraic solutions.

Chapter IV

1. Complex multiplication and singular moduli. Let Λ be a lattice in $\mathbf{C}$. The only holomorphic maps $f : \mathbf{C}/\Lambda \to \mathbf{C}/\Lambda$ are those lifting to a map $f : \mathbf{C} \to \mathbf{C}$ satisfying $f(z + \omega) - f(z) \in \Lambda$ whenever $\omega \in \Lambda$. This implies that $f(z + \omega) - f(z)$ is a constant independent of z and so $\dfrac{df}{dz}$ is constant and $f(z) = az + b$. By suitably renormalizing one can always rewrite f as $z \to az$, so the only maps of a lattice Λ to itself are of the form

$$\left. \begin{aligned} a\omega &= \alpha\omega + \beta\omega' \\ a\omega' &= \gamma\omega + \delta\omega' \end{aligned} \right\} , \alpha, \beta, \gamma, \delta \in Z,$$

where ω and ω' are generators of the lattice. Either $a = \alpha = \delta$, $\beta = \gamma = 0$, or, setting $\tau = \omega'/\omega$, $\beta\tau^2 + (\alpha - \delta)\tau - \gamma - 0$, so τ is a quadratic imaginary. In this case the lattice is said to possess a complex multiplication, by a. The theory of singular moduli was first developed by Abel [1828 2nd ed. p. 426] in terms of elliptic integrals. Each transformation of the lattice gives a holomorphic map from $\mathbf{C}/\Lambda$ to $\mathbf{C}/\Lambda$, Abel sought values of a for which the differential equation

$$\frac{dy}{\sqrt{[(1 - y^2)(1 + \mu y^2)]}} = a \frac{dx}{\sqrt{[(1 - x^2)(1 + \mu x^2)]}}$$

has algebraic solutions, and found that a is either rational or a quadratic imaginary $m + \sqrt{n}i$, m, n rational. This occurs only for certain μ which he called the singular moduli, and which he conjectured satisfied an algebraic equation solvable by radicals. The conjecture was proved by Kronecker [1857].

2. Hadamard [1954, 109] said of Hermite: "Methods always seemed to be born in his mind in some mysterious way".

3. See also Dugac [1976], Mehrtens [1979].

4. Compare Lang [1976], Serre [1973], or Schoeneberg [1974].

5. I cannot find that he ever took an opportunity to do this.

6. The group of matrices $\begin{pmatrix} a & b \\ c & d \end{pmatrix} \equiv \begin{pmatrix} 1 & 0 \\ 0 & 1 \end{pmatrix}$ mod 2 was connected with modular equations at arbitrary N, not merely the primes, by H. J. S. Smith [1877], in a paper much admired by Klein. Smith wanted to connect the theory of binary quadratic forms $ax^2 + 2bx + cy^2$, $N = b^2 - ac > 0$, with the modular equation at N. Results of Hermite and Kronecker concerning the case when N is negative were well known, he said, but the positive case was more difficult and little had been done beyond Kronecker [1863] which discussed the solution of Pell's equation by elliptic functions.

Smith's paper dealt with the study of the reduced forms equivalent to a given form $Q(x, y) = ax^2 + 2bxy + cy^2$, a, b, and c integers. Reduced forms are those $a'x^2 + 2b'xy + c'y^2$ for which the quadratic equation $a' + 2b't + c't^2 = 0$ has real roots of opposite sign such that the absolute values of one root is greater, and the other less, than unity. They are equivalent to $Q(x, y)$ if there is $g = \begin{pmatrix} \alpha & \beta \\ \gamma & \beta \end{pmatrix} \in SL(2; \mathbf{Z})$ such that $Q(\alpha x + \beta y, \gamma x + \delta y) = a'x^2 + 2b'xy + c'y^2$. They are associated with the continued fraction expansion of $\sqrt{N}$ in a way first described by Dirichlet [1854] (for the history of this matter see Smith [1861, III, §93]). Smith found it convenient to work with the weaker notion of equivalence where g satisfies $\begin{pmatrix} \alpha & \beta \\ \gamma & \delta \end{pmatrix} \equiv \begin{pmatrix} 1 & 0 \\ 0 & 1 \end{pmatrix}$ mod 2, $\alpha \equiv \delta \equiv 1$ mod 4, presumably to avoid the fixed points i, ρ of the Γ-action. He remarked of his restriction only that "c'est uniquement pour abréger le discours que nous l'admettons ici". This group acts on the upper half plane and a set of inequivalent points, $\sum$, is obtained by taking the region bounded by $P = x - 1 = 0$, $P^{-1} = x + 1 = 0$, and the two semicircles $Q = x^2 + y^2 - x = 0$, $Q^{-1} = x^2 + y^2 + x = 0$, including P and Q but not P^{-1} and Q^{-1}. To each quadratic form $[a, b, c] := ax^2 + 2bxy + cy^2$ he associated a circle $(a, b, c) := a + 2bx + c(x^2 + y^2) = 0$, and he studied the image of this circle reduced mod $\Gamma(2)$, a set of arcs in $\sum$ ("en apparence sera brisée, mais dont on mettra en évidence la continuité". The way in which these arcs went from one boundary of $\sum$ to another was, he showed, described by the periodic continued fraction expansion of $\sqrt{N}$, and thus to the production of the reduced forms equivalent to (a, b, c) (see Smith [1861, Part V, §93]).

The connection with the modular equation was established by introducing Hermite's functions $\phi^8(\omega) = k^2 = \frac{1}{2} + X + iY$, $\psi^8(\omega) = k'^2 = \frac{1}{2} + X - iY$ and mapping from the upper half ω-plane to the X, Y plane. Under this map $\omega = i$ goes to $X = 0$, $Y = 0$, the imaginary axis of $\sum$ is mapped onto the X-axis (Smith incorrectly said the real points of $\sum$ are sent to points with $Y = 0$) and the map is conformal even at $\omega = 0$, $\omega = 1 \equiv -1$, and $\omega = i\infty$. Each circle (a, b, c) is sent to an algebraic curve in the X, Y plane whose equation is $F(\phi^8(\omega), \psi^8(\omega)) = 0$, where $F = 0$ is the modular equation at N between k^2 and λ^2. If $k^2 = \Phi^8(\omega)$ then $\lambda^2 = \Phi^8\left(\dfrac{\gamma\omega + 2\delta}{\gamma'}\right)$, where $\gamma\gamma' = N$, $\delta \equiv 0, 1, \ldots, \gamma'^{-1}$ mod γ and $(\gamma, \gamma', \delta) = 1$. F is symmetric with respect to k^2 and λ^2 and is invariant under the six permutations of k^2 (and λ^2) which form the cross-ratio group $(k^2, 1 - k^2, \frac{1}{k^2}$ etc.), so there are 6 modular curves $F(k^2, \lambda^2) = 0$, $F(k^2, 1 - \lambda^2) = 0$ etc. These curves spiral around the images $A_1 = (+\frac{1}{2}, 0)$ of $\omega = 0$ and $A_2 = (-\frac{1}{2}, 0)$ of $\omega = i\infty$ like interlacing lemniscates symmetrically situated about the X-axis. If the (X, Y)-plane is cut along the X-axis from A_1 to $+\infty$ and A_2 to $-\infty$ the spirals are cut into whorls ('*spires*') which correspond to the individual arcs of the reduced image of (a, b, c) in $\sum$, and consequently to the continued fraction expansion of N.

7. The implicit isomorphism between $PSL(2, \mathbf{Z}/5\mathbf{Z})$ and A_5 was made explicit by Hermite [1866 = *Oeuvres*, II, 386–3871].

8. See Dugac [1976, 73] for a discussion of priorities, and Dauben [1978, 142] for Kronecker's tardiness. The matter is fully discussed in H. M. Edwards [1980, 370–371].

9. Klein [1926, I, 366] – strictly, a description of work on automorphic function theory.

10. For more information on this work of Betti's, see J. C. Nicholson's Oxford D. Phil Thesis, *Otto Hölder and the Development of Group Theory and Galois Theory*, 1993, and for the growing appreciation the concept of quotient group, see her article [1993].

11. For an account of Bring and his work, see Gärding [1997], 6–7.

12. This highly transitive group inspired Mathieu to look for others, and so led to his discovery of the simple, sporadic groups which bear his name, Mathieu [1860, 1861].

13. The multiplier equation associated to a modular transformation describes M (see Chapter I, p. 13) as a function of k.

14. Bottazzini is editing some unpublished correspondence between Italian and other mathematicians on this topic.

15. Gordan made a study of quintics in his [1878], discussed in Klein [1922, 380–4].

16. Notes of Weierstrass' lectures were presumably available, see Chapter VI n. 4. Klein based his approach to the modulus of elliptic functions on the treatments in Müller [1867, 1872 a, b] which I have not seen. Hamburger's report on them in *Fortschritte*, V, 1873, 256-257, indicates that they are based on Weierstrass' theory of elliptic functions and the invariants of biquadratic forms.

17. While recommending readers to consult these essays, I should also like to take this opportunity of thanking Peter Slodowy for the detailed criticisms he sent me of the first edition. I have tried to see that they produced improvements in this edition; these pages are a particular case in point.

Chapter V

1. The plane is generally the complex rather than the real plane, but early writers e.g., Cayley are ambiguous. Later writers, like Klein, are more careful.

2. If 4 or more inflection points were real they would all have to be, since the line joining two inflection points meets the curve again in a third. But then they would all lie on the same line, which is absurd. The configuration of inflection points on a cubic coincides with the points and lines of the affine plane over the field of 3 elements, an observation first made, in its essentials, by Jordan, *Traité* 302.

3. H. J. S. Smith [1877] pointed out that Eisenstein used the Hessian earlier, in 1844. Eisenstein's 'Hessian' is the Hessian of the cubic form $ax^3 + 3bx^2y + 3cxy^2 + dy^3$, i.e., $(b^2 - ac)x^2 + (bc - ad)xy + (c^2 - bd)y^2$, see Eisenstein [1975, I, 10].

4. It takes two to tango.

5. In a subsequent paper on this topic, [1853], Steiner was led to invent the Steiner triple system. However, these configurations had already been discovered and published by Kirkman [1847], as Steiner may have known, see the sympathetic remarks of Klein, *Entwicklung*, 129.

6. It has proved impossible to survey the literature of these wonderful configurations. For some historical comments see Henderson [1911, 1972], who also describes how models of the 27 lines may be constructed. For modern mathematical treatments see Griffiths and Harris [1978], Hartshorne [1977] and Mumford [1976]. For their connection with the Weyl group of E_7 see Manin [1974]. The existence of a finite number of lines on a cubic surface was discovered by Cayley [1849], and their enumeration is due to Salmon [1849]. The American mathematician A. B. Coble [1908] and [1913] seems to have been the first to illuminate the 27 lines and 28 bitangents with the elementary

theory of geometries over finite fields. I am grateful to J. W. P. Hirschfeld for drawing his work to my attention.

7. See Weierstrass *Werke*, IV, 9, quoted and discussed in Neuenschwander [1981b, 94].

8. Riemann tacitly assumed that the genus was finite. His proof that it was well-defined was imprecise and was improved by Tonelli [1875].

9. See [Gray 1984b] and later essays. Good accounts of it have been given in Scholz [1980] and Laugwitz [1996]; a lucid treatment of Riemann's theta function is given in Chapter VI of the book by Farkas and Kra.

10. Jacobi [1829 = 1969, I, 249–275] had introduced, in connection with the multiplier equation at the prime p, a class of equations of degree $p+1$ whose roots depend linearly on $\dfrac{p+1}{2}$ quantities $A_0, \ldots, A_{\frac{p+1}{2}}$. The $p+1$ roots take the form

$$\sqrt{z_\infty} = (-1)^{\frac{p-1}{2}}.p.A_0$$

$$\sqrt{z_i} = A_0 + \epsilon^\nu A_1 + \epsilon^{4\nu} A_2 + \cdots + \epsilon^{\left(\frac{p-1}{2}\right)^2}\nu\, \nu A_{\frac{p-1}{2}}$$

where $\epsilon = e^{2\pi i/p}$. Brioschi [1858] was able to show what the general form of an equation of degree 6 is if it is a Jacobian equation, and to exhibit explicitly the corresponding quintic. The multiplier equation was, he showed, a special Jacobian equation for which the corresponding quintic took Bring's form, so the solution of the quintic by elliptic functions was again accomplished. Kronecker [1858] showed that the general quintic yields rational resolvents which are Jacobian equations of degree 6, and conjectured that equations of degree 7 which were solvable by radicals would likewise be associated with Jacobian equations of degree 8.

Klein [1879 = 1922, 390–425] argued that, since the group of the modular equation of the prime 7 is $G_{168} = PSL(2; \mathbf{Z}/7\mathbf{Z})$ it is only possible to solve those equations of degree 7 whose Galois group is a quotient of G_{168}, i.e., is G_{168} itself. The general Jacobian equation of degree 8 has Galois group $SL(2 : \mathbf{Z}/7\mathbf{Z})$ so that task of solving every equation of degree 7 by modular functions is impossible. In fact $SL(2 : \mathbf{Z}/7\mathbf{Z})$ can also send each $\sqrt{z_i}$ to $-\sqrt{z_i}$ and each $\sqrt{z_i}(i = 0, 1, \ldots, 6, \infty)$ is linear in the A's. Klein took $[A_0 : A_1 : A_3 : A_4]$ as coordinates in $\mathbb{C}P^3$, so each $\sqrt{z_i} = 0$ defines a plane and (dually) a point. These eight points define a net of quadrics (after Hesse). If the Jacobian equation has Galois group G_{168} then the net contains a degenerate family of conics whose vertices lie on a space sextic (K, above, p. 243) which is the Hessian of a quartic [1879, §9]. Klein also showed that any polynomial equation whose Galois group was G_{168} could be reduced to the modular equation at the prime 7, thus answering affirmatively Kronecker's conjecture about the '*Affect*' of such equations.

11. The simplicity of the group emerges from Klein's description of its subgroups without Klein remarking on it explicitly. Silvestri's comments [1979, 336] in this connection are misleading, and he is wrong to say the group is the trans-formation group of a seventh order modular function: G_{168} is the quotient of $PSL(2, \mathbf{Z})$ by such a group.

12. In particular, the inflection points of this curve are its Weierstrass points. For example, when the equation of the curve is obtained in the form $\lambda^3\mu + \mu^3\nu + \nu^3\lambda = 0$ the line $z = 0$ is an inflection tangent at $[0, 1, 0]$ and meets the curve again at $[1, 0, 0]$. The function $g(\lambda, \mu, \nu) = \mu/\nu$ is regular at $[1, 0, 0]$ but has a triple pole at $[0, 1, 0]$.

13. Gordan [1880a,b] derived a complete system of covariants for $f := x_1^3x_2 + x_2^3x_3 + x_3^3x_1 = 0$ from a systematic application of transvection and convo-lution (*Faltung*). Convolution replaces a product $\alpha_x\beta_x$ in a covariant by the factor $(\alpha\beta)$, where $(\alpha\beta) := \alpha_1\beta_2 - \alpha_2\beta_1$, and $\alpha_x = (\alpha_1x_1 + \alpha_2x_2)^n = \sum \binom{n}{r} a_r x_1^{n-r} x_2^r$. These papers and other related ones by Gordan were summarized by Klein in his [1922], 426–438, where it is shown that they lead to a solution of the 'form problem' for $PSL(2, \mathbf{Z}/7\mathbf{Z})$ just as Klein's work had solved the 'form problem' for the Icosahedron. Klein also showed how Gordan's work enables one to resolve Kronecker's conjecture. The surviving correspondence in Göttingen between Klein and Gordan throws no extra light on the working relationship between the two men in this period, and Professor K. Jacobs has kindly informed me that there is no correspondence between them kept in Erlangen.

14. Halphen's solution appeared as a letter to Klein in [Halphen 1884 = *Oeu-vres* IV 112 – 5]. Hurwitz's more general solution was published as [Hurwitz 1886]. He argued that three functions λ, μ, ν of J such that $\lambda^3\mu + \mu^3\nu + \nu^3\lambda = 0$ could be found which are the solutions of a 3rd order linear differen-tial equation as follows. Pick 3 linear independent everywhere finite integrals J_1, J_2, J_3 on $\lambda^3\mu + \mu^3\nu + \nu^3\lambda = 0$ such that $dJ_1 : dJ_2 : dJ_3 = \lambda : \mu : \nu$. Then Fuchs' methods [Fuchs 1866, esp. 139 – 148] imply that, if $y_i = \dfrac{dJ_i}{dJ}$, $1 \leq i \leq 3$, then y_1, y_2, and y_3 are the solutions of an equation of the form

$$\frac{d^3y}{dJ^3} + \frac{aJ + b}{J(J-1)}\frac{d^2y}{dJ^2} + \frac{a'J^2 + b'J + c'}{J^2(J-1)^2}\frac{dy}{dJ} + \frac{a''J^3 + b''J^2 + c''J + d''}{J^3(J-1)^3}y = 0$$

since the only possible branch points are $J = 0$, 1, and ∞. Klein's [1878/79] establishes that the branching is 7-fold at $J = \infty$, 3-fold at $J = 0$, and 2-fold at $J = 1$, whence Fuchs' equation for the sum of the exponents yields values for most of $a, b, \ldots, d''$. Since the exponent differences are integers at $J = 1$, log terms could occur in the solution but they do not, so the differential equation can be found explicitly. It is

$$J^2(J-1)^2\frac{d^3y}{dJ^3} + (7J-4)J(J-1)\frac{d^2y}{dJ^2} + \left[\frac{72}{7}J(J-1) - \frac{20}{9}(J-1)\right.$$

$$+ \left. \frac{3J}{4} \right] \frac{dy}{dJ} + \left[\frac{72.11}{73}(J - 1) + \frac{5}{8} + \frac{2}{63} \right] y = 0.$$

A study of the invariants associated to the curve $\lambda^3 \mu + \mu^3 \nu + \nu^3 \lambda = 0$ enabled Hurwitz to show that his differential equation took Halphen's form when the substitution $y = (J(J - 1))^{-2/3} y'$ is made.

15. Two of Klein's students at this time deserve particular mention: Walther Dyck, and Adolf Hurwitz. Dyck presented his thesis at Munich in 1879 on the theory of regularly branched Riemann surfaces, and published two papers on this topic in the *Mathematische Annalen* the next year [1880a, b]. He called a surface regularly branched if each leaf was mapped by the self-transformations of the surface into every other leaf, and was particularly interested in the case where the group (supposed to be of order N) has a normal ("*ausgezeichneten*") subgroup of order N_1. He then interested himself in the surfaces corresponding to the subgroup and the quotient group, and in this way analysed surfaces until simple groups were reached. The map of a subgroup into a bigger group corresponds to a map of one Riemann surface onto another, and he gave explicit forms for these in the cases he looked at. In the example which most attracted him, the surface was composed of triangles with angles $\pi/2$, $\pi/3$, and $\pi/8$, and its genus is 3. The principal group has order 96, so Dyck regarded the surface as having 96 leaves. There is a normal subgroup of order 48, and so the surface is also made up of triangles with angles $\pi/3$, $\pi/3$, and $\pi/4$ in this case. This in turn sits over a 16-leaved surface made up of 2-sided leaves with vertical angles of $\frac{\pi}{3}$ and $\frac{\pi}{3}$ and so on until a double cover of the sphere is reached. He also noted that other decompositions are possible and showed that the corresponding algebraic curve is the 'Fermat' curve $x^4 + y^4 + z^4 = 0$.

Dyck was in some ways very like Klein as a mathematician, attracted to '*anschauliche*' geometry and the role of groups in describing the symmetry of figures. He also became very interested in constructing models of various mathematical objects, and collaborated with Klein in the design of these for the firm, owned by A. Brill's brother in Munich, which made them.

Klein's other, and younger, student was a more independent character. Hurwitz's *Inauguraldissertation* at Leipzig was published in the *Mathematische Annalen* for 1881 [Hurwitz 1881]. It recalls both Klein and Dedekind's work on modular functions, but it develops the theory entirely independently of the theory of elliptic functions. This had been Dedekind's aim as well, but he had not been able to carry it through completely. Hurwitz succeeded because he had read and understood Eisenstein, a debt he fully acknowledged. He did not say he was the first person to have read Eisenstein since the latter's death in 1852, although he might well have been. Weil points out [Weil 1976, 4] that only Hurwitz and Kronecker (in 1891) seem to mention their illustrious predecessor in the whole of the later nineteenth century. In Hurwitz's case the reference to Eisenstein is a little more substantial that Weil's comment might suggest, but even so it was not sufficient to rescue him from the shades. With some reluctance I have chosen to suppress the details of Hurwitz's important treatment of modular forms and modular quotients and the invariance group for Δ^{112}, although it well illustrates how Hurwitz grasped the connections between group theory, geometry, and function theory which I am tracing historically.

Chapter VI

1. The Lamé equation is discussed in Whittaker and Watson, Ch. XXIII.

2. The applications of elliptic functions are discussed in Houzel [1978, §15].

3. Halphen defined a single-valued function (*fonction uniforme*) thus (p. 4n): "Par ce mot *fonction uniforme*, j'entends ici une fonction ayant l'aspect d'une fonction rationnelle, c'est-à-dire développable sous la forme

$$(\alpha - \alpha_0)^n [A + B(\alpha - \alpha_0) + C(\alpha - \alpha_0)^2 + \cdots],$$

dans laquelle n est un nombre entier, positif on négatif."

4. See [Mittag-Leffler, 1876] in Swedish, translated by Hille in 1922. *A propos* of the Weierstrassian theory Halphen commented [69n]: "Il existe des tableaux de formules lithographiées qui ont été rédigées d'après les leçns de M. Weierstrass, et qui sont entre les mains de presque tous les géométrès allemands."

5. This information is taken from Darboux, Éloge Historique d'Henri Poincaré, printed in Poincaré, *Oeuvres*, II, vii-lxxi.

6. 7 letters between Poincare and Fuchs were printed in *Acta Mathematica* (38) 1921 175–187, and reprinted in Poincaré *Oeuvres*, XI, 13–25. An eighth letter is given in photographs in *Oeuvres*, XI, 275–276.

7. When the essay was published in 1923 this figure was incorrectly printed as figure 6, before the one depicting the situation which Fuchs had shown to be impossible. The text refers to the annular case as the second one, which it is in Poincaré's original essay as deposited in the Académie.

8. Accounts of the supplements appeared in Gray [1982a,b] and Dieudonné [1983]. I would like to thank J. Dieudonné for his help during my work on these supplements. The supplements themselves, with an introductory essay, have now been published as Poincaré [1997].

9. Poincaré presented his ideas on non-Euclidean geometry and arithmetic to l'Association Franąise pour l'avancement des Sciences at its 10th session, Algiers, 16 April 1881, [*Oeuvres* V, 267–274]. The ternary indefinite form $\zeta^2 + \eta^2 - \xi^2 = -1$ is preserved by a group of 3×3 matrices ($SO(2, 1)$)). If

new variables $X := \dfrac{\zeta}{\xi + 1}$ and $Y = \dfrac{\eta}{\xi + 1}$ are introduced and ζ is taken to be positive (a condition Poincaré forgot to state) then the point (X, Y) must lie inside the unit circle. The linear transformations in $SO(2, 1)$ induce transformations of the unit disc, planes through $(\xi, \eta, \zeta) = (0, 0, 0)$ cut the upper half of the hyperboloid of 2 sheets $(\xi^2 + \eta^2 - \zeta^2 = -1)$ in lines which correspond to arcs of circles in the (X, Y) domain meeting the unit circle at right angles, and the induced transformations map these arcs to other similar arcs. Poincaré showed that the concepts of non-Euclidean distance and angle may be imposed on the disc to yield a conformal map of the geometry-and that the induced transformations are then non-Euclidean isometries. His interest in the group $SO(2, 1)$ derived from Hermite's number theoretical work [1854]. I presume that Poincaré's paper [1881, 11 July] appeared in the *Comptes Rendus* before the account of this talk. There is no reason to suppose Poincaré knew of the Weierstrass-Killing interpretation of non-Euclidean geometry in terms of the hyperboloid of 2 sheets [Hawkins, 1980, 297].

10. See [Bottazzini 1977, 32].

11. c.f. Riemann, "Vorlesungen Uber die hypergeometrische Reihe", *Nachträge*, 77.

12. The correspondence is printed in Klein, *Gesammelte Mathematische Abhandlungen*, III, 1923, 587–626, in *Acta Mathematica* (38) 1922, and again in Poincaré *Oeuvres*, XI. The simplest way to refer to the letters is to give their date.

13. Published somewhat accidentally in Poincaré, *Oeuvres* XI, 275–276.

14. "Name ist Schall und Rauch" comes from the key scene in Goethe's Faust (Act I, Scene 16) where Gretchen asks Faust if he still believes in God. Unable, because of his bargain, to say that he does, he gives a long reply to the effect that no-one, not even philosophers and priests, can answer that question. His euphemistic speech ends

> "... Gefühl ist alles
> Name ist Schall und Rauch
> Umnebelend Himmelsglut",

which might be translated as

> "... Feeling is everything
> Name is sound and smoke
> obscuring the glow of heaven".

Gretchen sees through Faust's obscurantism almost at once and replies "Then you are not a Christian". To this day educated Germans speak of the Gretchen Question where English speakers might inquire "Where's the rub?" or, more simplistically ask "The 64,000 dollar question". Klein would undoubtedly have picked up the reference, but he might have thought it a little excessive.

15. The same approach was taken by Picard in proving his celebrated 'little' theorem that an entire function which fails to take two values is constant, Picard [1879a]. He argued that, if there is an entire function which is never equal to a, b, or ∞ (being entire), then a Möbius transformation produces an entire function, G, say, which is never 0, 1, or ∞. So G never maps a loop in the complex plane onto a loop enclosing 0, 1, or ∞. Now, G maps a small patch into a region of, say, the upper half plane, and if $k^2 : H \rightarrow \mathbf{C}$ is the familiar modular function, then composing G with a branch of the inverse of k^2 maps the patch into half a fundamental domain for k^2. Analytic continuation of this composite function cannot proceed on loops enclosing 0, 1, or ∞ (by the remark just made) so the composite $(k^2)^{-1} \circ G$ is single-valued. But it is also bounded, so, by Liouville's theorem, it is constant, and the 'little' theorem is proved. To see that the function is bounded replace the domain of k^2 by the conformally equivalent unit disc; Picard missed that trick and proved it directly. For a nearly complete English translation see Birkhoff and Merzbach, 79–80.

16. Many years later, Mittag–Leffler wrote in an editorial in *Acta Mathematica* ((39) 1923, iii) that Poincaré's work was not at first appreciated. "Kronecker, for example, expressed his regret to me via a mutual friend that the journal seemed bound to fail without help on the publication of a work so incomplete, so immature and so obscure."

17. I hope to discuss the work of Ritter, Fricke, and Fubini and its implications for function theory in a forthcoming book on the history of complex function theory, jointly with U. Bottazzini.

Appendix 1

1. For a recent look at the history of potential theory see Archibald [1996].

Appendix 2

1. There is an extensive and growing literature on the Riemann–Hilbert problem, entrance to which may be gained via Yoshida [1987] and Varadarajan [1991].

Appendix 4

1. This presentation therefore contributes to the misrepresentation of Riemann's ideas against which Radner, Scholz and others have fought. There is much more to this paper of Riemann's than a sketch of a theory of differential geometry, and the treatment of non-Euclidean geometry is but a speck. The influence of Herbart's ideas on Riemann has been given an interesting and surprising examination by Scholz [1982], which aquires renewed interest in the light of the study by Bottazzini and Tazzioli [1995].

2. 17 July, 1867 is the date when Dedekind handed the paper over to Weber for final editing [Dugac 1976, 171], and it is the date given in the French edition. The bound edition of the *Göttingen Abhandlungen*, volume 13, covers the years 1866 and 1867 and was published in 1868. Sommerville [1911, 38] gives 1866 and *Fortschritte* 1868.

3. I wish this account to replace the misleading one in Gray [1979a]. In particular the reference (p. 247) to geodesics appearing as circular arcs in the disc is a totally wrong summary of Beltrami's ideas.

4. A brief reference was also made to Gauss' views on non-Euclidean geometry in the biography [von Waltershausen, 1856, 80–81]. Gauss is a difficult case to understand, for he never brought his ideas together in one place, nor did he seem to like the idea. For a detailed discussion see [Reichardt 1976].

5. For a truly stupid piece of philosophizing in a book well received at the time, consider e.g. "The foregoing discussion has brought us to the point where the reader is in a condition, I hope, to realize the great fundamental absurdity of Riemann's endeavour to draw inferences respecting the nature of space and the extension of its concept from algebraic representations of "multiplicities". An algebraic multiple and a spatial magnitude are totally disparate. That no conclusion about forms of extension or spatial magnitudes are derivable from the forms of algebraic functions is evident upon the most elementary considerations". [Stallo, 1888 – 1960, 278]. On Frege, see Toth [1984].

6. Cayley solved his problem himself in [1872].

7. Weierstrass' seminar of 1872 seems to be the source of Killing's interest in non-Euclidean geometry. See [Hawkins 1979, Ch. 2].

8. Both Klein and Cayley made an attempt to define distance by (i) saying that if the cross ratio of A, B, C, and D is -1 and D is at infinity, then B is the midpoint of AC, and (ii) somehow passing to the infinitesimal case. The passage (ii) is not clear, nor was it generally accepted, The whole process recalls von Staudt's attempts [1856 – 1860] to define cross-ratio without the concept of distance and then to define distance in terms of cross-ratio. The difficulty is the introduction of continuity into the projective space.

Appendix 5

1. For a fuller discussion of this material, upon which this account is based, see Gray [1994].

Appendix 6

1. I have replaced Picard's n-tuples of functions with a vector notation.

2. Pépin's obituary by Augusto Statuti, Cenno Necrologico del Rev. Prof. P. Teofilo Pépin, can be found in *Atti della Pontifica Accademia Romana dei Nuovi Lincei*, Anno LVIII 1904–1905, 210–216. It is not listed in K.O. May's *Bibliography*.

Serials:
List of Abbreviations

The following abbreviations have been used:

Abh Berlin	Abhandlungen den Deutschen Akademie der Wissenschaften zu Berlin
A. H. E. S.	Archive for History of Exact Sciences
Ann. di Mat.	Annali di matematiche pura e applicata
Ann. Sci. Ec. Norm.	Annales des Sciences de l'Ecole Normale Supèrieure
Ann. Sci. Mat. Fis.	Annali di scienze matematiche e fisiche, (ed.Tortolini)
Astr. Nachr.	Astronomische Nachrichten
Bull. A. M. S.	Bulletin of the American Mathematical Society
Comm. Soc. Reg. Gött.	Commentationes (recentiores) societatis regiae scientiarum Göttingensis
C. R.	Comptes Rendus de l'Académie des Sciences, Paris
D. S. B.	Dictionary of Scientific Biography, Scribners, New York
Enc. Math. Wiss.	Encyklopädie der Mathematischen Wissenschaften
Fortschritte	Jahrbuch über die Fortschritte der Mathematik
Giornale di Mat.	Giornale di Matematiche (ed. Battaglini)
H. M.	Historia Mathematica
J. D. M. V.	Jahrsbericht den Deutschen mathematiker Vereinigung
J. de Math	Journal des mathématiques pures et appliquées
J. Ec. Poly	Journal de l'École Polytechnique, Paris
JfM.	Journal für die reine und angewandte Mathematik
Math. Ann.	Mathematische Annalen
Mem. Caen	Mémoires de l'Académie nationale des Sciences, *Arts et Belles –* Lettres de Caen
Monatsber. Königl. Preuss. Akad. der Wiss. zu Berlin	Monatsbericht der Königlich Preussichen
Akademie der Wissenschaften zu Berlin	
Nachr. König. Ges. Wiss. zu Gött	Nachrichten Königlichen Gesellschaft der Wissenschaften zu Göttingen
Nouv. Ann. Math	Nouvelles Annales des Mathématiques

Ouetelet Corr. Math	Correspondence mathématique et physique (ed. Quetelet)
Rep. f. Math.	Repertorium für die reine und angewandte Mathematik
Rev. phil	Revue philosophique
Schlömilch's Zeitschrift	Zeitschrift für Mathematik und Physik …
Tortolini's Ann	Ann. Sci. Mat. Fis.

Bibliography

Abel, N. H., 1828a, Ueber einige bestimmte Integrale, *JfM*, **2**, 22–30, tr. as Sur quelques intégrales définies, in *Oeuvres*, 2nd ed. (no. 15) I, 251–262

Abel, N. H., 1828b, Solution d'un problème général concernant la transformation des fonctions elliptiques, *Astr. Nachr.*, VI, col. 365–388, and Addition au Mémoire précédent, *Astr. Nachr*, VII, **147**, in *Oeuvres*, 2nd ed. (nos. 19, 20) I, 403–428 and 429–443

Abel, N. H., 1881, *Oeuvres complètes de Niels Hendrik Abel*, 2 vols, 2nd ed. L. Sylow and S. Lie, Christiania

Abikoff, W., 1981, The Uniformization Theorem, *American Mathematical Monthly*, **88.8**, 574–592

Anosov, D. V. and Bolibruch, A. A., 1994, *The Riemann-Hilbert Problem*, Steklov Institute of Mathematics, Vieweg

Appell, P., 1880, Sur la série $F_3(\alpha, \beta, \beta', \gamma, x, y)$, *Comptes Rendus*, **90**, 296–9, 731–5, 977–980

Appell, P., 1882, Sur les fonctions hypergéométriques de deux variables, *J. de Math*, **8**(3), 173–217

Appell, P., 1926, *Fonctions hypergéométriques*, Paris, Gauthier-Villars

Arbogast, L-F. A., 1791, *Mémoire sur la nature des fonctions arbitraires ...* St. Petersburg

Archibald, T., 1996, From attraction theory to existence proofs: the evolution of potential-theoretic methods in the study of boundary-value problems, 1860–1890, *Revue d'histoire des mathématiques* **2**, 67–93

Aronhold, S. H., 1872, Partial French translation of: Ueber den gegenseitigen Zusammenhang der 28 Doppeltangenten einer allgemeinen Curve 4-ten Grades (*Berlin Monatsber*. 1864, 499–523), in *Nouv. Ann. Math.*, XI, 438–443

Baldassarri, F. and Dwork, B., 1979, On second order linear differential equations with algebraic solutions, *American Journal of Mathematics*, **101**, 42–76

Baldassarri, F., 1980, On second order linear differential equations with algebraic solutions on algebraic curves, Seminario Matematico, Universita di Padova, Preprint

Barrow-Green, J.E., 1997, *Poincaré and the Three Body Problem*, American and London Mathematical Societies, History of Mathematics Series **11**, Providence, RI

Beardon, A.F., 1983, *The Geometry of Discrete Groups*, Springer-Verlag, New York

Beltrami, E., 1868, Saggio di interpretazione della geometria non Euclidea, *Giornale di Mat*, **6**, 284–312, tr. in Stillwell [1996] 7–34

Beltrami, E., 1869, Essai d'interprétation de la géométrie non-Euclidienne, *Annales Ecole Norm. Sup.*, **6**, 251–288 (translation of previous by Houël)

Bernoulli, Jakob, 1694, Curvatura laminae elasticae, *Acta erud*, 262–276, in *Opera*, II, 576–600

Betti, E., 1853, Sopra l'abbassamento delle equazioni modulari delle funzioni ellittiche, *Tortolini, Annali*, **4**, 81–100

Bieberbach, L., 1918, Über die Einordnung des Hauptsatzes der Uniformisierung in die Weierstrassische Funktionentheorie, *M. Ann.*, **78**, 312–331

Biermann, K. R., 1971a, (Julius Wilhelm) Richard Dedekind, *D. S. B.*, IV, 1–5

Biermann, K. R., 1971b, Ferdinand Gotthold Max Eisenstein, *D. S. B.*, IV, 340–343

Biermann, K. R., 1973a, Ernst Eduard Kummer, *D. S. B.*, V, 521–524

Biermann, K. R., 1973b, *Die Mathematik und ihre Dozenten an der Berliner Universität 1810–1920*, Akademie Verlag, Berlin

Biermann, K. R., 1976, Karl Theodor Wilhelm Weierstrass, *D. S. B.*, XIV, 219–224

Birkhoff, G. D., 1913a, A Theorem on Matrices of Analytic Functions, *Math Ann* **74**, 122–133, and Berichtigung dazu, ibid, 161

Birkhoff, G. D., 1913b, Equivalent singular points of ordinary linear differential equations, *Math Ann*, **74**, 134–139

Birkhoff, G. and Merzbach, U., 1973, *A Source Book in Classical Analysis*, Harvard University Press

Bolyai, J., 1832, Science Absolute of Space, trans. G. B. Halsted, Appendix in Bonola [1912]

Bonola, R., 1912, *Non-Euclidean Geometry*, tr. H. S. Carslaw from original Italian edition of 1906, Dover reprint 1955

Bottazzini, U., 1981, *Il calcolo sublime: storia dell' analisi matematica da Euler a Weierstrass*, Editore Boringhieri, Turin, tr. as *The Higher Calculus, a history of Real and Complex Analysis from Euler to Weierstrass*, Springer-Verlag, New York, 1986

Bottazzini, U., and Tazzioli, R., 1995, *Naturphilosophie* and its role in Riemanns mathematics, *Revue d'histoire des Mathématiques*, **1.1**, 3–38

Boulanger, A., 1898, Contribution à l'étude des équations différentielles linéaires homogènes intégrables algébriquement, *J. Ec Poly*, **4**(2), 1–122

Brill, A. and Noether, M., 1874, Ueber die algebraischer Functionen und ihre Anwendung in der Geometrie, *Math. Ann.*, **7**, 269–310

Brill, A. and Noether, M., 1892–93, Die Entwicklung der Theorie der algebraischen Functionen in alterer und neuerer Zeit, *J. D. M. V.*, **3**, 107–566

Bring, S., 1786, Promotionschrift, Lund

Brioschi, F., 1858a, Sulla equazioni del moltiplicatore per la transformazione delle Funzioni ellittiche, *Ann. di Mat.*, (I), **1**, 175–177, in *Opere*, I, (no. XLIX), 321–324

Brioschi, F., 1858b, Sulla risoluzione delle equazioni del quinto grado, *Ann. di Mat.*, (I), **1**, 256–259 and 326–328 in *Opere*, I, (no. LII), 335–341

Brioschi, F., 1877a, La théorie des formes dans l'intégration des équations différentielles linéaires du second ordre, *Math. Ann.*, **11**, 401–411 in *Opere*, II, (no.
CCXXXIV), 211–223

Brioschi, F., 1877b, Sopra una classe di forme binarie, *Ann. di. Mat.*, (II), **8**, 24–42
in *Opere*, II (no. LXII), 157–175

Brioschi, F., 1877c, Extrait d'une lettre de M. F. Brioschi a M. F. Klein, *M. Ann*, **11**,
111–114

Brioschi, F., 1878/79a, Sopra una classe di equazioni differenziali lineari del se-
condo ordine, *Ann. di Mat.*, (II), **9**, 11–20, in *Opere*, II (no. LXXIII) 177–187

Brioschi, F., 1878/79b, Nota alla memoria del sig. L. Kiepert "Über die Auflosung
der Gleichungen fünften Grades", *Ann. di Mat.*, (II) IX, 124–125, in *Opere*, II
(no. LXXXIV) 189–191

Brioschi, F., 1883/84, Sulla teoria delle funzioni ellittiche, *Ann. di Mat.*, (II), **12**,
49–72, in *Opere*, II (no. LXXXVI), 295–318

Brioschi, F., 1901, *Opere Matematiche*, 5 vols

Briot, C. and Bouquet, J., 1856a, Étude des fonctions d'une variable imaginaire,
J. Ec. Poly, **21**, 85–132

Briot, C. and Bouquet, J., 1856b, Recherches sur les propriétes des fonctions
définies par des équations différentielles, *J. Ec. Poly*, **21**, 133–198

Briot, C. and Bouquet, J., 1856c, Mémoire sur l'intégration des équations différentielles au moyen des fonctions elliptiques, *J. Ec. Poly*, **21**, 199–254

Briot, C. and Bouquet, J., 1859, *Théorie des fonctions doublement périodiques et,
en particulier, des fonctions elliptiques*, Paris

Bühler, W. K., 1981, *Gauss, a biographical study*, Springer-Verlag, New York

Bühler, W. K. 1985, The Life of the hypergeometric function – a biographical
sketch, *Mathematical Intelligencer*, **7.2**, 35-40

Burnside, W., 1891, On a Class of Automorphic Functions, *Proc L. M. S.*, **33**,
49–88

Burnside, W., 1892, Further Note on Automorphic Functions, *Proc L. M. S.*, **34**,
281–295

Cauchy, A. L., 1840, Mémoire sur l'intégration des équations différentielles in *Exercises d'Analyse et de physique mathématiques*, II, Bachelier, Paris, in *Oeuvres*,
11(2), 399–465

Cauchy, A. L., 1841, Note sur le nature des problèmes que presente le calcul intégrale, in *Exercises d'Analyse et de physique mathématiques*, I, Bachelier, Paris,
in *Oeuvres*, **12**(2), 263–271

Cauchy, A. L., 1843, Rapport sur un Mémoire de M. Laurent qui a pour titre: Extension du théorème de M. Cauchy relatif à la convergence du développment d'une
fonction suivant les puissances ascendantes de la variable x, *C. R.*, **17**, 938–940,
in *Oeuvres*, (I), **8**, (no. 233) 115–117

Cauchy, A. L., 1845, Sur le nombre des valeurs égales ou inégales que peut acquérir
une fonction, etc. *C. R.*, 21, 779–797, in *Oeuvres*, **9**(1), 323–34

Cauchy, A. L., 1851, Rapport sur un Mémoire presente à l'Académie par M. Puiseux
C. R., **32**, 276, 493, in *Oeuvres*, **11**(1), 380–382

Cauchy, A. L., 1981, *Équations différentielles ordinaires. Cours inedit. Fragment.* Introduction by Ch. Gilain. Johnson Reprint, New York

Cayley, A., 1859, A sixth memoir upon quantics, *Phil. Trans. Roy. Soc. London,* **149**, 61–90, in *Coll. Math. Papers,* **2**(158) 561–592

Cayley, A., 1859b, On the conic of 5 point contact at any point of a plane curve, Phil. Trans. Roy. Soc. CXLIX 371–400, in *Coll. Math. Papers,* **4**(261) 207–239

Cayley, A., 1865, Note on Lobachevskii's imaginary geometry, *Phil. Mag.,* **29**, 231–233, in *Coll. Math. Papers,* **5**(362) 471–472

Cayley, A., 1883, On the Schwarzian derivative and the polyhedral functions, *Trans. Camb. Phil. Soc.,* **13.1**, 5–6, in *Coll. Math. Papers,* **11**(745) 148–216

Chasles, M., 1837, *Aperçu historique sur l'origine et le développement des méthodes en Géométrie,* Bruxelles, Mem. Couronn. XI

Christoffel, B. E., 1867, Über einige allgemeine Eigenschaften der Minimumsflächen, *JfM,* **67**, 218–228

Christoffel, B. E., 1868, Allgemeine Theorie der geodetischen Dreiecke, *Abh. Berlin,* 119–176

Christoffel, B. E., 1869a, Über die Transformation der homogenen Differentialausdrücke Zweiten Grades, *JfM,* **70**, 46–70

Christoffel, B. E., 1869b, Über ein die Transformation homogener Differentialausdrücke Zweiten Grades betreffendes Theorem, *JfM,* **70**, 241–245

Clebsch, R. F. A., 1864, Ueber die Anwendung der Abelschen Functionen in der Geometrie, *JfM,* **63**, 189–243

Clebsch, R. F. A., 1876, *Vorlesungen über Geometrie,* ed. F. Lindemann, Teubner, Leipzig

Clebsch, R. F. A. and Gordan, P., 1866, *Theorie der Abelschen Functionen.* Teubner, Leipzig

Clemens, C. H., 1980, *A Scrapbook of Complex Curve Theory,* Plenum, New York

Coble, A., 1908, A configuration in finite geometry isomorphic with that of the 27 lines on a cubic surface, Johns Hopkins University Circular, no. 7, 80–88, 736–744

Coble, A., 1913, An application of finite geometry to the characteristic theory of the odd and even theta functions, *Trans. A. M. S.,* **14**, 241–276

Codazzi, D., 1857, Intorno alle superficie le quali hanno costante il prodotto de' due raggi di curvatura, *Ann. Sci. Mat. Fis,* **8**, 346–355

Courant, R., 1912, Über die Anwendung des Dirichlets Prinzipes auf die Probleme der konformen Abbildung, *M. Ann,* **71**, 145–183

Darboux, G., 1888, *Leçons sur la Théorie genérales des Surfaces,* **2**, 2nd ed. 1915, Gauthier-Villars, Paris

Dedekind, R., 1872, *Stetigkeit und irrationale Zahlen,* tr. Beman as *Continuity and irrational numbers,* Dover 1963

Dedekind, R., 1877, Schreiben an Herrn Borchardt über die Theorie der elliptischen Modulfunctionen, *JfM,* **83**, 265–292, in *Werke,* I, 174–201

Dedekind, R., 1882, (with H. Weber) Theorie der algebraischen Functionen einer Veränderlichen, *JfM,* **92**, 181–290, in *Werke,* I 238–350

Dedekind, R., 1969, *Gesammelte mathematische Werke*, I, New York

Dickson, L. E., 1900, *Linear Groups*, reprinted Dover 1958

Dieudonné, J., 1974, *Cours de géométrie algébrique*, I, (Collection SUP) Presses Univ. de France, Paris

Dieudonné, J., 1978, *Abrégé d'histoire des mathématiques*, (editor) Hermann, Paris

Dieudonné, J., 1983, La découverte des fonctions fuchsiennes, C. R. Paris

Dieudonné, J. and Carroll, J., 1971, *Invariant Theory, old and new*, New York

Dirichlet, P. G. L., 1829, Sur la convergence des séries trigonométriques, *JfM*, **4**, 157–169, in *Werke*, I, 117–132

Dirichlet, P. G. L., 1850, Die reduction der positiven quadratischen Formen mit drei unbestimmten ganzen Zahlen, *JfM*, **40**, 213–220, in *Werke*, II, 34–41

Dirichlet, P. G. L., 1889–97, *Gesammelte Werke*, 2 vols, ed. L. Fuchs and L. Kronecker, Berlin

Dubrovin, B. A., 1981, Theta functions and non-linear equations, English tr. in *Russian Math. Surveys*, **36**(2), 11–92

Dugac, P., 1976, *Richard Dedekind et les fondements des mathématiques*, Vrin, Paris

Dunnington, G. W., 1955, *Carl Friedrich Gauss, Titan of Science*, New York

Dutka, J., 1984, The Early History of the Hypergeometric Function, *A. H. E. S.*, **31.1**, 15–34

Dyck, W. von, 1880a, Ueber Ausstellung und Untersuchung von Gruppe und Irrationalität regular Riemann'scher Flächen, *M. Ann.*, **17**, 473–509

Dyck, W. von, 1880b, Notiz über eine reguläre Riemann'scher Flächen vom Geschlechte drei ..., *M. Ann.*, **17**, 510–516

Edwards, H. M., 1977, *Fermat's Last Theorem*, Springer-Verlag, New York

Edwards, H. M., 1978 On the Kronecker Nachlass, *H. M.*, **5.4**, 419–426

Eisenstein, F. G. M., 1847, Genaue Untersuchungen der unendlichen Doppelproducte ..., *JfM*, **35**, 137–274, in *Mathematische Werke*, I, 1975, (no. 28) 357–478

Euler, L., 1769, *Institutiones Calculi Integralis*, **2**, in *Opera Omnia*, ser. 1, **12**

Euler, L., 1770, *Institutiones Calculi Integralis*, **3**, in *Opera Omnia*, ser. 1, **13**

Euler, L., 1794, Specimen Transformationis singularis serierum, *Nova Acta Acad Sci Petrop*, 58–70, in *Opera Omnia*, ser. 2, **16**, 41 -55 (E710)

Farkas, H. M. and Kra, I., 1980, *Riemann Surfaces*, Springer Verlag, New York

Fisher, C., 1966, The death of a mathematical theory, *A. H. E. S.*, **3**, 137–159

Ford, L. R., 1929, *Automorphic functions*, reprinted Chelsea, 1972

Forsyth, A. R., 1900–06, *Theory of Differential Equations*, 6 vols. Cambridge University Press

Freudenthal, H., 1954, Poincaré et les fonctions automorphes, in Poincaré *Oeuvres*, **11**, 213–219

Freudenthal, H., 1970, L'Algèbre topologique en particulier les groupes topologiques et de Lie, XIIe *Congrès International d'Histoire des Sciences, Paris, Actes*, I, 223–243

Freudenthal, H., 1971, Augustin – Louis Cauchy, *D. S. B.*, **3**, 131–148

Freudenthal, H., 1975, Bernhard Riemann, *D. S. B.*, **11**, 447–456

Fricke, R., 1913, Automorphe Funktionen mit Einschluss der elliptischen Modulfunktionen, *Enc. Math. Wiss.*, II B 4

Fricke, R. and Klein, C. F., 1897, 1912 *Vorlesungen über die Theorie der Automorphen Functionen*, 2 vols, Teubner, Leipzig and Berlin

Frobenius, F. G., 1873a, Über die Integration der linearen Differentialgleichungen durch Reihen, *JfM*, **76**, 214–235, in *Ges. Abh.*, I, 84–105

Frobenius, F. G., 1873b, Über den Begriff der Irreducibilität in der Theorie der linearen Differentialgleichungen, *JfM*, **76**, 236–270, in *Ges. Abh.*, I, 106–140

Frobenius, F. G., 1875a, Über algebraisch integrirbare lineare Differentialgleichungen, *JfM*, **80**, 183–193, in *Ges. Abh.*, I, 221–231

Frobenius, F. G., 1875b, Über die reguläre Integrale der linearen Differentialgleichungen, *JfM*, **80**, 317–333, in *Ges. Abh.*, I, 232–248

Frobenius, F. G., 1968, *Gesammelte Abhandlungen*, ed. J. P. Serre, Springer Verlag

Fubini, G., 1908, *Introduzione alla Teoria dei Gruppi discontinui e delle Funzioni automorfe*, Enrico Spoerri, Pisa

Fuchs, L. I., 1865, Zur Theorie der linearen Differentialgleichungen mit veränderlichen Coefficienten, *Jahrsber. Gewerbeschule Berlin*, Ostern, in *Werke*, I, 111–158

Fuchs, L. I., 1866, Zur Theorie der linearen Differentialgleichungen mit veränderlichen Coefficienten, *JfM*, **66**, 121–160, in *Werke*, I, 159–204

Fuchs, L. I., 1868, Zur Theorie der linearen Differentialgleichungen mit veränderlichen Coefficienten, Ergänzungen zu der im 66-sten Bände dieses Journal enthaltenen Abhandlung, *JfM*, **68**, 354–385, in *Werke*, I, 205–240

Fuchs, L. I., 1870a, Die Periodicitätsmoduln der hyperelliptischen Integrale als Functionen eines Parameters aufgefasst, *JfM*, **71**, 91–127, in *Werke*, I, 241–282

Fuchs, L. I., 1870b, Über eine rationale Verbindung der Periodicitätsmoduln der hyperelliptischen Integrale, *JfM*, **71**, 128–136, in *Werke*, I, 283–294

Fuchs, L. I., 1870c, Bemerkungen zu der Abhandlungen "Über hypergeometrische Functionen $n$ter Ordnung in diesem Journal", **71**, 316, in *JfM*, **72**, 255–262, in *Werke*, I, 311–320

Fuchs, L. I., 1871a, Über die Form der Argumente der Thetafunctionen und über die Bestimmung von $\theta(0, 0, \dots, 0)$ als function der Klassenmoduln, *JfM*, **73**, 305–323, in *Werke*, I, 321–342

Fuchs, L. I., 1871b, Über die linearen Differentialgleichungen, welchen die Periodicitatsmoduln der Abelschen Integrale genügnen, und über verschiedene Arten von Differentialgleichungen fur $\theta(0, 0, \dots, 0)$, *JfM*, **73**, 324–339, in *Werke*, I, 343–360

Fuchs, L. I., 1875, Über die linearen Differentialgleichungen zweiter Ordnung, welche algebraische Integrale besitzen, und eine neue Anwendung der Invariantentheorie ..., *Nachr. Königl. Ges. der Wiss Gött*, 568–581, 612–613, in *Werke*, II, 1–10

Fuchs, L. I., 1876a, Über die linearen Differentialgleichungen zweiter Ordnung, welche algebraische Integrale besitzen, und eine neue Anwendung der Invariantentheorie, *JfM*, **81**, 97–142, in *Werke*, II, 11–62

Fuchs, L. I., 1876b, Extrait d'une lettre adressée à M. Hermite, *J. de Math*, **2**(3), 158–160, in *Werke*, II, 63–66

Fuchs, L. I., 1876c, Sur les équations différentielles linéaires du second ordre, *C. R.*, **82**, 1494–1497 and **83**, 46–47, in *Werke*, II, 67–72

Fuchs, L. I., 1877a, Selbstanzeige der Abhandlung: "Über die linearen Differentialgleichungen zweiter Ordnung, welche algebraische Integrale besitzen und eine neue Anwendung der Invariententheorie" *Borchardts Journal*, Bd. 81, p. 97 sqq., *Rep. f. Math*, **1**, 1–9, in *Werke*, II, 73–86

Fuchs, L. I., 1877b, Sur quelques proprietes des intégrales des équations différentielles, auxquelles satisfont les modules de périodicité des intégrales elliptiques des deux premières especes. Extrait d'une lettre adressée à M. Hermite, *JfM*, **83**, 13–37, in *Werke*, II, 87–114

Fuchs, L. I., 1877c, Extrait d'une lettre adressée à M. Hermite, *C. R.*, **85**, 947–950, in *Werke*, II, 145–150

Fuchs, L. I., 1878a, Über die linearen Differentialgleichungen zweiter Ordnung, welche algebraische Integrale besitzen. Zweite Abhandlung, *JfM*, **85**, 1–25, in *Werke*, II, 115–144

Fuchs, L. I., 1878b, Über eine Klasse von Differentialgleichungen, welche durch Abelsche oder elliptische Functionen integrirbar sind, *Nachrichten von der Königl. Gesellschaft der Wissenschaften*, Göttingen 19–32; *Ann di mat*, (II), **9**, 25–35, in *Werke*, II, 151–160

Fuchs, L. I., 1878c, Sur les équations différentielles linéaires qui admettent des intégrales dont les différentielles logarithmiques sont des fonctions doublement périodiques. Extrait d'une lettre à M. Hermite, *J. de Math.*, **4**(3), 125–140, in *Werke*, II, 161–17

Fuchs, L. I., 1879a, Selbstanzeige der Abhandlung: "Sur quelques propriétés des intégrales des équations différentielles, auxquelles satisfont les modules de périodicité des intégrales elliptiques des deux premières especes. Extrait d'une lettre adressée à M. Hermite". *Borchardts Journal*, **83**, 13, *Rep. Math*, **2**, 235–240, in *Werke*, II, 177–184

Fuchs, L. I., 1880a, Über eine Klasse von Functionen mehrerer Variabeln, welche durch Umkehrung der Integrale von Lösungen der linearen Differentialgleichungen mit rationalen Coefficienten entstehen, *Nachr Königl. Ges. der Wiss, Göttingen*, 170–176, in *Werke*, II, 185–190

Fuchs, L. I., 1880b, Über eine Klasse von Functionen mehrerer Variabeln, welche durch Umkehrung der Integrale von Lösungen der linearen Differentialgleichungen mit rationalen Coefficienten entstehen, *JfM*, **89**, 151–169, in *Werke*, II, 191–212

Fuchs, L. I., 1880c, Sur une classe de fonctions de plusieurs variables tirées de l'inversion des intégrales de solutions des équations différentielles linéaires dont les coefficients sont des fonctions rationnelles. Extrait d'une lettre adressée à M. Hermite, *C. R.*, **90**, 678–680 and 735–736, in *Werke*, II, 213-218

Fuchs, L. I., 1880d, Über die Functionen, welche durch Umkehrung der Integrale von Lösungen der linearen Differentialgleichungen entstehen, *Nachr Königl Ges. der Wiss, Göttingen*, 445–453, in *Werke*, II, 219–224

Fuchs, L. I., 1880e, Sur les fonctions provenant de l'inversion des intégrales des solutions des équations différentielles linéaires . . . , *Bulletin des Sciences mathématiques et astronomiques*, **4**(2), 328–336, in *Werke*, II, 229–238

Fuchs, L. I., 1881a, Auszug aus einem Schreiben des Herrn L. Fuchs an C. W. Borchardt, *JfM*, **90**, 71–73, in *Werke*, II, 225–228

Fuchs, L. I., 1881b, Über Functionen zweier Variabeln, welche durch Umkehrung der Integrale zweier gegebener Functionen entstehen, *Abh. König. Ges. der Wiss. zu Göttingen*, **27**, 1–39, in *Werke*, II, 239–274

Fuchs, L. I., 1881c, Sur les fonctions de deux variables qui naissent de l'inversion des intégrales de deux fonctions données. Extrait d'une lettre adressée à M. Hermite, *C. R.*, **92**, 1330–1331, 1401–1403, in *Werke*, II, 275–289

Fuchs, L. I., 1881d, Sur une équation differentielle de la forme $f\left(u, \frac{du}{dz}\right) = 0$. Extrait d'une lettre adressée à M. Hermite, *C. R.*, **93**, 1063–1065, in *Werke*, II, 283–284

Fuchs, L. I., 1882a, Über Functionen, welche durch lineare Substitutionen unverändert bleiben, *Nachr. Königl. Ges. der Wiss. Göttingen*, 81–84, in *Werke*, II, 285–288

Fuchs, L. I., 1882b, Über lineare homogene Differentialgleichungen, zwischen deren Integralen homogene Relationen hoheren als ersten Grades bestehen, *Sitzungsberichte der K. Preuss Akademie der Wissenschaften zu Berlin*, 703–710, in *Werke*, II, 289–298

Fuchs, L. I., 1882c, Über lineare homogene Differentialgleichungen, zwischen deren Integralen homogene Relation hoheren als ersten Grades bestehen, *Acta Mathematica*, **1**, 321–362, in *Werke*, II, 299–340

Fuchs, L. I., 1884a, Über Differentialgleichungen, deren Integrale feste Verzeigungspunkte besitzen, *Sitzungsberichte der K. Preuss. Akademie der Wissenschaften zu Berlin*, 699–710, in *Werke*, II, 355–368

Fuchs, L. I., 1884b, Antrittsrede gehalten am 3 Juli 1884 in der offentlichen Sitzung zur Feier des Leibniztages der Königl. Preuss Akademie der Wissenschaften zu Berlin, *Sitzungsberichte der K. Preuss. Akademie der Wissenschaften zu Berlin*, 744–747, in *Werke*, II, 369–372

Fuchs, L. I., 1904, 1906, 1909, *Gesammelte Mathematische Werke*, 3 vols, ed. R. Fuchs and L. Schlesinger, vol I, Berlin

Galois E., 1846a, Lettre à M. A. Chevalier sur la théorie des équations et les fonctions intégrales, *J. de Math.*, **11**, 408–416, in *Oeuvres*, 25–32

Galois E., 1846b, Mémoire sur les conditions de la resolubilité des équations par radicaux, *J. de Math.*, **11**, 417–433.

Galois, E., 1897, *Oeuvres Mathématiques*, ed. E. Picard.

Gårding, L., 1997, *Mathematics and Mathematicians – Mathematics in Sweden before 1950*, History of Mathematics **13** American and London Mathematical Societies, Providence, R. I

Garnier, R., 1928, Le problème de Plateau, Annales de L'École Normale Supérieure **45**(3), 53–144

Gauss, C. F., 1801, *Disquistiones Arithmeticae*, in *Werke*, I.

Gauss, C. F., 1812a, Disquisitiones generales circa seriem infinitam, Pars prior, *Comm. Soc. reg Gött.*, II, in *Werke*, III, 123–162.

Gauss, C. F., 1812b, Determinatio seriei nostrae per aequationem differentialem secundi ordinis, Ms. in *Werke*, 207–230.

Gauss, C. F., 1822, Allgemeine Auflösung etc, *Astron. Abh.*, III, **1**, in *Werke*, IV, 189–216.

Gauss, C. F., 1827, Disquisitiones generales circa superficies curvas, *Comm. Soc. Reg. Gött.*, VI, in *Werke*, IV, 217–258

Gauss, C. F., 1839-40, Allgemeine Lehrsatz ein Beziehung auf die im verkehrten Verhältnisse des Quadrats der Entfernung, Resultate aus den Beobachtungen des magnetischen Vereins im Jahre 1839. Herausg. v. Gauss u. Weber, Leipzig, in *Werke*, V, 195–240

Gauss, C. F., 1860–65, *Briefwechsel zwischen C. F. Gauss and H. C. Schumacher*, 6 vols, Altona

Gauss, C. F., 1863–67, *Werke*, vols 1–V

Geiser, C. F., 1869, Ueber die Doppeltangenten einer ebenen Curve Vierten Grades, *M. Ann.*, **1**, 129–138

Geppert, H., 1927, *Bestimmung der Anziehung eines elliptischen Ringes, Nachlass zur Theorie des arithmetisch-geometrischen Mittels und der Modulfunktion von C. F. Gauss.* Ostwald Klassiker no. 225, Leipzig

Gilain, C., 1977, *La theorie géométrique des équations différentielles de Poincaré et l'histoire de l'analyse*, Thesis, Paris

Göpe1, A., 1847, Theorie transcendentium Abelianarum primi ordinis adumbratio levis, *JfM*, **35**, 277–312

Gordan, P., 1868, Beweis, dass jede Covariante und Invariante einer binaren Form eine ganze Function mit numerischen Coefficienten einer endlichen Anzahl solcher Formen ist, *JfM*, **69**, 323–354

Gordan, P., 1877a, Über endliche Gruppen linearer Transformationen einer Veranderlichen, *M. Ann.*, **12**, 23–46

Gordan, P., 1877b, Binare Formen mit verschwinden Covarianten, *M. Ann.*, **12**, 147–166

Gordan, P., 1878, Über die Auflösung der Gleichungen vom fünften Grade, *M. Ann.*, **13**, 375–404

Gordan, P., 1880, Ueber das volle Formen system der terneren biquadratischen Form $f = x_1^3 x_2 + x_2^3 x_3 + x_3^3 x_1$, *M. Ann.*, **17**, 217–233

Goursat, E., 1882, Extension de problème de Riemann, *Comptes Rendus*, **94**, 903–904, 1044–1047

Goursat, E., 1883a, Sur les fonctions hypergéométriques de deux variable *C. R.*, **95**, 717–9

Goursat, E., 1883b, Sur une classe des fonctions, *Acta Mathematica*, **2**, 1–70

Gray, J. J., 1979a, Non-Euclidean Geometry – a re-interpretation, *H. M.*, **6**, 236–258

Gray, J. J., 1981, Les trois suppléments au Mémoire de Poincaré, écrit en 1880, sur les fonctions fuchsiennes et les équations différentielles. *C. R.*, **293**, Vie Académique, 87–90

Gray, J. J., 1982, The three supplements to Poincaré's prize essay of 1880 on Fuchsian functions and differential equations, *Arch. Int. Hist. Sci.*, **32**, No. 109, 221–235

Gray, J. J., 1984a, Fuchs and the theory of differential equations, *Bull. A. M. S.*, **10.1**, 1–26, and corrigendum

Gray, J. J., 1984b, A commentary on Gauss's mathematical diary, 1796–1814, with an English translation, *Expositiones Mathematicae*, **2**, 97–130, with corrigenda

Gray, J. J., 1987, The Riemann-Roch Theorem: the acceptance and rejection of geometric ideas, 1857–74, *Cahiers d'histoire et de philosophie des sciences*, **20**, 139–151

Gray, J. J., 1989, *Ideas of Space*, 2nd ed. Oxford Univ. Press

Gray, J. J., 1994, On the history of the Riemann Mapping Problem, *Supplemento ai Rendiconti del Circolo Matematico di Palermo*, **34**(2), 47–94

Green, G., 1835, On the Determination of the Exterior and Interior Attractions of Ellipsoids of Variable Densities, *Trans Camb Phil Soc*, **5**(3), 395–430, in *Mathematical Papers* 1883, 187–222

Hadamard, J., 1920, *The Early Scientific Work of Poincaré*, Rice Institute Pamphlet, no. 9.3, 111–183

Hadamard, J., 1954, *The Psychology of Invention in the Mathematical Field*, Dover

Haedenkamp, H., 1841, Ueber Transformation vielfacher Integrale, *JfM*, **22**, 184–192

Halphen, G. H., 1881a, Sur des fonctions qui proviennent de l'équation de Gauss, *C. R.*, **92**, 856, in *Oeuvres*, II, 471–474

Halphen, G. H., 1881b, Sur un système d'équations différentielles, *C. R.*, **92**, 1101, in *Oeuvres*, II, 475–477

Halphen, G. H., 1884a, Sur une équation différentielle linéaires du troisième ordre, *M. Ann.*, **21**, 461, in *Oeuvres*, IV, 112–115

Halphen, G. H., 1884b, Sur le reduction des équations différentielles linéaires aux formes intégrables, *Mémoires presentés par divers savants à l'Académie des Sciences*, **28**, 1–260, in *Oeuvres*, III, 1–260

Halphen, G. H., 1921, *Oeuvres*, 4 vols, ed. C. Jordan, H. Poincaré, E. Picard, E. Vessiot, Gauthier-Villars, Paris

Hamburger, M., 1873, Bemerkungen über die Form der Integrale der linearen Differentialgleichungen mit veränderlichen Coefficienten, *JfM*, **76**, 113–125

Harkness, J., Wirtinger, W., and Fricke, R., 1913, Elliptische functionen, *Enc. Math. Wiss.*, II, B3., **13**, 1–52

Harnack, A., 1887, *Grundlagen der Theorie des logarithmischen Potentiales*, etc., Teubner, Leipzig

Hartshorne, R., 1977, *Algebraic Geometry*, Springer-Verlag, New York

Haskell, M. W., 1890, Über die zu der Curve $\lambda^3\mu + \mu^3 v + v^3\lambda = 0$ ... *American Journal of Mathematics*

Hawkins, T., 1975, *Lebesgue's Theory of Integration*, Chelsea (2nd edition)

Hawkins, T., 1977, Weierstrass and the Theory of Matrices, *A. H. E. S.*, **17.2**, 119–163

Hawkins, T., 1980, Non-Euclidean Geometry and Weierstrassian Mathematics: The background to Killing's work on Lie Algebras, *H. M.*, **7.**3, 289–342

Heaviside, O., 1876, On duplex telegraphy *Philosophical Magazine*, **1**(5), 32–43, in *Electrical papers*, **1**, 53–64

Heine, H. E., 1845, Beitrag zur Theorie der Anziehung und der Wärme, *JfM*, **29**, 185–208

Heine, H. E., 1846, Über die Reihe $1 + (q - l)(q - 1)x+$ etc., *JfM*, **32**, 210–212

Heine, H. E., 1847, Untersuchungen über die Reihe ... etc., *JfM*, **34**, 285–328

Helmholtz, H., 1868, Über die tatsachlichen Grundlagen der Geometrie, *Nachr. König. Ges. Wiss. zu Göttingen*, **15**, 193–221, in *Abh.*, **2**, 1883, 618–639

Hermite, C., 1851, Sur les fonctions algébriques, *C. R.*, **32**, 458–461, in *Oeuvres*, I, 276–280

Hermite, C., 1858, Sur la résolution de l'équation du cinquième degré, *C. R.*, **46**, 508–515, in *Oeuvres*, II, 5–12

Hermite, C., 1859a, Sur la théorie des équations modulaires, *C. R.*, **48**, 940–947, 1079–1084, 1095–1102 and *C. R.*, **49**, 16–24, 110–118, 141–144, in *Oeuvres*, II, 38–82

Hermite, C., 1859b, Sur l'abaissement de l'équation modulaire du huitième degré, *Ann. di Mat.*, II, 59–61, in *Oeuvres*, II, 83–86

Hermite, C., 1862, Note sur la théorie des fonctions elliptiques, in Lacroix: *Calcul différentiel et Calcul intégral*, Paris 6th ed. 1862, in *Oeuvres*, II, 125–238

Hermite, C., 1863, Sur les fonctions de sept lettres, *C. R.*, **57**, 750–757, in *Oeuvres*, II, 280–288

Hermite, C., 1865–66, Sur l'équation du cinquième degré, *C. R.*, **66** and *C. R.*, **67** passim, in *Oeuvres*, II, 347–424

Hermite, C., 1877–82, Sur quelques applications des fonctions elliptiques, *C. R.*, **85–94** passim, in *Oeuvres*, III, 266–418

Hermite, C., 1900, Lettre de Charles Hermite à M. Jules Tannery sur les fonctions modulaires, ms. in *Oeuvres*, III, 13–21

Hermite, C., 1905–17, *Oeuvres*, 4 vols, Paris

Hesse, L. O., 1844, Ueber die Wendepunkte der Curven dritter Ordnung, *JfM*, **28**, 97–106, in *Werke*, (no. 9), 123–136

Hesse, L. O., 1855a, Ueber Determinanten und ihre Anwendung in der Geometrie, insbesondere auf Curven Vierter Ordnung, *JfM*, **49**(24), 243–264, in *Werke*, 319–343

Hesse, L. O., 1855b, Ueber die Doppeltangenten der Curven Vierter Ordnung, *JfM*, **49**, 279–332, in *Werke*, (no. 25) 345–404

Hesse, L. O., 1897, *Gesammelte Werke*, 1st pub. 1897, reprinted Chelsea, 1972,

Hilb, E., 1913, Lineare Differentialgleichungen im komplexen Gebiet, *Enc Math Wiss*, IIB5, II.2, 471–562

Hilbert, D., 1900a, Ueber das Dirichlet'sche Prinzip. *J. D. M. V.*, VIII, 184–188, tr. in Birkhoff Merzbach, 399–402

Hilbert, D., 1900b, Mathematische Probleme, *Göttinger Nachrichten*, 253–297 (translated in *Bull A. M. S.* 1902, 437–479 reprinted in *Mathematical develop-*

ments arising from Hilbert Problems, Proc. Symp. Pure Math. A. M. S. 1972, vol. I, 1–34.)

Hilbert, D., 1905, Über eine Anwendung der Integralgleichungen auf ein Problem der Funktionentheorie, *Verhandlungen 3. Internationaler Mathematiker-Kongresses, Heidelberg*, Teubner, Leipzig

Hille, E., 1976, *Ordinary Differential Equations in the Complex Domain*, Wiley

Hölder, E., 1981, Historischer Überblick zur mathematischen Theorie von Unstetigkeitswellen seit Riemann und Christoffel, in *E. B. Christoffel*, Basel 1981, 412–434

Holzapfel, R. P., 1986a, A Voyage with three Balloons, *Mathematical Intelligencer*, **12.1**, 33–39

Holzapfel, R. P., 1986b, *Geometry and Arithmetic around Euler Partial Differential Equations*, Reidel, Dordrecht

Hurwitz, A., 1881, Grundlagen einer independenten Theorie der elliptischen Modulfunktionen …, *M. Ann.*, **18**, 528–592, in *Math. Werke*, Birkhäuser, Basel, 1932, I, 1–66

Hurwitz, A., 1886, Über einige besondere homogene linearen Differentialgleichungen, *M. Ann.*, **26**, 117–126, in *Math. Werke*, I (no. 9) 153–162

Jacobi, C. G. J., 1829, *Fundamenta Nova Theoriae Functionum Ellipticarum*, in *Ges. Werke*, I, (2nd ed.) 49–239

Jacobi, C. G. J., 1834, De functionibus duarum variabilium quadrupliciter periodicis, quibis theoria transcendentium Abelianarum innititur, *JfM*, **13**, *Ges. Werke*, II, (2nd ed.) 23–50

Jacobi, C. G. J., 1850, Beweis des Satzes, dass ein Curve n^{ter} Grades im Allgemein $\frac{1}{2}n(n - 2)(n^2 - 9)$ Doppeltangenten hat, *JfM*, **40**, 237–260, in *Ges. Werke*, III, 517–542

Jacobi, C. G. J., 1969, *Gesammelte Werke*, 2nd edition, 8 vols. Chelsea

Jevons, W. S., 1871, Helmholtz on the Axioms of Geometry, *Nature*, **4**, 481–482

Johnson, D. M., 1979, The Problem of Invariance of Dimension in the Growth of Modern Topology, Part I, *A. H. E. S.*, **20.2**, 97–188

Johnson, D. M., 1981, The Problem of Invariance of Dimension in the Growth of Modern Topology, Part II, *A. H. E. S.*, **20.2/3**, 85–267

Jordan, C., 1868a, Note sur les équations modulaires, *C. R.*, **66**, 308–312, in *Oeuvres*, I, 159–163

Jordan, C., 1868b, Mémoire sur les groupes de mouvement, Ann. di Mat., II, 167–215, 322–345, in *Oeuvres*, IV, 231–302

Jordan, C., 1869, Commentaire sur Galois, *M. Ann.*, **1**, 142–160, in *Oeuvres*, I, 211–230

Jordan, C., 1870, *Traité des substitutions et des équations algébriques*, Paris

Jordan, C., 1871, Sur la résolution des équations différentielles linéaires, *C. R.*, **73**, 787–791, in *Oeuvres*, I, 313–317

Jordan, C., 1876–77, Sur une classe de groupes d'ordre fini contenus dans les groupes linéaires, *Bull. Math. Soc. France*, **5**, 175–177, in *Oeuvres*, II, 9–12

Jordan, C., 1876a, Sur les équations du second ordre dont les intégrales sont algébriques, *C. R.*, **82**, 605–607, in *Oeuvres*, II, 1–4

Jordan, C., 1876b, Sur la détermination des groupes formé d'un nombre fini des substitutions linéaires, *C. R.*, **83**, 1035–1037, in *Oeuvres*, II, 5–6

Jordan, C., 1876c, Mémoire sur les covariants des formes binaires, I, *J. de Math.*, **2**(3), 177–232, in *Oeuvres*, III, 153–208

Jordan, C., 1877, Détermination des groupes formé d'un nombre fini de substitutions, *C. R.*, **84**, 1446–1448, in *Oeuvres*, II, 7–8

Jordan, C., 1878, Mémoire sur les équations différentielles linéaires à intégrale algébrique, *JfM*, **84**, 89–215, in *Oeuvres*, II, 13–140

Jordan, C., 1879, Mémoire sur les covariants des formes binaires, II, *J. de Math.*, **5**(3), 345–378, in *Oeuvres*, III, 213–246

Jordan, C., 1880, Sur la détermination des groupes d'ordre fini contenus dans le groupe linéaire, *Atti Accad. Nap*, **8**, no. 11, in *Oeuvres*, II, 177–218

Jordan, C., 1915, *Cours d'Analyse* III, 3rd ed. Paris

Jordan, C., 1961, *Oeuvres de Camille Jordan*, 4 vols, ed. J. Dieudonné, Paris.

Kaplansky, I., 1976, *An Introduction to Differential Algebra*, 2nd ed. Hermann, Paris

Katz, N. M., 1976, An overview of Deligne's work on Hilbert's Twenty First Problem, in *Mathematical Developments arising from Hilbert Problems*, Proc. Symp. Pure Math AMS XXVIII, vol. 2, 537–557

Kiepert, L., 1876, Wirkliche Ausführung der ganzzähligen Multiplication der elliptischen Functionen, *JfM*, **76**, 21–33

Kiernan, B. M., 1972, The Development of Galois Theory from Lagrange to Artin, *A. H. E. S.*, **8**, 40–154

Kirchhoff , 1857, Ueber die Bewegung der Elektricität in Drählen, *Annalen der Physik und Chemie*, **100**(4), 193–217, in *Gesammelte Abhandlungen*, 131–155

Klein, C. F., 1871, Über die sogenannte Nicht – Euklidische Geometrie, I, *M. Ann.*, **4**, in *Ges. Math. Abh.* I, (no. XVI) 254–305

Klein, C. F., 1872, *Vergleichende Betrachtungen über neuere geometrische Forschungen* (Erlanger Programm) 1st pub. Deichert, Erlangen, in *Ges. Math. Abh.*, I (no. XXVII) 460–497

Klein, C. F., 1873, Über die sogenannte Nicht-Euklidische Geometrie, II, *M. Ann.*, **6**, in *Ges. Math. Abh.*, I (no. XVIII) 311–343

Klein, C. F., 1875–76, Über binare Formen mit linearen Transformationen in sich selbst, *M. Ann.*, **9**, in *Ges. Math. Abh.*, II (no. LI) 275–301

Klein, C. F., 1876, Über [algebraisch integrirbare] lineare Differentialgleichungen, I, *M. Ann.*, **11**, in *Ges. Math. Abh.*, II (no. LII) 302–306

Klein, C. F., 1877a, Über [algebraisch integrirbare] lineare Differential gleichungen, II, *M. Ann.*, **12**, in *Ges. Math. Abh.*, II (no. LIII) 307–320

Klein, C. F., 1877b, Weitere Untersuchungen über das Ikosaeder, *M. Ann.*, **12**, in *Ges. Math. Abh.*, II (no. LIV) 321–379

Klein, C. F., 1878–79a, Über die Transformationen der elliptischen Funktionen und die Auflosung der Gleichungen Fünften Grade, *M. Ann.*, **14**, in *Ges. Math. Abh.*, III (no. LXXXII) 13–75

Klein, C. F., 1878–79b, Über die Erniedrigung der Modulargleichungen, *M. Ann.*, **14**, in *Ges. Math. Abh.*, III (no. LXXXIII) 76–89

Klein, C. F., 1879a, Über die Transformation siebenter Ordnung der elliptischen Funktionen, *M. Ann.*, **14**, in *Ges. Math. Abh.*, III (no. LXXXIV) 90–134

Klein, C. F., 1879b, Über die Auflosung gewisser Gleichungen vom siebenten und achten Grades, *M. Ann.*, **15**, in *Ges. Math. Abh.*, II (no. LVII) 390–425

Klein, C. F., 1880–81, Zur Theorie der elliptischen Modulfunktionen, *M. Ann.*, **17**, in *Ges. Math. Abh.*, III (no. LXXXVII) 169–178

Klein, C. F., 1881, Über Lamé'sche Funktionen, *M. Ann.*, **18**, in *Ges. Math. Abh.*, II (no. LXII) 512–520

Klein, C. F., 1882a, *Über Riemanns Theorie der algebraischen Funktionen und ihrer Integrale*, Teubner, Leipzig, in *Ges. Math. Abh.*, III (no. XCIX) 499–573

Klein, C. F., 1882b, Über eindeutige Funktionen mit linearen Transformationen in sich, I, *M. Ann.*, **19**, in *Ges. Math. Abh.*, III (no. CI) 622–626

Klein, C. F., 1882c, Über eindeutige Funktionen mit linearen Transformationen in sich, II, *M. Ann.*, **20**, in *Ges. Math. Abh.*, III (no. CII) 627–629

Klein, C. F., 1883, Neue Beiträge zur Riemannschen Funktionentheorie, *M. Ann.*, **21**, in *Ges. Math. Abh.*, III (no. CIII) 630–710

Klein, C. F., 1884, *Vorlesungen über das Ikosaeder und die Auflösung der Gleichungen vom fünften Grade*. Teubner, Leipzig (English translation *Lectures on the Icosahedron* by G. G. Morrice, C. F. Klein, 1888, Dover reprint 1956), reprinted 1993, mit einer Einführung und mit Kommentaren, P. Slodowy, Birkhäuser, Basel

Klein, C. F., 1894a, *Vorlesungen über die hypergeometrische Funktion*, Göttingen

Klein, C. F., 1894b, *Höhere Geometrie*, **2**, Teubner, Leipzig

Klein, C. F., 1921, 1922, 1923a, *Gesammelte Mathematische Abhandlungen*, 3 vols Springer-Verlag, Berlin

Klein, C. F., 1923b, Göttinger Professoren (Lebensbilder von eigener Hand): Felix Klein, *Mitteilung des Universitätsbundes Göttingen*, **5.1**, 11–36, tr. as Klein (forthcoming)

Klein, C. F., (Forthcoming), *Vorlesungen über die Entwicklung der Mathematik im 19. Jahrhundert*. Chelsea reprint, 2 vols. in 1

Klein, C.F., (Forthcoming), *The Evanston Colloquium and other Lectures on Mathematics*, edited with introductory essays, by Rowe D.E., and Gray J.J., Springer-Verlag, New York

Kline, M., 1972, *Mathematical Thought from Ancient to Modern Times*. Oxford University Press

Koebe, P., 1907a, Über die Uniformisierung reeller algebraischer Kurven, *Göttingen Nachrichten*, 177–190

Koebe, P., 1907b, Über die Uniformisierung beliebiger analytischer Kurven, *Göttingen Nachrichten*, 191–210

Koebe, P., 1907c, Über die Uniformisierung beliebiger analytischer Kurven, (2), *Göttingen Nachrichten*, 633–669

Koebe, P., 1908, Über die Uniformisierung beliebiger analytischer Kurven, (3), *Göttingen Nachrichten*, 337–358

Koebe, P., 1909a, Über die Uniformisierung der algebraischen Kurven, I, *M. Ann*, **67**, 145–224

Koebe, P., 1909b, Über die Uniformisierung beliebiger analytischer Kurven, *Göttinger Nachrichten*, (4), 337–358

Koebe, P., 1910a, Über die Uniformisierung der algebraischen Kurven, II, *M. Ann*, **69**, 1–81

Koebe, P., 1910b, Über die Uniformisierung beliebiger analytischer Kurven, I, *JfM*, **138**, 192–253

Koebe, P., 1911, Über die Uniformisierung beliebiger analytischer Kurven, II, *JfM*, **139**, 251–292

Koebe, P., 1912, Ueber eine neue Methode der konformen Abbildung und Uniformisierung, *Göttingen Nachrichten*, 844–848

Koebe, P., 1913, Ränderzuordnung bei konformer Abbildung, *Göttingen Nachrichten*, 286–288

Koenigsberger, L., 1868, *Die Multiplication und die Modulargleichungen der elliptischen Functionen*, Teubner, Leipzig

Koenigsberger, L., 1871, Die linearen transformationen der Hermite'schen ϕ-Function, *M. Ann.*, **3**, 1–10

Kolchin, E., 1973, *Differential algebra and algebraic groups*, Academic Press, New York

Krazer, A., 1909, Zur Geschichte des Umkehrproblems der Integrale, *J. D.M.V.*, **18**, 44–75

Krazer, A. and Wirtinger, W., 1913, Abelsche Funktionen und allegemeine Thetafunktionen, *Enc. Math. Wiss*, II. B. 7, 604–873

Kronecker, L., 1857, Über die elliptischen Functionen, für welche complexe Multiplication stattfindet, *Monatsber. Königl. Preuss. Akad. der Wiss. zu Berlin*, 455–460, in *Werke*, II, (no. XVII) 177–184

Kronecker, L., 1858a, Über Gleichungen des siebenten Grades, *Monatsber, etc. Berlin*, 287–289, in *Werke*, II, (no. V), 39–42

Kronecker, L., 1858b, Sur la résolution de l'équation du cinquième degré, (Extrait d'une lettre adressée à M. Hermite), *C. R.*, **46**, 1150–1152, in *Werke*, II, (no. VI), 43–48

Kronecker, L., 1859, Sur la théorie des substitutions (Extrait d'une lettre adressée à M. Brioschi) *Ann. di mat*, II, **131**, in *Werke*, II (no. VII) 51–52

Kronecker, L., 1863, Über die Auflösung der Pell'schen Gleichung mittels elliptischer Functionen, *Monatsber. etc. Berlin*, 44–50, in *Werke*, II (no. XXI), 219–226

Kronecker, L., 1881–82, Grundzüge einer arithmetischen Theorie der algebraischen Grossen, Festschrift Kummer, *JFM*, **92**, 1–122, in *Werke*, II (no. XI), 237–388

Kronecker, L., 1883, Über bilineare Formen mit vier Variabeln, *Abh. Königl. Preuss. Akad. des Wiss. zu Berlin* II, 1–60, in *Werke*, II, (no. XVII), 425–496

Kronecker, L., 1897, *Werke*, II

Kronecker, L., 1929, *Werke*, IV

Kronecker, L., 1929, *Werke*, 5 vols, ed. K. Hensel, Teubner, Leipzig, reprinted Chelsea 1968

Kummer, E. E., 1834, De generali quadam aequatione differentiali tertii ordinis, *Programme of the Liegnitz Gymnasium*, in *JfM*, **100**, 1–9 (1887), in *Coll. Papers*, II, 33–39

Kummer, E. E., 1836, Über die hypergeometrische Reihe ..., *JfM*, **15**, 39–83 and 127–172, in *Coll. Papers*, II, 75–166

Kummer, E. E., 1975, *Collected Papers*, 2 vols, edited A. Weil, Springer-Verlag, New York

Kung, J. P. S. and Rota, G. C., 1984, The invariant theory of binary forms, *Bull. A. M. S.*, **10.1**, 27–85

Lamé, G., 1837, Mémoire sur les surfaces isothermes, etc., *J. de Math.*, **2**, 147–183

Lampe, E., 1892–3, Nachruf für Ernst Eduard Kummer, *J. D. M. V.*, **3**, 13–28, in Kummer, *Coll. Papers*, I, 1975

Lang, S., 1976, *Modular Functions*, Springer-Verlag, New York

Laurent, P. A., 1843, Extension du théorème de M. Cauchy, relatif à la convergence du développement d'une fonction suivant les puissances ascendantes de la variable, in J. Peiffer, *Les premiers exposes globaux de la théorie des fonctions de Cauchy, 1840–1860*, Thesis, Paris

Lauricella, G., 1893, Sulle funzioni ipergeometriche a piu variabile, *Rendiconti Palermo*, **7**, 111–158

Legendre, A. M., 1786, Mémoire sur les intégrations par les arcs d'ellipse, *Hist. Acad. Paris*, 616–644

Legendre, A. M., 1814, *Exercises de calcul intégral*, vol. II

Legendre, A. M., 1823, *Éléments de Géométrie* (12th ed.) Paris

Legendre, A. M., 1825, *Traité des fonctions elliptiques et des Intégrales euleriennes*, 3 vols. Paris

Lehner, J., 1966, *A short course in automorphic functions*, Holt, Rinehart, and Winston, New York

Lie, M. S., 1888, *Theorie der Transformationsgruppen*, **1**, Teubner, Leipzig

Lie, M. S., 1891, Die linearen homogenen gewöhnlichen Differentialgleichungen *Leipziger Berichte*, 253–270

Lindemann, F., 1899, Zur Theorie der automorphen Functionen, *Sitzberichte der Math-Physik Classe München*, **29**, 423–454

Liouville, J., 1845, Sur diverses questions d'analyse et de physique mathématique, *J. de Math.*, **10**, 222–228

Lobachevskii, N. I., 1829, Ueber die Anfangsgründe der Geometrie, Kasan Bulletin

Lobachevskii, N. I., 1837, Géométrie Imaginaire, *JfM*, **17**, 295–320

Lobachevskii, N. I., 1840, *Geometrische Untersuchungen etc.* Berlin, tr. G.B. Halsted as Geometrical researches on the theory of parallels, appendix in Bonola [1912]

Lützen, J., 1990, *Joseph Liouville, 1809–1882, Master of Pure and Applied Mathematics*, Springer-Verlag, New York

Magid, A. R., 1994, *Lectures on Differential Galois Theory*, American Mathematical Society University Lecture Series, Providence RI

Manin, Ju., 1974, *Cubic Forms*, trans. M. Hazewinkel, North Holland, Amsterdam

Mathieu, E., 1860, Le nombre de valeurs que peut acquérir une fonction quand on y permute ses variables de toutes les manières possibles, *J. de Math.*, **5**, 9–42

Mathieu, E., 1861, Mémoire sur l'étude des fonctions des plusieurs quantitées, *J. de Math.*, **6**, 241–323

Mathieu, E., 1875, Sur le fonction cinq fois transitive de 24 quantitées, *J. de Math.*, **18**, 25–46

Matsuda, M., 1980, *First Order Algebraic Differential Equations —A Differential Algebraic Approach*, Springer-Verlag, Heidelberg, New York, L. N. I. M. 804

May, K. O., 1972, Gauss, Carl Friedrich, *D. S. B.*, V

Mehrtens, H., 1979, Das Skelett der modernen Algebra, *Disciplinae Novae* ed. C. J. Scriba, Joachim Jungius Gesellschaft, Hamburg

Meyer, F., 1890–91, Bericht über den gegenwartigen Stand der Invarianten Theorie, *J. D. M. V.*, **1**, 79–288

Miller, G. A., Blichfeldt, H. F. and Dickson, L. E., 1916, *The Theory and Application of Finite Groups*, 1st ed. New York, Dover reprint 1962

Minding, F., 1839, Wie sich entscheiden lässt, ob zwei gegebener Krummen Flächen, etc., *JfM*, **19**, 370–387

Mitschi, C. and Singer, M.F., 1996, Connected Linear Groups as Differential Galois Groups, *Journal of Algebra*, **184**, 333–361

Mittag-Leffler, G., , 1876, En Metod alt kommer i analytiske berittning af de Elliptiska functionerna, Helsingfors, translated by E. Hille as An Introduction to the Theory of Elliptic functions, *Acta Mathematica*, **24**(2), 271–351

Mittag-Leffler, G., 1880, Sur les équations différentielles linéaires à coefficients doublement périodiques, *C R.*, **90**, 299–300

Monna, A. F., 1975, *Dirichlet's Principle. A mathematical comedy of errors and its influence on the development of analysis.* Oosthoek, Scheltema and Holkema, Utrecht

Montel, P., 1907, Sur les suites infinies de fonctions, *Annales de l'Ecole Normale* **4**(3), 233–304

Muller, F., 1872, Ueber die Transformation vierten Grades der elliptischen functionen, Berlin

Muller, F., 1873, Beziehungen zwischen dem Moduln der elliptischen functionen, etc. *Schlömilch's Zeitschrift*, **18**, 280–288

Mumford, D. and others, 1983, 1984, 1991, *Tata Lectures on Theta*, 3 vols. Birkhäuser Boston

Netto, E., 1882, *Die Substitutionentheorie und ihre Anwendung auf die Algebra*, Teubner, Leipzig

Neuenschwander, E., 1978a, The Casorati-Weierstrass Theorem, *H. M.*, **5**, 139–166

Neuenschwander, E., 1978b, Der Nachlass von Casorati (1835–1890) in Pavia, *A. H. E. S.*, **19.1**, 1–89

Neuenschwander, E., 1980, Riemann und das "Weierstrassche" Prinzip der Analytischen Fortsetzung durch Potenzreihen, *J. D. M. V.*, **82**, 1–11

Neuenschwander, E., 1981a, Lettres de Bernhard Riemann à sa famille, *Cahiers du Seminaire d'Historie des Mathématiques*, 85–131

Neuenschwander, E., 1981b, Studies in the History of Complex Function Theory, II, *Bull. A. M. S.*, **5.2**, 87–105

Neumann, C. A., 1865a, *Vorlesungen über Riemann's Theorie der Abel'schen Integrale*, Teubner, Leipzig

Neumann, C. A., 1865b, *Das Dirichlet'sche Princip in seiner Andwendung auf die Riemanns'chen Flächen*, Teubner, Leipzig

Newton, I., 1968, *The Mathematical Papers of Isaac Newton*, ed. D. T. Whiteside, II, 10–89, Cambridge University Press

Nicolson, J.C., 1993, The development and understanding of the concept of quotient group, *Historia Mathematica* **20**, 68–88

Osgood, W. F., 1900, On the existence of the Green's function for the most general simply connected plane region, *Trans AMS*, **1**, 310–314

Osgood, W. F., 1912, *Lehrbuch der Funktionentheorie*, 2nd ed. vol. 1, Teubner, Leipzig

Osgood, W. F. and Taylor, E. H., 1913, Conformal transformations on the boundary of their regions of definition, *Trans AMS*, **14**, 277

Painlevé, P., 1887, Sur les équations différentielles linéaires du troisième ordre, *C.R.* **104**, 1829–1832

Papperitz, E., 1889, Ueber die Darstellung der hypergeometrischen Transcendenten durch eindeutige Functionen, *M. Ann.*, **34**, 247–296

Pépin P. Th., 1863, Mémoire sur l'intégration sous forme finie de l'équation différentielle du second ordre à coefficients rationnels, *Ann. di Mat.* **5**(1) , 185–224

Pépin P. Th., 1878, Sur les équations différentielles du second ordre, *Ann. di Mat.* **9** (2), 1

Pépin P. Th., 1881, Méthode pour obtenir les intégrales algbriques des équations différentielles linéaires du second ordre, *Atti dell'Accademia Pontificia de' Nuovi Lincei*, **34**, 243–388

Pfaff, J. F., 1797, *Disquisitiones Analyticae*, I, Helmstadt

Piaggio, H. T. H., 1962, *An Elementary Treatise on Differential Equations and their applications*, London

Picard, E., 1887, Sur les équations différentielles linéaires et les groupes algébriques de transformations, *Annales de la Faculté des Sciences de Toulouse*, **1**, A.1–A.15

Picard, E., 1879a, Sur une propriété des fonctions entières, *C. R.*, **88**, 1024–1027

Picard, E., 1879b, Sur une généralisation des fonctions périodiques et sur certaines équations différentielles linéaires, *C. R.*, **89**, 140–144

Picard, E., 1880a, Sur une classe d'equations différentielles linéaires, *C. R.*, **90**, 128–131

Picard, E., 1880b, Sur les équations différentielles linéaires à coefficients doublement périodiques, *C. R.*, **90**, 293–295

Picard, E., 1880c, Sur une extension aux fonctions de deux variables du problème de Poincaré relatif aux fonctions hypergéométriques, *C. R.*, **91**, 1267–9

Picard, E., 1881a, Sur les équations différentielles linéaires à coefficients doublement périodiques, *JfM*, **90**, 281–302

Picard, E., 1881b, Sur une extension aux fonctions de deux variables du problème de Riemann relatif aux fonctions hypergéométriques, *Ann. Ec Norm*, **10**(2), 305–322

Picard, E., 1885, Sur les formes quadratiques ternaires indéfinies à indeterminées conjugées, *Acta Mathematica*, **5**, 121–182

Picard, E., 1895, *Traité d'analyse* Gauthier-Villars, Paris

Pincherle, S., 1888, Sulle funzione ipergeometriche generalizzate, *Rend Accad Lincei Rome*. **4.1**, 694–700 and 792–9

Plemelj, J., 1908, Riemannsche Funktionenscharen mit gegebener Monodromiegruppe, *Monatshefte für Mathematik und Physik*, **19**, 211–246

Plemelj, J., 1909, Über Schlesingers "Beweis" der Existenz Riemannscher Funktionenscharen mit gegebener Monodromiegruppe, *J. D. M. V.*, **18**, 15–20

Plemelj, J., 1964, *Problems in the sense of Riemann and Klein*, Interscience Tract no. 16, Wiley

Plücker, J., 1834, *System der analytischen Geometrie*, Berlin

Plücker, J., 1839, *Theorie der algebraischen Curven*, Berlin

Pochhammer, L., 1870, Ueber hypergeometrische Functionen n^{ter} Ordnung, *JfM*, **71**, 316–352

Pochhammer, L., 1889, Ueber ein Integral mit doppeltem Umlauf, *M. Ann.*, **35**, 470–494

Poincaré, H., 1880, Extrait d'un Mémoire inédit de Henri Poincaré sur les fonctions fuchsiennes, *Acta Mathematica.*, **39** (1923), 58–93, in *Oeuvres*, I, 578–613

Poincaré, H., 1881a, Sur les fonctions fuchsiennes, *C. R.*, **92**, 333–335, in *Oeuvres* II, 1–4

Poincaré, H., 1881b, Sur les fonctions fuchsiennes, *C. R.*, **92**, 395–398, in *Oeuvres*, II, 5–7

Poincaré, H., 1881c, Sur une nouvelle application et quelques propriétés importantes des fonctions fuchsiennes, *C. R.*, **92**, 859–861, in *Oeuvres*, V, 8–10

Poincaré, H., 1881d, Sur les applications de la géométrie non-Euclidienne à la théorie des formes quadratiques, *Association Française pour l'avancement des sciences*, 10th session, Algiers, 16, in *Oeuvres*, V, 267–276 April

Poincaré, H., 1881e, Sur les fonctions fuchsiennes, *C. R.*, **92**, 957, in *Oeuvres*, II, 11

Poincaré, H., 1881f, Sur les fonctions abeliennes, *C. R.*, **92**, 958–959, in *Oeuvres*, IV, 299–301

Poincaré, H., 1881g, Sur les fonctions abeliennes, *C. R.*, **92**, 1198–1200, in *Oeuvres*, II, 12–15

Poincaré, H., 1881h, Sur les fonctions abeliennes, *C. R.*, **92**, 1214–1216, in *Oeuvres*, II, 16–18

Poincaré, H., 1881i, Sur les fonctions abeliennes, *C. R.*, **92**, 1484–1487, in *Oeuvres*, II, 19–22

Poincaré, H., 1881j, Sur les groupes kleinéens, *C. R.*, **93**, 44–46, in *Oeuvres*, II, 23–25

Poincaré, H., 1881k, Sur une fonction analogue aux fonctions modulaires, *C. R.*, **93**, 138–140, in *Oeuvres*, II, 26–28

Poincaré, H., 1881l, Sur les fonctions fuchsiennes, *C. R.*, **93**, 301–3, in *Oeuvres*, II, 29–31

Poincaré, H., 1881m, Sur une fonction analogue aux fonctions modulaires, *C. R.*, **93**, 581–582, in *Oeuvres*, II, 32–34

Poincaré, H., 1881n, Sur une fonction analogue aux fonctions modulaires, *Mem. Caen* for 1882, 3–29, in *Oeuvres*, II, 75–91

Poincaré, H., 1882a, Sur une fonction analogue aux fonctions modulaires, *C. R.*, **94**, 163–166, in *Oeuvres*, II, 35–37

Poincaré, H., 1882b, Sur les groupes discontinus, *C. R.*, **94**, 840–843, in *Oeuvres*, II, 38–40

Poincaré, H., 1882c, Sur les fonctions fuchsiennes, *C. R.*, **94**, 1038–1040, in *Oeuvres*, II, 41–43

Poincaré, H., 1882d, Sur les fonctions fuchsiennes, *C. R.*, **94**, 1166–1167, in *Oeuvres*, II, 44–46

Poincaré, H., 1882e, Sur une classe d'invariants relatifs aux équations linéaires, *C. R.*, **94**, 1042–1045, in *Oeuvres*, II, 47–49

Poincaré, H., 1882f, Sur les fonctions fuchsiennes, *C. R.*, **95**, 626–628, in *Oeuvres*, II, 50–52

Poincaré, H., 1882g, Sur les fonctions uniformes qui se réproduisent par des substitutions linéaires, *M. Ann.*, **19**, 553–564, in *Oeuvres* II, 92–105

Poincaré, H., 1882h, Sur les fonctions uniformes qui se réproduisent par des substitutions linéaires (Extrait d'une lettre adressée à M. F. Klein.), *M. Ann.*, **20**, 52–53, in *Oeuvres*, II, 106–107

Poincaré H., 1882i, Théorie des Groupes Fuchsiens, *Acta Mathematica*, **1**, 1–62, in *Oeuvres*, II, 108–168

Poincaré H., 1882j, Sur les Fonctions fuchsiennes, *Acta Mathematica*, **1**, 193–294, in *Oeuvres*, II, 169–257

Poincaré, H., 1883, Sur l'intégration algébrique des équations linéaires, *Comptes Rendus*, **97**, 984–5, and 1189–1191, in *Oeuvres*, 3, 101–2, 103–5

Poincaré, H., 1883, Sur un théorème de la théorie générale des fonctions, *Bulletin Société Mathématique de France*, **11**, 112–125, in *Oeuvres*, 4, 57–69

Poincaré H., 1883a, Sur les groupes des équations linéaires, *C. R.*, **96**, 691–694, in *Oeuvres*, II, 53–55

Poincaré, H., 1883c, Sur les fonctions fuchsiennes, *C. R.*, **96**, 1485–1487, in *Oeuvres* II, 59–61

Poincaré, H., 1883d, Mémoire sur les Groupes kleinéens, *Acta Mathematica*, **3**, 49–92, in *Oeuvres*, II, 258–299

Poincaré, H., 1883h, Mémoire sur les Groupes kleinéens, *C. R.*, **96**, 1302–1304, in *Oeuvres*, II, 56–58

Poincaré, H., 1884a, Sur les groupes des équations linéaires, *Acta Mathematica*, **4**, 201–311, in *Oeuvres*, 2, 300–401, tr. Stillwell, [1985] 357–483

Poincaré, H., 1884b, Mémoire sur les fonctions zétafuchsiennes, *Acta Mathematica*, **5**, 209–278, in *Oeuvres*, II, 402–462

Poincaré, H., 1886, Sur les intégrales irrégulières des équations linéaires, *Acta Mathematica*, **8**, 295–344, in *Oeuvres*, I, 290–332

Poincaré, H., 1893, Sur la propagation de l'électricité, *C. R.*, **117**, p. 1027 in *Oeuvres*, **9**, 278–283

Poincaré, H., 1894, *Cours sur les oscillations électriques*, Paris

Poincaré, H., 1907, Sur l'uniformisation des fonctions analytiques, *Acta Mathematica*, **31**, 1–63, in *Oeuvres*, **4**, 70–139

Poincaré, H., 1907, Sur l'uniformisation des fonctions analytiques, *Acta Mathematica*, **31**, 1–63, in *Oeuvres*, IV, 70–139

Poincaré, H., 1908, *Science et méthode*, Paris

Poincaré, H., 1916–54, *Oeuvres*, 11 vols. Paris

Poincaré, H., 1985, *Papers on Fuchsian Functions*, tr. J. Stillwell, Springer-Verlag, New York

Poincaré, H., 1997, *Three Supplementary Essays on the Discovery of Fuchsian Functions*, ed. Gray, J. J. and Walter, S. A. with an introductory essay, Akademie Verlag, Berlin, Blanchard, Paris

Poncelet, V., 1832, Théorèmes et problémes sur les lignes du troisième ordre, *Quetelet, Corr. Math.*, **7**, 79–84

Pont, J. C., 1974, *La Topologie algébrique des origines à Poincaré*, P. U. F. Paris

Poole, E. G. C., 1936, *Introduction to the Theory of Linear Differential Equations*, Oxford University Press, Dover reprint 1960

Prosser, R.T., 1994, On the Kummer Solutions of the Hypergeometric Equation, *American Mathematical Monthly*, 535–543

Puiseux, V., 1850, Recherches sur les fonctions algébriques, *J. de Math.*, **15**

Puiseux, V., 1851, Recherches sur les fonctions algébriques, Suite, *J. de Math*, **16**

Purkert, W., 1971, Zur Genesis des abstrakten Korperbegriffs, I, *N. T. M.*, **8.1**, 23–37

Purkert, W., 1973, Zur Genesis des abstrakten Korperbegriffs II, *N. T. M.*, **10.2**, 8–26

Purkert, W., 1976, Ein Manuskript Dedekinds über Galoistheorie, *N. T. M.*, **13.2**, 1–16

Radner, J. M., 1979, Bernhard Riemann's probationary lecture, Cambridge M. Phil. Thesis

Reichardt, H., 1976, *Gauss und die nicht-euklidische Geometrie*, Teubner, Leipzig

Richards, J. L., 1979, The Reception of a Mathematical Theory: Non Euclidean Geometry in England 1865–1883 in B. Barnes, S. Shapin (eds.) *Natural Order: Historical Studies of Scientific Culture*, Sage Publications, Beverley Hills

Riemann, B., 1851, Grundlagen für eine allgemeine Theorie der Functionen einer veränderlichen complexen Grösse (Inaugural dissertation), Göttingen, in *Werke*, 3–45

Riemann, B., 1854a, Über die Darstellbarkeit einer Function durch einer trigonometrische Reihe, *K. Ges. Wiss. Göttingen*, **13**, 87–132, in *Werke*, 227–271

Riemann, B., 1854b, Ueber die Hypothesen welche der Geometrie zu Grunde liegen, *K. Ges. Wiss. Göttingen*, **13**, 1–20, in *Werke*, 272–287

Riemann, B., 1857a, Beiträge zur Theorie der durch Gauss'sche Reihe $F(\alpha, \beta,$

γ, x) darstellbaren Functionen, *K. Ges. Wiss. Göttingen*, in *Werke*, 67–83

Riemann, B., 1857b, Selbstanzeige der vorstehenden Abhandlung, *Göttingen Nachr*. no. 1, in *Werke*, 84–87

Riemann, B., 1857c, Theorie der Abelschen Functionen, *JfM*, **54**, 115–155, in *Werke*, 88–144

Riemann, B., 1859, Ueber die Anzahl der Primzahlen unter einer gegebene Grösse, *Monatsberichte Berlin Akademie*, 671–680, in *Werke*, 145–153

Riemann, B., 1860, Ueber die Fortpflanzung ebener Luftwellen von endlicher Schwingungsweite, *Abh. König. Ges Wiss Gött*, **8**, in *Werke*, 3rd ed, 188–207

Riemann, B., 1865, Ueber das Verschwinden der Theta-functionen, *JfM*, **65**, in *Werke*, 212–224

Riemann, B., 1867, Ueber die Fläche vom kleinsten Inhalt bei gegebener Begrenzung, K. Ges. Wiss. Göttingen, **13**, 21–57, in *Werke*, 301–337

Riemann, B., 1990, *Bernhard Riemann's Gesammelte Mathematische Werke und Wissenschaftliche Nachlass*, ed. R. Dedekind and H. Weber, with *Nachträge*, ed. M. Noether and W. Wirtinger. 3rd ed. R. Narasimhan

Riemann, B., 1990a, Zwei allgemeine Lehrstäze über lineare Differentialgleichungen mit algebraischen Coefficienten, ms. in *Werke*, 379–390

Riemann, B., 1990b, Fragmente über die Grenzfälle der elliptischen Modulfunctionen, ms. in *Werke*, 455–465

Riemann, B., 1990c, Convergenz der p-fach unendlichen Theta Reihe, ms. in *Werke*, 483–486

Riemann, B., 1990d, Zur Theorie der Abel'schen Functionen, ms. in *Werke*, 487–504

Riemann, B., 1990e, Nachträge, in *Werke* after 559

Ritt, J.F. , 1950, *Differential Algebra*, American Mathematical Society Colloquium Publications, **33**, Dover reprint New York 1966

Ritter, E., 1892, Die eindeutigen automorphen Formen vom Geschlechte Null, *Math. Ann*, **41**, 1–82

Roch, G., 1864, Ueber die Anzahl der willkurlichen Constanten in algebraischen Functionen, *JfM*, **64**, 372–376

Roch, G., 1866, Ueber die Doppeltangenten an Curven vierter Ordnung, *JfM*, **66**, 97–120

Röhrl, H., 1957, Das Riemann–Hilbertsche Problem der Theorie der linearen Differentialgleichungen, *M. Ann*, **133**, 1–25

Rosenhain, G., 1851, Mémoire sur les fonctions de deux variables et à quatre périodes, qui sont les inverses des intégrales ultraelliptiques de la première classe, *Paris, Mem. Savans. Etrang.*, **11**, 361–468

Rowe, D. E., 1983, A forgotten chapter in the History of Klein's Erlanger Programm, *H. M.*, **10.4**, 448–454

Salmon, G., 1879, *A Treatise on Higher Plane Curves*, 3rd ed. Chelsea reprint

Schläfli, L., 1870, Über die Gauss'sche hypergeometrische Reihe, *M. Ann.*, **3**, 286–295, in *Ges. Math. Abh.*, III, 153–162

Schläfli, L., 1956, *Gesammelte Mathematische Abhandlungen.*, III, Basel

Schlesinger, L., 1895, 1897, 1898, *Handbuch der Theorie der Linearen Differentialgleichungen*, **3**, vols. Teubner, Leipzig

Schlesinger, L., 1904, Über das Riemannsche Fragment zur Theorie der Linearen Differentialgleichungen und daran anschliessende neuere Arbeiten, in *Verhandlungen des Dritten Internationalen Mathematiker Kongresses in Heidelberg*, ed. A. Krazer, Teubner, Leipzig

Schlesinger, L., 1906, Bemerkung zu dem Kontinuitätsbeweise des Riemannschen Problems, *Math. Ann*, **63**, 273–276

Schlesinger, L., 1909, Bericht über die Entwicklung der Theorie der linearen Differentialgleichungen seit 1865, *J. D. M. V.*, XVIII, 133–267

Schlesinger, L., 1909, Bemerkungen zum Kontinuitätsbeweise für die Lösbarkeit des Riemannschen Problems, *J. D. M. V.*, **18**, 21–25

Schlesinger, L., 1922, *Einführung in die Theorie der gewöhnlichen Differentialgleichungen auf funktiontheoretischer Grundlage*, 3rd ed. De Gruyter, Berlin and Leipzig

Schlissel, A., 1976–77, The Development of Asymptotic Solutions of Linear Ordinary Differential Equations, 1817–1920, *A. H. E. S.*, **16**, 307–378

Schoeneberg, B., 1974, *The Theory of Elliptic Modular Functions*, Springer-Verlag Heidelberg and New York

Scholz, E., 1980, *Geschichte des Mannigfaltigkeitsbegriffs von Riemann bis Poincaré*, Birkhäuser, Boston

Scholz, E., 1982, Herbart's Influence on Bernhard Riemann, *H. M.*, **9.4**, 413–440

Schottky, F., 1877, Ueber die conforme Abbildung mehrfach zusammenhängender ebener Flächen, *JfM*, **83**, 300–351

Schottky, F., 1887, Ueber eine specielle Function, welcher bei einer bestimmten linearen Transformationen ihres Argument unverändert bleiben, *JfM*, **101**, 227

Schwarz, H. A., 1869a, Ueber einige Abbildungsaufgaben, *JfM*, **70**, 105–120, in *Abh.*, II, 65–83

Schwarz, H. A., 1869b, Conforme Abbildung der Oberflache eines Tetraeders auf die Oberflache einer Kugel, *JfM*, **70**, 121–136, in *Abh.*, II, 84–101

Schwarz, H. A., 1869c, Zur Theorie der Abbildung, *Programm ETH Zürich*, in *Abh.*, II, 108–132

Schwarz, H. A., 1870a, Ueber einen Grenzübergang durch altenirendes Verfahren, *Vierteljahrschrift Natur. Gesellschaft Zürich*, **15**, 272–286, in *Abh.*, II, 133–143

Schwarz, H. A., 1870b, Ueber die Integration der partiellen Differentialgleichung $\frac{\partial^2 u}{\partial x^2} + \frac{\partial^2 u}{\partial y^2} = 0$ unter vorgeschriebenen Grenz – und Unstetigkeits-bedingungen, *Monatsber. K. A. der Wiss. Berlin*, 767–795, in *Abh.*, II, 144–171

Schwarz, H. A., 1872, Ueber diejenigen Falle, in welchen die Gaussische hypergeometrische Reihe eine algebraische Function ihres vierten Elementes darstellt, *JfM*, **75**, 292–335, in *Abh.*, II, 211–259

Schwarz, H. A., 1880, Auszug aus einem Briefe an Herrn F. Klein, *M. Ann.*, **21**, 157–160, in *Abh.*, II, 303–306

Schwarz, H. A., 1890, *Gesammelte Mathematische Abhandlungen*, 2 vols. 1st ed.

Berlin 1890, 2nd ed. reprinted in 1 vol. Chelsea, 1972

Serre, J. P., 1973, *A Course in Arithmetic*, Springer-Verlag, New York

Serre, J. P., 1980, Extensions Icosaédriques, Seminaire de Theories des Nombres de Bordeaux, Année 1979–80, no. 19, 19–01, 19–07, in *Oeuvres*, III, 550–554, Springer-Verlag, New York

Serret, J. A., 1879, *Cours d'Algèbre Superiéure*, 4th ed. Paris

Silvestri, A., 1979, Simple Groups of Finite Order in the Nineteenth Century, *A. H. E. S.*, **20.3**, 313–357

Singer, M. F., 1981, Liouvillian Solutions of nth order homogeneous linear differential equations, *Amer. J. Math*, **103.4**, 661–682

Singer, M. F. and Ulmer, F., 1993 Liouvillian and Algebraic Solutions of Second and Third Order Linear Differential Equations, *J. Symbolic Computation*, **16**, 37–73

Smith, H. J. S., 1859–65, Report on the Theory of Numbers, Chelsea reprint 1965

Smith, H. J. S., 1877, Mémoire sur les équations modulaires, *Atti della R. Accad. Lincei*, in *Coll. Math. Papers*, II, (no. XXXV) 224–241

Smith, H. J. S., 1894, *Collected Mathematical Papers*, Oxford University Press 2 vols, Chelsea reprint 1965

Sohnke, L. A., 1834, Aequationes modulares pro transformatione functionum ..., *JfM*, **12**, 178

Sommerville, D. M. Y., 1911, *Bibliography of non-Euclidean Geometry*, reprint Chelsea

Springer, T. A., 1977, *Invariant Theory*, L. N. I. M. 585 Springer Verlag, Heidelberg and New York

Stallo, J. B., 1888, *The Concepts and Theories of Modern Physics*, Belknap Press, Harvard University, reprint 1960

Staudt, K. G. C. von, 1856–60, *Beiträge zur Geometrie der Lage*, 3 vols, Nurnberg

Steiner, J., 1848, Ueber allgemeine Eigenschaften der algebraischen Curven, *Berlin Bericht*, 310–316

Steiner, J., 1852, Ueber solche algebraische Curven, welche einen Mittelpunkt haben, *JfM*, **47**, 7–105

Stillwell, J., 1996, *Sources of Hyperbolic Geometry*, History of Mathematics vol. 10, American and London Mathematical Societies, Providence, R. I

Sylow, L., 1872, Théorèmes sur les groupes de substitutions, *M. Ann.*, **5**, 584–594

Tannery, J., 1875, Propriété des intégrales des équations différentielles linéaires à coefficients variables, *Ann. Sci. de l'Ecole Normale Superieure*, (2), **4.1**, 113–182

Tannery, J., 1876, La géométrie imaginaire et la notion d'espace, *Rev. Phil.*, **2**, 433–451, 553–575

Thomae, J., 1870, Über die höheren hypergeometrischen Reihen, insbesondere über die Reihe ..., *M. Ann.*, **2**, 427–444

Thomae, J., 1874, Integration einer linearen Differentialgleichungen zweiter Ordnung durch Gauss'sche Reihen, *Schlömilch's Zeitschrift*, **19**, 273–285

Thomé, L. W., 1841–1910, 1872, Zur Theorie der linearen Differentialgleichungen, *JfM*, **74**, 193–217

Thomé, L. W., 1873, Zur Theorie der linearen Differentialgleichungen, *JfM*, **75**, 265–291

Thomé, L. W., 1884, Zur Theorie der linearen Differentialgleichungen, *JfM*, **96**, 185–281

Tissot, J., 1852, Sur un determinant d'intégrales définies, *J. de. Math.*, **17**, 177–185

Tobies, R., 1981, *Felix Klein, Biographien . . .*, Teubner, Leipzig

Toth, I., 1977, La révolution non-euclidienne, *La Recherche*, **75.8**, 143–151

Toth, I., 1984, Three errors in the Grundlagen of 1884: Frege and non-Euclidean Geometry, *Frege Conference 1984*, ed. G. Werschung, Akademie-Verlag, Berlin

Tretkoff, C. and Tretkoff, M., 1979, Solution of the inverse problem of differential Galois theory in the classical case, *Amer J. Math*, **101**, 1327–1332

Valentiner, H., 1889, De endelige Transformations – Gruppers Theori, Danish Academy publications, series V, vol. 6

Varadarajan, V. S., 1991, Meromorphic Differential Equations, *Expositiones Mathematicae*, **9.2**, 97–188

Vessiot, 1892, Sur l'integration des équations différentielles linéaires, *Annales de l'École Normale*, **9**(3), 199–264

Wallis, J., 1656, *Arithmetica Infinitorum*, Oxford

Waltershausen, S. von, 1856, *Gauss zum Gedächtniss*, Leipzig

Waterhouse, W. C., 1980, The early proofs of Sylow's Theorem, *A. H. E. S.*, **21**, 279–290

Weber, H., 1875, Neuer Beweis des Abel'schen Theorems, *M. Ann.*, **8**, 49–53

Weber, H., 1876, *Theorie der Abel'schen Functionen vom Geschlecht 3*, Reimer, Berlin

Weber, H., 1886, Ein Beitrag zu Poincarés Theorie der Fuchsschen Functionen, *Göttingen Nachrichten*, 359–370

Wedekind, L., 1876, Beiträge zur geometrischen Interpretation binärer Formen, *M. Ann.*, **9**, 209–217

Wedekind, L., 1880, Das Doppelverhaltnis und die absolute Invariante binarer biquadratischer Formen, *M. Ann.*, **17**, 1–20

Weierstrass, K. T. W., 1841, Zur Theorie der Potenzreihen, ms. in *Werke*, I, 67–74

Weierstrass, K. T. W., 1842, Definition analytischer Functionen einer Veränderlichen vermittelst algebraischer Differentialgleichungen, ms. in *Werke*, I, 75–84

Weierstrass, K. T. W., 1854, Zur Theorie der Abel'schen functionen, *JfM*, **47**, 289–306, in *Werke*, I, 133–152

Weierstrass, K. T. W., 1856a, Über die Theorie der analytischen Facultäten, *JfM*, **51**, 1–60, in *Werke*, I, 153–221

Weierstrass, K. T. W., 1856b, Theorie der Abel'schen functionen, *JfM*, **52**, 285–339, in *Werke*, I, 297–355

Weierstrass, K. T. W., 1894, *Mathematische Werke*, 7 vols. Olms, Hildesheim

Weierstrass, K. T. W., 1902, Vorlesungen über die Theorie der Abelschen Transcendenten, in *Werke*, IV

Weil, A., 1974, Two lectures on number theory, past and present, *L'Enseignement Mathématique*, **20**(2), 87–110, in *Oeuvres*, III, 279–302

Weil, A., 1976, *Elliptic functions according to Eisenstein and Kronecker*, Springer

Weil, A., 1979, *Oeuvres Scientifiques, Collected Papers*, Springer-Verlag, New York

Weyl, H., 1913, reprinted 1923 *Die Idee der Riemannschen Fläche*, Teubner, Leipzig and Berlin

Whittaker, E. T. and Watson, G. N., 1973, *A Course of Modern Analysis* (1st edition 1902, 4th 1927, reprinted). Cambridge University Press, Cambridge

Wiman, A., 1913, Endliche Gruppen Linearen Substitutionen, *Enc. Math. Wiss.*, IB 3f

Wirtinger, W., 1904, Riemanns Vorlesungen über die hypergeometrische Reihe und ihre Bedeutung, in *Verhandlungen des dritten internationalen Kongresses in Heidelberg*, ed. A. Krazer, Teubner, Leipzig

Wirtinger, W., 1913, Algebraische Functionen und ihre Integrale, *Enc. Math. Wiss.*, II B2

Wussing, H., 1969, *Die Genesis des abstrakten Gruppenbegriffes*, Berlin. 1979 C. F. Gauss, Teubner, Leipzig, tr. *The Genesis of the Abstract Group Concept*, MIT Press 1984

Yavetz, I., 1995, *From Obscurity to Enigma: The Work of Oliver Heaviside, 1872–1889*, Birkhäuser Verlag, Basel

Yoshida, M., 1987, *Fuchsian Differential Equations*, Max-Planck Institut für Mathematik, Bonn and Vieweg

Youschkevitch, P., 1976–77, The Concept of Function up to the Middle of the 19th Century, *A. H. E. S.*, **16**, 37–85.

Historical List of Names

Some principal figures, showing their age in 1870

Niels Hendrik Abel, 1802–1829 (d.).
Enrico Betti, 1823–1892 (47).
Jean-Claude Bouquet, 1819 –1885 (51).
Francesco Brioschi, 1824 -1897 (46).
Charles-Auguste Briot, 1817-1882 (53).
Augustin Louis Cauchy, 1789 –1857 (d.).
Arthur Cayley 1821 –1895 (49).
Rudolf Friedrich Alfred Clebsch, 1833 –1872 (37).
Richard Dedekind, 1831 –1916 (39).
Peter Gustav Lejeune Dirichlet, 1805 –1859 (d.).
Ferdinand Gotthold Max Eisenstein, 1823 –1852 (d.).
Leonhard Euler, 1707 –1783 (d.).
Ferdinand Georg Frobenius, 1849 –1917 (21).
Immanuel Lazarus Fuchs, 1833 –1902 (37).
Evariste Galois, 1811 –1832 (d.).
Carl Friedrich Gauss, 1777 –1855 (d.).
Paul Gordan, 1837 –1912 (33).
Georges Henri Halphen 1844 -1889 (26).
Charles Hermite, 1822 –1901 (48).
Ludwig Otto Hesse, 1811 –1874 (59).
Adolph Hurwitz, 1859 –1919 (11).
Carl Custav Jacob Jacobi, 1804 –1851 (d.).
Camille Jordan, 1838 –1922 (32).
Christian Felix Klein, 1849 –1925 (21).
Leopold Kronecker, 1823 –1891 (47).
Ernst Eduard Kummer, 1810 –1893 (60).
Adrien Marie Legendre, 1752 –1833 (d.).
Carl Neumann, 1832 –1925 (38).
Charles Emile Picard, 1856 –1941 (14).
Julius Plücker, 1801 –1868 (d.).
Henri Poincaré, 1854 –1912 (16).
Victor Alexandre Puiseux, 1829 –1883 (50).

Georg Friedrich Bernhard Riemann, 1826 –1866 (d.).
Gustav Roch, 1836 –1866 (d.).
Ludwig Schläfli, 1814 –1895 (56.).
Hermann Amandus Schwarz, 1843 –1921 (27).
Jacob Steiner, 1796 –1863 (d.).
Heinrich Weber, 1843 –1913 (27).
Karl Theodor Wilhelm Weierstrass, 1815 –1897 (55).

Index